# PLASTIC-ENCAPSULATED MICROELECTRONICS

# PLASTIC-ENCAPSULATED MICROELECTRONICS

## Materials, Processes, Quality, Reliability, and Applications

*Edited by*

**Michael G. Pecht**
**Luu T. Nguyen**
**Edward B. Hakim**

A Wiley-Interscience Publication

**JOHN WILEY & SONS, INC.**

New York / Chichester / Brisbane / Toronto / Singapore

This text is printed on acid-free paper.



***Library of Congress Cataloging-in-Publication Data:***

Pecht, Michael.
Plastic encapsulated microelectronics: materials, processes, quality, reliability, and applications / Michael G. Pecht, Luu T. Nguyen, Edward B. Hakim.
p. cm.
Includes index.
ISBN 0-471-30625-8 (cloth ; alk. paper)
1. Microelectronic packaging--Materials. 2. Microencapsulation. 3. Plastics in packaging. I. Nguyen, Luu T. II. Hakim, Edward B. III. Title.
TK7874.P428 1995
621.381'046--dc20 94-46528

Printed in the United States of America

10 9 8 7 6 5 4 3 2 1

to the students and the staff at the University of Maryland
CALCE Electronic Packaging Research Center
who helped in this effort, and to those who
will follow in their footsteps

# CONTRIBUTORS

**Rakesh Agarwal** is a project engineer at GM Hughes, Delco Electronics where he is working on mechanical and thermal reliability of electronic modules used in under-the-hood automotives. His research experience includes mechanics of materials, damage mechanics, fracture mechanics, and mechanics of fatigue damage in heterogeneous materials. He received a Ph.D. in Mechanical Engineering from the University of Maryland. Dr. Agarwal contributed to Chapters 5, 6, 7, and 8.

**Ajay Arora** is a project engineer at Hughes Network Systems. He received an M.S. in Mechanical Engineering from the University of Maryland. His research focus is on the design of electronic packages. He contributed to Chapters 2 and 3.

**Vikram Chandra** is a engineer at Hughes Network Systems. He received his M.S. in Mechanical Engineering at the University of Maryland. His research interests include reliability analysis based on physics-of-failure principles for microelectronics. He is a member of ISHM, IEEE, and ASME. He contributed to Chapter 2.

**Lloyd W. Condra** is an engineer at Boeing Commercial Airplane Group. He has over twenty-five years experience in R&D, design, and manufacturing of high-reliability electronics, having worked as an engineer and manager at ELDEC, Bell Telephone Laboratories and Medtronic. Mr. Condra is a Senior Member of IEEE. He is a winner of the 1987 Taguchi Quality Award, and has won two ELDEC Quality Medals. He holds an M.S. from Lehigh University, in Materials Engineering. He contributed to Chapter 4.

**Abhijit Dasgupta** is a faculty member and researcher in the CALCE Electronic Packaging Research Center, University of Maryland. He received his Ph.D. in Theoretical and Applied Mechanics from the University of Illinois. He conducts research in the area of micromechanical modeling of constitutive and damage behavior of heterogenous materials and structures, with particular emphasis on fatigue and creep-fatigue interactions. His research also includes associated stress analysis techniques under combined thermomechanical loading, formulating physics-of-failure models to evolve guidelines for design, validation testing, screening, and derating for reliable electronic packages. He contributed to Chapter 7.

**Gerard Durback** is a mechanical engineer with NASA, Goddard Space Flight Center in Greenbelt, Maryland. He has been the lead mechanical engineer on space shuttle payload projects for Goddard's Special Payloads Division. He received his M.S. in Mechanical Engineering from the University of Maryland. He contributed to Chapter 5.

**Rathindra N. Ghoshtagore** received an Sc.D. in 1965 from Massachusetts Institute of Technology, Cambridge. He worked as a consultant at the CALCE Electronic Packaging Research Center, University of Maryland at College Park (1992-1993). He also worked as an Advisory Engineer of VLSI/ULSI Process Technology at Westinghouse Electric Corp. (1980-1992). He is a fellow of the American Physical Society and a senior member of IEEE. He contributed to Chapters 1, 2, 3, 4, and 10.

**Edward B. Hakim** is chief of the Reliability, Testability, and Quality Assurance Branch, Electronics Technology and Devices Laboratory, U.S. Army Laboratory

Command, at Fort Monmouth, New Jersey. For more than 30 years he has been involved in the Army microcircuit and semiconductor reliability program. He has served on numerous committees, including the Defense Science Board Task Force on Commercial Components — Electronic Devices, the Department of Defense (DoD) Industry Task Force on Electrostatic Discharge, the DoD Microcircuit Planning Group, and the DoD/Industry Coordination Working Group on Generic Qualification. He contributed to Chapter 1 and is a book editor.

**Qazi Ilyas** received a M.S. in Electrical Engineering from the University of Ilinois. In 1979, he joined AT&T Bell Laboratories as a member of the technical staff. He worked as an ASIC designer for over 5 years. His current responsibilities include reliability and qualification of integrated circuits in plastic and ceramic packages. Previously, he worked for Sperry Corporation at Goddard Space Flight Center, Maryland as a senior Engineer/Group leader for over 5 years. He is a senior member of ASME, and a professional engineer. He contributed to Chapter 8.

**Lawrence W. Kessler** is the President of SONOSCAN, INC. (formerly Sonoscope, Inc.), a high technology-based company he founded in 1973, which pioneered the development and applications of acoustic microscopy in the commercial marketplace. Through 1973 he was a Research Scientist at Zenith Radio Corporation and was instrumental in the development of new ultrasonic imaging systems. He was an Adjunct Professor of Information Engineering at the University of Illinois, Chicago Circle, for 3 years and now teaches ultrasonic imaging at short courses and workshops. Dr. Kessler is a Fellow of the Acoustical Society of America and was the National Lecturer of the IEEE-UFFC Society. Dr. Kessler received his Ph.D. degree in Electrical Engineering from the University of Illinois, Urbana, in 1968. He contributed to Chapter 9.

**Pradeep Lall** is an assistant research scientist at Motorola Radio Products Group. His research involves: development of physics of failure based software tools for design, reliability assessment, testing, and screening of microelectronics; temperature dependence of microelectronic device failure mechanisms and performance parameters, and guidelines for derating. He is a member of ASME, IEEE, SAMPE, and ISHM. He received his Ph.D. in Mechanical Engineering from the University of Maryland. He contributed to Chapters 2 and 3.

**Junhui Li** is a research scientist at the CALCE Electronic Packaging Research Center. Since 1982, she has been involved in research on microstructure relationships, failure mechanisms and reliability models for electronic packages. Her Ph.D. (1992) is in Materials Science and Engineering from the University of Maryland. She contributed to Chapter 9.

**Anupam Malhotra** is an engineer at GM Corp. He received an M.S. in Mechanical Engineering at the University of Maryland. He is a contributing author to the book, *Soldering Processes and Equipment*, published by John Wiley and Sons. His interests include physics-of-failure based methodologies for reliable product development and reliability issues in printed wiring board assembly. He is a member of the ASME, ISHM, and IEEE. He contributed to Chapters 4 and 10.

**Steven R. Martell** graduated with a B.S. in Mechanical/Ocean Engineering from the University of Rhode Island in 1981. He is an active member of ASME, ASNT, SEMI, and ICP and is chairman of the ASME Nondestructive Task Group for TAB Interconnections. He received an "Outstanding Performance" award for participation in writing of IPC-SM-786. Recommended Procedures for Handling of Moisture Sensitive Plastic IC Packages". His current position at Sonoscan allows him to work with companies and standards organizations on an international basis. He contributed to Chapter 9.

**Luu T. Nguyen** received his Ph.D. in Mechanical Engineering from MIT in 1984. He is currently a Program Manager with the Package Technology Group at National Semiconductor Corp., Santa Clara, working on several ARPA - and DoD-funded contracts on MCMs, single chip package assembly reliability, and design for manufacturability issues. He also serves as a Packaging Technical Advisory Board member for the Semiconductor Research Corporation. Previously, he was with the Philips R&D Center in Sunnyvale, CA, and the IBM T. J. Watson Research Center in Yorktown Heights, NY. He contributed to Chapter 6 and is a book editor.

**Tsutomu Nishioka** graduated in the factory of technology from Hiroshima University in 1985, where he received a bachelor of engineering. His special study is in applied chemistry. Tsutomo joined Nitto Denko Corporation in 1985. From 1985 to 1992, he was in charge of the development of molding compounds. Since 1993, he has been in charge of technical service for semiconductor companies in United States. He contributed to chapter 10.

**Thomas E. Paquette** joined Ford Microelectronics, Inc. (Colorado Springs, CO) in 1984 and is currently the Reliability Physics Manager. He is responsible for the failure analysis and reliability labs, and the assessment of advanced integrated circuits for use in automotive applications. Mr. Paquette also provides consulting services in reliability and failure analyses through his company, Insight Analytical Labs. He is a member of the IEEE and has a patent for high-frequency ESD protection for GaAs devices. He contributed to Chapter 7.

**Michael Pecht** is a professor at the University of Maryland and Director of CALCE Electronic Packaging Research Center. He is the Chief Editor of the *IEEE Transactions on Reliability*, and is on the Editorial Board of the *Journal of Electronics Manufacturing* and the IEEE SPECTRUM. Dr. Pecht is a Professional Engineer and an IEEE Fellow with an M.S. in Electrical Engineering, and an M.S. and Ph.D. in Engineering Mechanics from the University of Wisconsin. He contributed to Chapters 1, 3, 4, 5, 8, 10, and is a book editor.

**Ashok S. Prabhu** is an engineer at Nitto Denko. His interests focus on high-density interconnects and stress analysis using finite element methods. He received an M.S. in Mechanical Engineering at the University of Maryland. He contributed to Chapter 2.

**Dan Quearry** is a lead electronic technician working at the Craine Division, Naval Surface Warfare Center. In 1992 Mr. Quearry began working for the Standard Hardware Acquisition and Reliability Program (SHARP) and is the SHARP project lead concerning use of plastic encapsulated microcircuits in military applications. Mr. Quearry worked as a first line supervisor over the component Automatic Test Equipment (ATE) test development section from 1988 to 1992. From 1983 to 1988 he worked as a test development technician on ATE, developing programs to perform electrical tests on microcircuits. Mr. Quearry began his carrier as an Electronic Technician at the Crane Division in 1980. He contributed to Chapter 1.

**Janet E. Semmens** is a graduate of Western Illinois University and has been employed by Sonoscan since 1981. She is currently the Manager of the Technology Support Group. She was Project Manager on a government contract to develop a nondestructive test method for TAB (Tape Automated Bonding). As a senior scientist, she is consulted to apply Acoustic Microscopy to solve the problems in the fields of microelectronics and materials science for client companies. She contributed to Chapter 9.

**Jack Stein** received an Honours Bachelor of Applied Science in Electrical Engineering in 1982 and a Bachelor of Commerce in Business Administration in 1984 from the University of Windsor, Ontario, Canada. He is currently employed with Ford Motor Company Automotive Components Electronics Division and is Chairman of the Society of Automotive Engineers (SAE) International Reliability Standards Committee Subcommittee on Environmental factors. Mr. Stein has worked in the area of automotive electronics reliability and test engineering since 1984 and is presently completing a graduate study program in Reliability Engineering. He contributed to Chapter 8.

# CONTENTS

# PREFACE

Since the 1980s, plastic-encapsulated microcircuits (PEMs), or noncavity packages, have made significantly gains in lowering cost, rising the availability of functions, achieving high reliability, and increasing applications. Today, enhanced encapsulant materials, robust design, and improved process control have made PEMs the package of choice for most applications.

## Motivation

As part of their research charter, CALCE Electronics Packaging Research Center personnel have surveyed the available literature, held discussions with numerous professionals, and conducted research on the qualification, screening, and reliability assessment of PEMs. With such commercial companies, military and government agencies, and professional societies as the Joint Electronic Devices Engineering Council, the Semiconductor Industry Association, the Department of Defense, Boeing, Airbus, Motorola's Iridium, and the French military all interested in a quality and reliability approach for selecting and using PEMs, demand was high for a book discussing the state of the art in PEMs. It was also apparent that no standards and procedures based on scientific and cost-effective considerations have been generated for specific PEM issues.

## What This Book Is About

This book addresses issues relevant to PEM devices and to assemblies that incorporate PEMs. The materials, manufacturing processes, screens, tests, failure mechanisms, reliability, and applications of PEMs are presented in ten chapters organized to present the science and technology behind PEMs.

Chapter 1 provides a historical perspective on PEMs, and discusses the advantages of using plastic packages, the state of the current technology, and the primary package styles.

Chapter 2 discusses the materials employed in plastic packages, including their composition and characteristics. Significant recent developments in plastic packaging materials technology are included.

Chapter 3 describes the fabrication technology and process steps used in modern plastic packages. Manufacturing equipment for fabricating PEMs is only briefly addressed; this information is available in other books.

Chapter 4 is a basic description of the manufacturing of electronic systems using plastic encapsulated components. Particular attention is paid to the processes that can affect the reliability of PEMs on circuit cards.

Chapter 5 focuses on aspects of shipping containers and handling that are of special significance for PEMs during storage, transportation, and pre-assembly including information pertaining to moisture and electrostatic discharge protection in containers.

Chapters 6 through 9 present issues related to the quality and reliability of PEMs. Chapter 6 discusses the failure mechanisms, sites, modes, and modeling techniques unique to plastic-encapsulated microcircuits, using science-based explanations and equations.

Chapter 7 outlines the physics-of-failure approach to PEM quality. PEM defects are presented and screening is examined with respect to the defects that may be introduced, with the aim of moving away from mandatory screens towards defect-specific in-line process control methods.

Chapter 8 examines PEM qualification, with emphasis on its effectiveness in simulating field failures. Current commercial practices and relevant tests for specific use environments are discussed.

Chapter 9 covers the principles, strengths, and limitations of failure analysis techniques. The techniques discussed are applicable to field failures as well as to components that fail during screening and qualification.

Chapter 10 presents a perspective on future trends in plastic encapsulation, especially in chip technology, packaging, design, materials selection, manufacturing processes, device integration, and application-specific reliability.

## Who This Book Is For

Since this book describes the materials, manufacturing processes, failure mechanisms, accelerated tests, and science-based procedures for qualification and quality conformance of PEMs, it will prove most useful to those who manufacture, supply, or procure PEMs; those involved in developing guidelines and standards; reliability engineers; and package design teams.

While the needs of the electronics industry have been carefully considered, the book is also for students of electronic packaging, our future leaders, scientists, and engineers.

## Acknowledgments and Confessions

This book attempts to document state-of-the-art knowledge of PEMs to support current and future activities. The inspiration for the book came from various industry- and government-sponsored research projects completed by the CALCE Electronic Packaging Research Center and its members. We hope that this book will expedite the scientific understanding of PEMs, generate much-needed procurement standards, facilitate reliability assessment of application and use environments, and provide a guide for the manufacture and application of PEM devices. This work contains nothing that has not been discussed before at some technical conference or R&D laboratory. However, we hope that our attempt to report the state of the art in plastic-encapsulated microcircuits and future trends will prove useful.

Sincere thanks is given to experts in the electronic packaging field who reviewed this document and offered valuable comments and suggestions, especially Charles Packard of Paramax Systems Corporation, Nicholas A. Rounds of Plaskon Electronic Materials, George Perreault of Cypress Semiconductor, Dan Quearry of the Naval Surface Weapons Center, John A. Negrych of MINCO, Inc., and Mark Bahmueller of Hull Corporation. We also thank Kirk Bonner from NASA-JPL, Norman Owens from Motorola, My Nguyen from Johnson Matthey Electronics, Anthony Gallo from Dexter Electronic Materials, Leon Lantz from the Department of Defense at Fort Meade, Kenneth R. White, Andrea Chen from National Semiconductors and Alexander Teverovsky for their contributions to the chapters.

We also thank the CALCE EPRC members for their support in the development of this book. They include Allied Signal Aerospace Company, AMP Incorporated, Armstrong Laboratories, Bell Communications Research Incorporated, Boeing Commercial Airplane Group, Cetar Ltd., Collins

Commercial Avionics, David Taylor Laboratories, the Department of Defense, Eldec Corporation, Flomerics Incorporated, Ford Motor Company, Lockheed Aerospace, Martin Marietta, Hamilton Standard, United Technologies, Honeywell Incorporated, IBM Corporation (Federal Systems Division), IBM Corporation (Technology Lab.), Loral, McDonnel Douglas Aerospace, the NASA Goddard Space Flight Center, the National Science Foundation, the Naval Weapons Support Center, Rockwell International, Simmonds Precision Products, Sonoscan Incorporated, the State of Maryland, Texas Instruments Incorporated, the U.S. Army Material Systems Analysis Activity, Westinghouse Electric Corporation, and WPAFB-Armstrong Laboratories. Special acknowledgments are due to ELDEC Corporation, Hamilton Standard, Texas Instruments, and Motorola for conducting and sharing some of the highly accelerated stress test experiments and data used in reliability modeling and model validation presented in this book. Finally, we thank University of Maryland CALCE Electronic Packaging Research Center staff Iuliana Roshou Bordelon, Joanyuan Lee, Amal Hammad, and Kim Jong, and our English reviewer, Dr. Lesley Northup, who helped prepare this book. Thanks are also due to my graduate students, M. A. Padmanabhan and Anupam Malhotra, who helped organize the effort and bring the book into its final format.

*Michael G. Pecht*

# LIST OF SYMBOLS

| | |
|---|---|
| $\alpha$ | coefficient of thermal expansion |
| $\delta$ | maximum clearance between chip pad and encapsulant |
| $\Delta K$ | stress intensity factor range |
| $\Delta_m$ | the quantity of absorbed water |
| $\Delta V_{TH}$ | threshold-voltage shift |
| $\varepsilon$ | real part of the dielectric constant |
| $\eta$ | viscosity of the molding compound |
| $\theta$ | diffraction angle |
| $\theta_{ca}$ | case-to-ambient thermal impedance |
| $\theta_{JC}$ | junction-to-case thermal impedance |
| $\lambda$ | wavelength of the x-ray |
| $\mu$ | channel mobility |
| $\nu$ | Poisson's ratio |
| $\nu_{filler}$ | volume part of the filler |
| $\rho$ | mass density |
| $\rho_c$ | specific density of the composite |
| $\rho_w$ | specific density of water |
| $\sigma$ | shrinkage stress due only to the molding compound |
| $\sigma(r)$ | stress component at distance r from the corner |
| $\sigma_p$ | maximum stress loading at the edge of the die pad |

| | |
|---|---|
| $\sigma_R$ | flexural strength of the encapsulant at the soldering temperature |
| $\tau_M$ | local von Mises stress |
| a | crack length |
| $A$ | area of the chip pad |
| $A_{bag}$ | area of the bag |
| $A_{elec}$ | area of the measuring electrode |
| $a_{max\ dp}$ | the longer dimension of the die pad |
| $b$ | fatigue constant |
| $B$ | pad width |
| $b_{beam}$ | width of the beam |
| $b_{shear}$ | shear fatigue constant of the material |
| $C$ | total capacitance |
| $C_0$ | gate oxide capacitance |
| $C_{0(x)}$ | initial distribution of moisture in the encapsulant |
| $c_1$ | constant |
| $c_2$ | constant |
| $c_3$ | material constant |
| $c_4$ | material constant |
| $c_5$ | geometric constant |
| $c_6$ | device constant |
| $c_7$ | geometric constant |
| $c_8$ | material constant |
| $C_{fatigue}$ | shear fatigue constant of the material |
| $C_L$ | loading capacitance |
| $C_{LL}$ | lead-to-lead capacitance |
| $d$ | depth of the beam |
| $D$ | additional amount of desiccant for dunnage |
| $D_0$ | diffusion coefficient |
| $DRF$ | the desiccant reduction factor |
| $D_{wire}$ | wire displacement |
| $e$ | charge carried by the electron |
| $E$ | modulus of elasticity |
| $E_a$ | activation energy |
| $E_B$ | flexural modulus |
| $E_d$ | activation energy for the diffusion |
| $E_{encapsulant}$ | flexural modulus of the encapsulant |
| $E_S$ | activation energy for solution |
| $E_T$ | temperature-dependent tensile modulus |

| | |
|---|---|
| $f$ | failure density |
| $F$ | deviation of deformed wire length |
| $f(t)$ | failure density |
| FIT | failure unit |
| $f_{load}$ | scaling factor for the load |
| $F_n$ | Fourier coefficients |
| $G$ | total fracture energy |
| $h$ | Planck's constant |
| $h_{bb}$ | ball-bond height |
| $h_{bn}$ | nail head ball-neck height |
| $h_{enc}$ | encapsulant thickness |
| $h_{package}$ | package thickness under the die pad |
| $I_s$ | stress index |
| $J$ | linear elastic component |
| $k$ | Boltzmann constant |
| $K$ | stress intensity factor |
| $k_D$ | diffusion coefficient |
| $K_I$ | characterize the severity of the crack-tip stress field |
| $K_{IC}$ | critical stress intensity factor |
| $K_{IIC}$ | critical stress intensity factor |
| $K_{max}$ | maximum stress intensity factor |
| $k_r$ | rate constants |
| $K_{stress}$ | stress intensity factor |
| $K_w$ | coefficient determined from experiments |
| $l$ | support span |
| $L$ | self-inductance of the leads |
| $m$ | slope of the tangent to the initial straight-line portion of the load-deflection curve |
| $M$ | mechanical accuracy of the wirebonder |
| $m_0$ | initial mass of water |
| $M_c$ | the mass of the composite |
| $M_L$ | multilayer substrate |
| $m_r$ | pseudo-reaction orders |
| $n$ | number of thermal shock cycles |
| $N$ | device population |
| $N_{cycles}$ | number of cycles |
| $N_{fs}$ | cycles to failure due to shear fatigue |
| $N_{ft}$ | cycles to failure due to tensile fatigue |

| | |
|---|---|
| $N_{IT}$ | radiation-induced interface charge |
| $N_{OT}$ | trapped oxide charge |
| $N_p$ | number of passivation cracks |
| $n_r$ | pseudo-reaction orders |
| $N_R$ | radiation-induced interface charge |
| $N_{Tc}$ | number of temperature cycles |
| $O_0$ | solubility coefficient at 0°C |
| $O_{conc}$ | moisture concentration |
| $O_{des}$ | minimum moisture capacity of the desiccant |
| $O_{diff}$ | moisture diffusion constant |
| $O_{sat}$ | encapsulant moisture saturation coefficient |
| $O_{sol}$ | solubility coefficient of moisture in plastic |
| $P$ | bonding pitch |
| $P_a$ | water steam pressure outside the package |
| $P_c$ | water steam pressure inside the gap |
| $p_{chip}$ | vapor pressure in the chip |
| $p_{pe}$ | vapor pressure in the pad-encapsulant space |
| $P_{rupture}$ | load at the rupture |
| $p_s$ | saturated vapor pressure |
| $P_{sat}$ | saturation vapor pressure |
| $p_{water}$ | vapor pressure of water at the soldering temperature |
| $q$ | electronic charge |
| $R$ | direct current resistance |
| $R_{acc}$ | recognition accuracy |
| $r_{ball}$ | ball radius |
| $r_{bb}$ | ball-bond radius |
| $r_{ch}$ | capillary hole radius |
| RH | relative humidity |
| $R_v$ | measured volume resistance |
| $r_{wire}$ | wire radius |
| $S$ | flexural strength |
| $S_c$ | sorption coefficient |
| $S_L$ | single-layer substrate |
| $t$ | thickness of the specimen |
| $T$ | absolute temperature |
| $T_a$ | ambient temperature |
| $T_D$ | signal propagation delay |
| $T_{desiccant}$ | total desiccant units |

| | |
|---|---|
| $T_F$ | time-to-failure |
| $T_g$ | glass transition temperature |
| $t_{gap}$ | gap thickness |
| $T_j$ | junction temperature |
| $T_m$ | mold compound temperature |
| $T_{max}$ | maximum temperature of the test |
| $T_{maxstor}$ | maximum storage time |
| $t_{package}$ | package thickness |
| $t_{pad}$ | thickness of the plastic below the chip pad |
| $T_T$ | steady-state temperature |
| $T_{wirebond}$ | teaching accuracy of the wirebonder |
| $U$ | amount of desiccant units |
| $V$ | accelerating potential |
| $V_A$ | applied voltage |
| $V_{chamf}$ | capillary chamfer |
| $V_{fab}$ | free air ball volume |
| $v_{light}$ | velocity of the light |
| $v_{mc}$ | velocity of the molding compound |
| $v_{quan}$ | velocity of the quantum |
| $v_{sound}$ | velocity of sound through the material that the wave is propagating |
| $W$ | deformed wire length |
| $W_a$ | reversible work of adhesion |
| $W_p$ | irreversible work of deformation |
| $w_{pad}$ | pad width |
| WVTR | the water vapor transmission rate |
| $x$ | distance from the die paddle to the bottom of the package |
| $X$ | fitting parameters for the conversion of epoxide groups water vapor transmission rate |
| $x_1$ | intermetallic layer thickness |
| $Z$ | acoustic impedance |

# LIST OF ACRONYMS

| | |
|---|---|
| AER | alpha emission rates |
| AES | Auger electron spectroscope |
| AFM | atomic force microscopy |
| AIPD | acoustic impedance polarity detection |
| AMI | acoustic microimaging |
| ANOVA | analysis of variance |
| ASIC | application specific integrated circuit |
| ASMAT | application-specific material |
| ASTM | American Society for Testing and Materials |
| C-SAM | C-mode scanning acoustic microscope |
| CADMP | computer-aided design of microelectronic packages |
| CERDIP | ceramic dual in-line package |
| CMOS | complementary metal oxide semiconductor |
| COB | chip-on-board |
| COF | chip-on-film |
| COG | chip-on-glass |

| | |
|---|---|
| CTE | coefficient of thermal expansion |
| CVD | chemically vapor-deposited |
| DIP | dual in-line package |
| DOE | design of experiment |
| DRAM | dynamic random access memory |
| DSC | dynamic scanning calorimetry |
| DTAB | demountable TAB |
| EDX | energy dispersion |
| EFO | electric flame-off |
| EIAJ | Electronics Industry Association of Japan |
| EM | emission microscopy |
| EMC | epoxy molding compound |
| ESD | electrostatic discharge |
| ESEM | environment scanning electron microscopy |
| FIE | front interface echo |
| HAST | highly accelerated stress testing |
| HIC | humidity indicator card |
| I/O | input and output |
| JEDEC | Joint Electron Devices Engineering Council |
| JQA | Joint Qualification Alliance |
| LCC | leadless chip carrier |
| LCD | liquid crystal display |
| LCP | liquid crystal polymers |
| LEEM | low-energy electron microscopy |
| MBB | moisture barrier bag |
| MCM-L | multichip modules in laminates |
| MC | molding compound |
| MOS | metal oxide semiconductor |
| MSMT | micro surface mount technology |
| MTBF | mean time between failures |
| MTTF | mean-time-to-failure |
| OM | optical microscopy |
| OMPAC | overmolded pad-array carrier |
| PCT | pressure-cooker test |
| PDIP | plastic dual in-line package |
| PECVD | plasma-enhanced chemically vapor-deposited |
| PEM | plastic-encapsulated microcircuit |
| PGA | pin-grid array |
| PLCC | plastic leaded chip carrier |

| | |
|---|---|
| PMC | post-mold cure |
| PPF | pre-plated frame |
| PPGA | plastic pin-grid array |
| PQFP | plastic quad flatpack package |
| PWB | printed wiring board |
| Q-BAM | quantitative B-scan analysis mode |
| QML | qualified manufacturing lines |
| QSSOP | quartersize small-outline package |
| RP | reverse polarity |
| RWoH | reliability without hermeticity |
| SAM | scanning acoustic microscopy |
| SDIP | shrink dip |
| SEM | scanning electron microscopy |
| SIP | single in-line package |
| SLAM | scanning laser acoustic microscope |
| SMT | surface mount technology |
| SNFOM | scanning near-field optical microscopy |
| SOIC | small-outline integrated circuit |
| SOJ | small-outline J-leaded |
| SOP | small-outline package |
| SP | straight polarity |
| SPC | statistical process control |
| SQFP | shrink quad flatpack |
| SSOP | skinny or shrink small outline package |
| STEM | scanning and transmission electron microscopy |
| STM | scanning tunneling microscopy |
| T/F | trim and form |
| TAB | tape automated bonding |
| TCP | tape-carrier package |
| TEM | transmission electron microscopy |
| TMA | thermomechanical analysis |
| TOF | time of flight |
| TOP | thin outline package |
| TQFP | thin quad flatpack package |
| TSOP | thin small-outline package |
| TSSOP | thin sealed small-outline |
| ULSI | ultra-large-scale integration integrated circuit |
| VLSI | very large scale integration |
| VSOP | very small-outline package |

| | |
|---|---|
| WDX | wavelength dispersion |
| WVTR | water vapor transmission rate |
| XM | x-ray microscopy |

# 1

# INTRODUCTION

A plastic-encapsulated microcircuit (PEM), often called a plastic package, consists of an integrated circuit chip physically attached to a leadframe, electrically interconnected to input-output leads, and molded in a plastic that is in direct contact with the chip, leadframe, and interconnects. In comparison, a hermetically sealed microcircuit (generally called a hermetic package) consists of an integrated circuit chip mounted in a metal or ceramic cavity, interconnected to the leads, and hermetically sealed to maintain a contact environment within the package.

Historically, PEMs have been used in commercial and telecommunications electronics and consequently have a large manufacturing base. With major advantages in cost, size, weight, performance, and availability, plastic packages have attracted 97% of the market share of worldwide microcircuit sales, although they encountered formidable challenges in gaining acceptance for use in government and military applications. In fact, it was only in the early 1990s that the industry dispelled the notion that hermetic packages were superior in reliability to plastic packages, in spite of their low production and procurement volumes and the outdated government and defense department standards and handbooks associated with their manufacture and use microcircuits.

Certainly, plastic-encapsulated microcircuits have not been free of problems. Prior to the 1980s, moisture-induced failure mechanisms, such as corrosion, cracking, and interfacial delamination, were significant [Flood 1972]. Early PEMs were also plagued by thermally induced intermittence problems, resulting in open-circuit failures at elevated temperatures [Motorola 1974].

However, the decade of the 1980s brought revolutionary changes in electronics technology in general, and in plastic packaging in particular. Early transistor and diode plastic encapsulation, accomplished by dispensing a small amount of room-temperature vulcanizing silicone or flexible epoxy material over the die and bond wires (glob-topping), was replaced by advanced molding techniques, including transfer molding, injection molding, and potting. Hundreds of molding materials — epoxies, silicones, and phenolics — were evaluated by manufacturers for cost, performance, producibility, shelf life, flammability, and reliability. Additives improved thermal conductivity, the coefficient of thermal expansion, adhesion, viscosity, mold release, flame retardance and appearance. Today, high-quality, high-reliability, high-performance, and low-cost plastic-encapsulated microcircuits are common. Hermetic packaging, on the other hand, has not kept pace.

## 1.1 WHAT IS A PLASTIC-ENCAPSULATED MICROCIRCUIT?

In general, a PEM consists of a silicon chip, a metal support or leadframe, wires that electrically attach the chip circuits to the leadframe and thus to the external leads, and a plastic epoxy encapsulating material to protect the chip and the wire interconnects. The leadframe is made of a copper alloy, Alloy 42 (42Ni/58Fe), or Alloy 50 (50Ni/50Fe), and is plated with gold and silver or palladium, either completely or in selected areas over nickel or nickel/cobalt. The silicon chip is usually mounted to the leadframe with an organic conductive formulation of epoxy. Wires, generally of gold but also of aluminum or copper, are bonded to the aluminum bonding pads on the chips and to the fingers of the leadframe. The assembly is then typically transfer-molded in epoxy. Following the molding operations, the external pins are plated with a lead-tin alloy, cut away from the strip, and formed [Cook and Servais 1983].

Plastic packages are either premolded or postmolded. In the former, a plastic base is molded; the chip is then placed on it and connected to an I/O fanout pattern with wire. The die and wirebonds are usually protected by an epoxy-attached lead which forms a cavity. Premolded packages are most often used for high-pin-count devices or pin-grid arrays that are not amenable to flat leadframes and simple fanout patterns. In the postmolded type, the die is attached to a leadframe, which is then loaded into a multicavity molding tool and encapsulated in a thermoset molding compound via the transfer molding process. Postmolded packages are less expensive than premolded ones because here are fewer parts and assembly steps. In the 1990s, about 90% of plastic packages were made using postmolding techniques.

PEMs are made in either surface-mount or through-hole configurations. The common families of surface-mounted PEMs are the small-outline package (SOP), the plastic-leaded chip carrier (PLCC), and the plastic quad flatpack (PQFP). The common families of through-hole-mounted PEMs are the plastic dual in-line (PDIP), the single in-line (SIP), and the plastic pin-grid array (PPGA). The various PEM configurations are given in Figures 1.1 and 1.2.

### 1.1.1 Through-Hole Mounted Devices

The common families of through-hole mounted PEMs are the plastic dual-in-line (PDIP), the single in-line (SIP), and the plastic pin-grid array (PPGA). The plastic dual in-line package, the most commonly used PEM in the 1980s, has a rectangular plastic body with two rows of leads, often on 2.45-mm (100-mil) centers on the long sides. The typical lead spreads at the lead tips between the two rows are 7.35 mm (300 mils), 9.80 mm (400 mils), or 14.7 mm (600 mils), with 4 to 20 leads, 22 leads, and 24 to 64 leads, respectively. Dimensional data for some common through-hole mounted PDIPs are given in Figure 1.3 and Table 1.1.

The design of the plastic dual in-line package is conducive to high-volume manufacturing at low cost. The dual in-line structure of the leads allows the packages to be shipped in plastic tubes end to end, without contacting the leads. The leads are bent up at a small angle from the package body for through-hole insertion mounting. The package layout allows automated board mounting and offers highly reliable joints.

Single in-line packages are rectangular with leads on one of the long sides. The leads are typically on 1.225-mm (50-mil) centers at the plastic body interface, formed into two staggered rows spaced 2.45 mm (100 mils) and 1.225 mm (50 mils) apart. These packages offer a high profile but small footprint on the board, maintaining a 2.45-mm (100-mil) hole-mounting standard. They offer all the advantages of dual in-line packages in ease and low manufacturing cost.

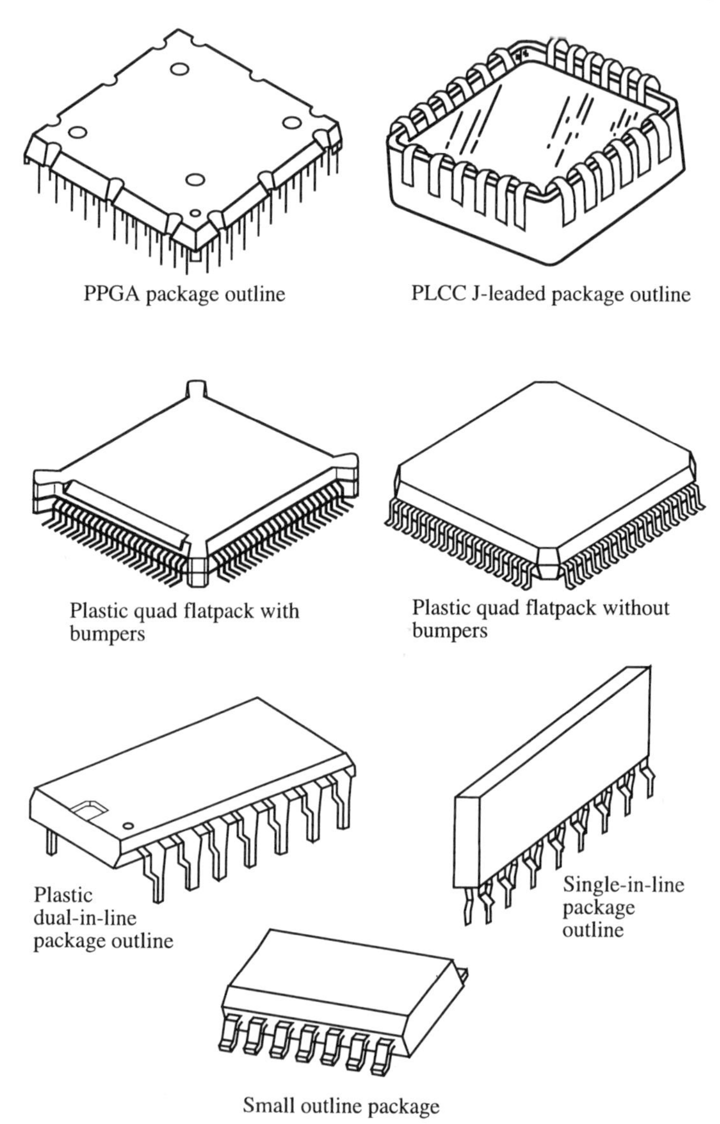


**Figure 1.1** *Plastic package configurations [Electronic Materials Handbook 1989]*

**Table 1.1 Typical dimensional data on dual in-line package**

| Pin count | Package length (mm [in]) | Package width (mm [in]) | Pin spacing (mm [in]) | Footprint width (min) (mm [in]) |
|---|---|---|---|---|
| 14 | 19.812(0.780) | 7.62±0.025(0.300±0.010) | 2.54(0.100) | 3.175(0.125) |
| 16 | 19.812(0.780) | 7.62±0.025(0.300±0.010) | 2.54(0.100) | 3.175(0.125) |
| 18 | 23.368(0.920) | 7.62±0.025(0.300±0.010) | 2.54(0.100) | 3.175(0.125) |
| 20 | 24.307(0.975) | 7.62±0.025(0.300±0.010) | 2.54(0.100) | 3.175(0.125) |
| 22 | 28.488(1.120) | 10.16±0.025(0.400±0.010) | 2.54(0.100) | 3.175(0.125) |
| 28 | 36.576(1.440) | 15.24±0.025(0.600±0.010) | 2.54(0.100) | 3.175(0.125) |
| 40 | 53.086(2.090) | 15.24±0.025(0.600±0.010) | 2.54(0.100) | 3.175(0.125) |
| 28 | 25.508(1.020) | 10.35±0.44(0.4075±0.018)7. | 1.78(0.070) | 2.032(0.080) |
| 24 | 28.575(1.125) | 62±0.025(0.300±0.010) | 2.54(0.100) | 3.175(0.125) |
| 24 | 32.766(1.290) | 15.24±0.025(0.600±0.010) | 2.54(0.100) | 3.175(0.125) |
| 8 | 10.160(0.400) | 7.62±0.025(0.300±0.010) | 2.54(0.100) | 3.175(0.125) |
| 28 | 33.985(1.338) | 7.87±0.025(0.310±0.010) | 2.54(0.100) | 3.175(0.125) |
| 48 | 62.230(2.450) | 15.24±0.025(0.600±0.010) | 2.54(0.100) | 3.175(0.125) |
| 52 | 67.310(2.650) | 15.24±0.025(0.600±0.010) | 2.54(0.100) | 3.175(0.125) |
| 64 | 81.280(3.200) | 22.86±0.025(0.900±0.010) | 2.54(0.100) | 3.175(0.125) |

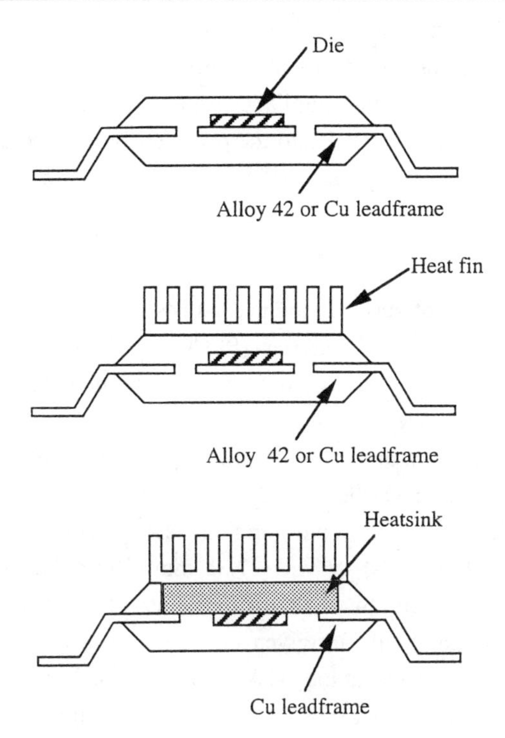


**Figure 1.2** *Plastic package configurations [Courtesy of OKI 1994]*

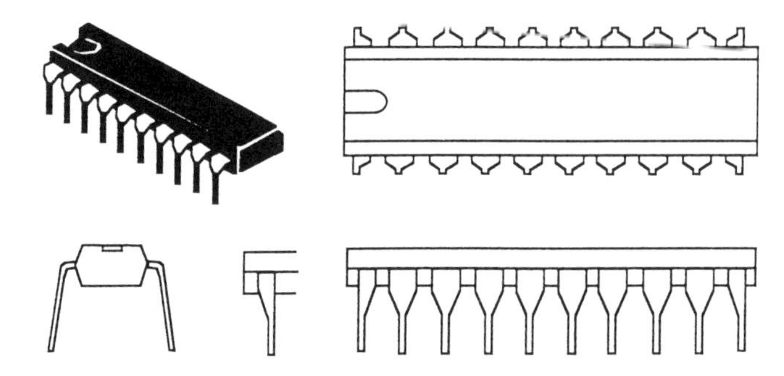

**Figure 1.3** *Mechanical data for a typical through-hole-mounted PDIP package*

Plastic pin-grid arrays are packages with leads located in a grid array under a plastic body. The pins are often located on 2.45-mm (100-mil) centers, although 1.715-mm (70-mil) centers are also available. The 2.45-mm (100-mil) centered pin matrix takes advantage of mature through-hole board mounting technology for high pin-count devices. Plastic pin-grid arrays are the highest density through-hole packages and offer the highest available pin counts for PEM. The grid-array layout of the interconnects allows impedance matching of the package to the chip. Using multilayer printed-board technology, ground and power planes can be incorporated into the package for high-performance devices. Heat sinks can also be incorporated, if needed.

### 1.1.2 Surface-mounted Devices

The common families of surface-mounted PEMs are the small-outline packages (SOP), the plastic-leaded chip carrier (PLCC), and the plastic quad flatpack (PQFP). These packages are designed for low-profile mounting on printed wiring boards for products smaller overall.

Small-outline packages are like dual in-line packages, with leads on two sides of the package body, but the leads are typically on 1.225-mm (50-mil) centers and are formed in a gull-wing configuration. Body widths are typically either narrow, 3.675 mm (150 mils), or wide, 7.35 mm (300 mils), with pin counts of 8 to 16 for the former and 14 to 28 for the latter. Dimensional data for some common surface-mount SOPs are given in Figure 1.4 and Table 1.2.

A variation on the small-outline package is the small-outline J-leaded (SOJ) package, in which the lead is formed in a J-bend configuration and folded under the body. The advantage is an even smaller footprint than that of a gull-wing, although solder joint inspection becomes more difficult.

Plastic-leaded chip carrier packages are molded PEMs with leads on all four

**Table 1.2 Typical dimensional data on small outline packages**

| Pin count | Package length (mm [in]) | Package width (max) (mm [in]) | Pin spacing (mm [in]) | Footprint width (max) (mm [in]) |
|---|---|---|---|---|
| 8 | 5.004(0.197) | 3.988(0.157) | 1.27(0.050) | 6.198(0.244) |
| 14 | 8.738(0.344) | 3.988(0.157) | 1.27(0.050) | 6.198(0.244) |
| 16 | 10.008(0.394) | 3.988(0.157) | 1.27(0.050) | 6.198(0.244) |

sides of the plastic body. Typically, from 18 to 124 leads are present on 1.225-mm (50-mil) centers and formed in the J-bend configuration. Since the leads are on all four sides of the plastic body, this package style offers the advantages of dense mounting on the board. Electrically, the interconnect leads of plastic-leaded chip carriers are shorter on average and more consistent in length than those of equivalent dual in-line packages, resulting in a better match of the impedances of the package leads.

Plastic quad flatpack packages (PQFPs) are square or rectangular plastic packages, typically with 40 to 240 leads distributed on all four sides. PQFPs are either bumpered, with leads on 0.6125-mm (25-mil) centers, or non-bumpered, with leads on centers of 0.98, 0.735, 0.6125, and 0.49 mm (40, 30, 25, and 20 mils).

## 1.2 WHY PLASTIC PACKAGES?

Plastic-encapsulated microcircuits (PEMs) have many advantages over hermetic packages in the areas of size, weight, performance, cost, reliability, and availability.

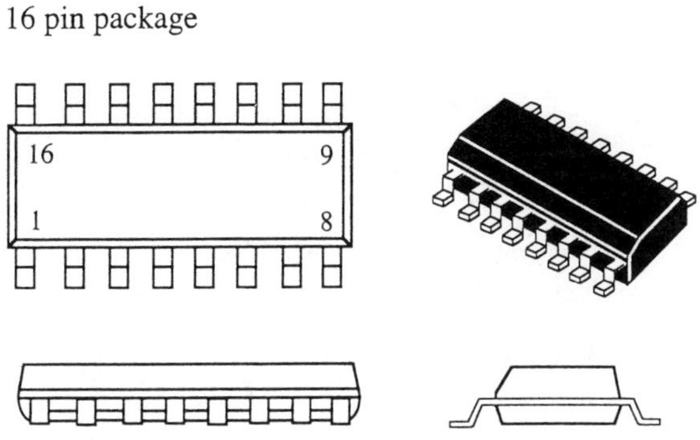

16 pin package

16 9

1 8

**Figure 1.4** *Mechanical data for a typical surface-mount SOP package*

### 1.2.1 Size and Weight

Commercial PEMs generally weigh about half as much as ceramic packages. For example, a fourteen-lead plastic dual in-line package (DIP) weighs about one gram, versus two grams for a fourteen-lead ceramic DIP. Although there is little difference in size between plastic and ceramic DIPs, smaller configurations, such as small-outline packages (SOPs), and thinner configurations, such as thin small-outline packages (TSOPs), are available only in plastic. The use of SOPs and TSOPs also enables better-performing circuit boards due to higher packing density and consequent reduced component propagation delays. Figure 1.5 shows size-versus-lead comparisons of various microcircuit packaging options. A smaller form factor naturally implies higher board density and more functionality packed into the same precious board real estate. Similarly, a lighter package results in a smaller overall payload for the same board functionality, a concern of critical importance for avionics equipment. Consumer and commercial electronics will also benefit from these advantages of plastics over ceramics.

### 1.2.2 Performance

Plastics have better dielectric properties than ceramics. Although plastic packages, such as DIPs, are not the most conducive to propagating high-frequency signals [Sinnadurai 1985], plastic quad flatpack (PQFP), pin-grid arrays, and ball-grid arrays are favored for minimizing propagation delays. For the typical applications encountered commercially, in which frequencies do not

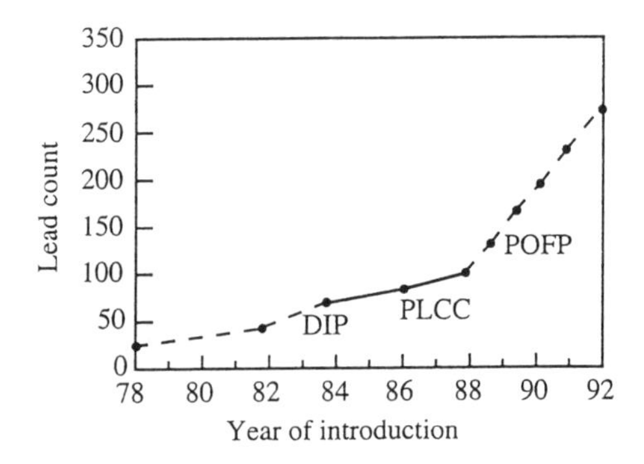


**Figure 1.5** *Package size versus lead count for various square packages and DIP*

exceed 2 to 3 GHz, plastic packages perform better than their ceramic counterparts in the same form factor. Two main features account for this characteristic: the lower dielectric constant of the epoxy compared with that of the standard cofired ceramic, and the smaller lead inductance of the copper lead frames next to the Kovar leads. When the application calls for much higher frequencies (up to 20 GHz), however, better and more predictable performance is obtained with ceramic packages. The dielectric constant of typical ceramics stays the same over a wider frequency range than that of plastic molding compounds. This is because the dielectric constant is dependent on the amount of moisture absorption.

Tables 1.3 to 1.5 provide a summary of typical electrical performance data measured or calculated for some PEMs. In these tables, $R$ is the direct current (DC) resistance (excluding the gold interconnection wire) of the leads, $C$ is the total capacitance, including lead-to-lead ($C_{LL}$) and loading capacitances ($C_L$), and $L$ is the self-inductance of the leads. The symbol $S_L$ indicates a single-layer substrate (i.e., no power and ground planes) and $M_L$ indicates a multi-layer substrate. A comparative evaluation indicates that for the same lead count, PEMs provide three to four times the lead self-inductance, half the loading capacitance, one hundred times less the total capacitance, and the same DC resistance of ceramic packages [Intel 1993].

**Table 1.3 Summary of electrical data on typical plastic packages for pin-grid arrays [Intel 1993]**

| Lead count | I/O lines | | | | Power and ground | | |
|---|---|---|---|---|---|---|---|
| | L (nH) | $C_L$ (pF) | $C_{LL}$ (pF) | R (mΩ) | L (nH) | C (pF) | R (mΩ) |
| 68(S) | 9 to 13 | 2 to 3 | 2 to 3 | 200 to 300 | 9 to 13 | 2 to 3 | 200 to 300 |
| 68(M) | 5 to 9 | 2 to 3 | ≤ 0.5 | 200 to 300 | 2 to 4 | ≥ 75 | 50 to 100 |
| 88(S) | 9 to 14 | 2.5 to 4 | 2.5 to 4 | 220 to 350 | 9 to 14 | 2.5 to 4 | 220 to 350 |
| 88(M) | 5.8 to 8.6 | 2.5 to 6 | ≤ 0.5 | 220 to 350 | 2 to 3 | ≥ 121 | 50 to 100 |
| 132(M) | 6 to 11 | 6 to 15 | ≤ 0.3 | 220 to 700 | 0.5 to 3.9 | 300 to 900 | 20 to 50 |
| 168(M) | 9 to 22 | 4 to 9 | ≤ 0.5 | 220 to 800 | 0.5 to 1.0 | 300 to 900 | 20 to 50 |

**Table 1.4 Summary of electrical data on typical plastic packages for plastic quad flatpack [Intel 1993]**

| Lead count | Electrical parameter | | | |
|---|---|---|---|---|
| | R (MΩ) | L (nH) | $C_L$ (pF) (loading) | $C_{LL}$ (pF) |
| 84 | 70 to 80 | 5.8 to 6.6 | < 1.0 | < 0.5 |
| 100 | 70 to 80 | 6.1 to 7.1 | < 1.5 | < 1.0 |
| 132 | 70 to 80 | 7.3 to 8.5 | < 2.0 | < 1.5 |
| 164 | 70 to 80 | 8.0 to 14.5 | < 2.2 | < 1.5 |
| 196 | 70 to 80 | 9.0 to 15.5 | < 1.5 | < 1.6 |

**Table 1.5 Summary of electrical data on typical plastic packages [Intel 1993] for plastic-leaded chip carrier**

| Lead count | L (nH) | Loading $C_L$ (pF) | Lead-to-lead $C_{LL}$ (pF) | R (mΩ) |
|---|---|---|---|---|
| 20 square | 3.5 to 6.3 | 0.8 to 0.9 | 0.3 to 0.4 | 68 to 78 |
| 28 square | 3.7 to 7.8 | 1.0 to 1.1 | 0.4 to 0.5 | 68 to 78 |
| 28 rectangular | 3.3 to 4.2 | 0.6 to 0.7 | 0.2 to 0.3 | 68 to 78 |
| 32 rectangular | 3.9 to 5.8 | 0.8 to 1.1 | 0.1 to 0.6 | 68 to 78 |
| 44 square | 4.3 to 6.1 | 1.1 to 1.4 | 0.2 to 0.8 | 68 to 78 |
| 52 square | 6.1 to 8.4 | 1.0 to 1.3 | 0.2 to 0.8 | 68 to 78 |
| 68 square | 5.3 to 8.9 | 1.4 to 2.0 | 0.2 to 1.0 | 68 to 78 |
| 84 square | 8.4 to 10.8 | 1.8 to 2.7 | 0.25 to 1.2 | 68 to 78 |

### 1.2.3 Cost

The cost of a packaged electronic part is determined by several factors: die, package, volume, size, assembly cost and yield, screening, pre-burn-in and its yield, burn-in, final testing and its yield, and the specified qualification-required tests. Table 1.6 shows an example of items that affect the total package cost. Table 1.7 presents the relative cost for various microcircuits packaging options. Tables 1.8 to 1.10 reports the relationship of the cost to its screening level and consequent availability from four semiconductor suppliers. Table 1.11 compares

**Table 1.6 Cost drivers in packaging; the die is approximately 50,000 square mils and is assembled in a 64-pin, $5 package; total costs and typical yields are shown for two categories: commercial and MIL-STD-833 Class B screened**

| Package element/process | Cost | |
|---|---|---|
| | Commercial | Screened |
| Die | 2.00 | 2.10 |
| Package | 5.00 | 5.00 |
| Assembly | 0.44 | 2.22 |
| Pre-cap visual | 0.00 | 0.15 |
| Pre-cap yield (%) | 84.00 | 71.40 |
| Assembled | 8.86 | 13.27 |
| Screening | 0.00 | 0.72 |
| Pre burn-in test | 0.00 | 0.23 |
| Pre burn-in yield (%) | 100.00 | 73.63 |
| Burn-in | 0.00 | 0.64 |
| Final test | 0.23 | 0.68 |
| Final test yield (%) | 83.84 | 88.47 |
| Qualification | 0.00 | 0.00 |
| Factory | 10.84 | 23.32 |

**Table 1.7 Relative cost of packaging options**

| Package type | Relative cost |
|---|---|
| Small-outline integrated circuit (SOIC) | 0.9 |
| Plastic dual in-line package (PDIP) | 1.0 |
| Plastic leadless chip carrier (PLCC) | 1.2 |
| Ceramic dual in-line package (CERDIP) | 4.0 |
| Plastic pin-grid array (PPGA) | 10.0 |
| Plastic quad flatpack (PQFP) | 50.0 |
| Pin-grid array (PGA) | 130.0 |
| Leadless chip carrier (LCC) | 150.0 |

**Table 1.8 Typical price, screening level, and availability [Westinghouse 1991]**

| | Part number | Price | Discriminator | Lead times (weeks) |
|---|---|---|---|---|
| Memory manufacturer: Integrated Device Technology | | | | |
| 28nsec LSI error detection and correction | IDT7140LA45FB | $152.00 | configuration control | 10 to 12 |
| | IDT7140LA45FB | $129.56 | 100% temperature cycle, burn-in, ambient, centrifuge, sample hot/cold AC/DC | 8 to 10 |
| | IDT7140LA45FB | $129.56 | same as above, except sampling at temperatures to internal specification | 8 to 10 |
| | IDT7140LA45FM | $88.70 | -55°C + 125°C usually 100% ambient, with guaranteed limits | 6 to 8 |
| | IDT7140LA45F | $49.00 | 0°C + 70°C usually 100% ambient | 6 |
| | IDT7140LA45J | $22.50 | 0°C + 70°C usually 100% ambient | 4 |

**Table 1.8 (cont.)**

| | Part number | Price | Discriminator | Lead times (weeks) |
|---|---|---|---|---|
| Logic manufacturer: Advanced Micro Devices | | | | |
| 45-nsec LSI dual port RAM | 1A20752H04 | $107.20 | configuration control | 8 to 10 |
| | 5962-8853304UX | $87.10 | 100% temperature cycle, burn-in, ambient, centrifuge, sample hot/cold AC/DC | 4 |
| | Am29C660B/BZC | $87.10 | same as above, except sampling at temperatures to internal specification | 4 |
| | Am29C660BGC | $46.55 | 0°C + 70°C usually 100% ambient | 4 |
| | Am29C660BJC | $34.47 | 0°C + 70°C usually 100% ambient | 2 |

**Table 1.9 Typical price, screening level, and availability**

| | Part number | Price | Discriminator | Lead times (weeks) |
|---|---|---|---|---|
| Analog manufacturer: Texas Instruments | | | | |
| SSI dual-line driver | 583R932H02 | $4.62 | configuration control | 16 |
| | 5962-8754701CA | $4.13 | 100% temperature cycle, burn-in, ambient, centrifuge, sample hot/cold AC/DC | stock to 10 |
| | SNJ55110AJ | $4.13 | same as above, except sampling at temp. to internal specification | stock to 10 |
| | SN55110AJ | $3.38 | -55°C + 125°C usually 100% ambient with guaranteed limits | stock to 8 |
| | SN75110AJ | $1.00 | 0°C + 70°C usually 100% ambient | stock to 8 |
| | SN75110AN | $.65 | 0°C + 70°C usually 100% ambient | 6 |

**Table 1.10 Typical price, screening level, and availability**

| | Part number | Price | Discriminator | Lead times (weeks) |
|---|---|---|---|---|
| Logic Array manufacturer: Cypress Semiconductor | | | | |
| 24-pin programmable logic array | 587R262H02 | $37.45 | configuration control | 8 |
| | 5962-8867002LX | $17.85 | 100% temperature cycle, burn-in, ambient, centrifuge, sample hot/cold AC/DC | 4 |
| | PALC22V10-30DMB | $17.85 | same as above, except sampling at temperatures to internal specification | 4 |
| | PALC22V10-30DM | $15.17 | -55°C + 125°C usually 100% ambient, with guaranteed limits | 2 to 4 |
| | PALC22V10-30DC | $12.50 | 0°C + 70°C usually 100% ambient | 2 to 4 |
| | PALC22V10-30PC | $6.25 | 0°C + 70°C usually 100% ambient | 1 to 2 |

the ceramic-to-plastic package cost for ceramic (hermetic) dual in-line packages, leadless chip carriers, and pin-grid arrays. Because more than 97% of the integrated circuit market is plastic-packaged, cost has been lowered by high yield and high quality automated volume manufacturing. Hermetic packages usually have a higher material cost and are fabricated with more labor-intensive manual processes. For example, Thomson-CSF reported a 45% purchase cost reduction for each of twelve printed wiring boards in a manpack transceiver application implemented with plastic, rather than ceramic, components [Brizoux et al. 1990].

A hermetically packaged integrated circuit may also cost up to ten times more than a plastic-packaged integrated circuit because of the rigorous testing and screening required for the low-volume hermetic parts [RAC 1985]. When both types were screened to customer requirements, ELDEC [Condra and Pecht 1990] estimated that purchased components for plastic packaging of integrated circuits cost 12% less than their hermetic counterparts, primarily due to the economics of high-volume production.

The cost benefits of PEMs decrease with higher integration levels and pin counts, because of the high price of the die in relation to the total cost of the packaged device. While these cost benefits may not be realized for complex monolithic very-large-scale integrated circuits, cost advantages may accrue for complex package styles, such as multichip modules, because of the ease of assembly. Indeed, the trend toward future multichip modules in laminates (MCM-L) packaged in form factors such as PQFPs or ball-grid arrays (BGAs) will make plastic packages even more popular. In the MCM-L, several dies and passive components can be combined on a printed circuit board substrate and integrated into a leadframe to enhance part functionality. This approach shortens the time to market (by using readily available dies to eliminate the need for die integration), increases the process yield (there are no large dies and no mixture of such different technologies as CMOS and bipolar), and lowers package cost (compared with a ceramic equivalent). MCM-L products have already been produced for some time and the cost savings achieved will impact the proliferation of plastic packages [Nachnani et al. 1993].

### 1.2.4 Reliability

The reliability of plastic-encapsulated microelectronics has increased tremendously since the 1970s, due largely to improved encapsulating materials, die passivation, and manufacturing processes. In particular, modern encapsulating materials have low ionic impurities, good adhesion to other packaging materials, a high glass transition temperature, high thermal conductivity, and coefficients of thermal expansion matched to the leadframe. Advances in passivation include better die adhesion, fewer pinholes or cracks, low ionic impurity, low moisture absorption, and thermal properties well matched to the substrate. The failure rate of plastic packages has decreased from about 100 per million device hours in 1978 to about 0.05 per million device hours in 1990 [Watson 1991]. Figure 1.6 summarizes typical published improvements in PEM reliability since 1976 [Condra and Pecht 1991].

Figure 1.7 presents comparative failure-rate data for PEMs and hermetically packaged devices from first-year warranty information on commercial equipment operating primarily in ground-based applications (office, laboratory, and transportable equipment) from 1978 to 1990; the rates are for the same part (or part function) over time [Priore and Farrell 1992]. As shown in Figure 1.6, during this period, both types of packaged devices improved by more than one order of magnitude in early-life failure rate. For PEMs, the early 1990s failure rate was between 0.3 to 3.0 failures per $10^6$ device-hours, with less variability between encapsulant materials and their vendors. This very closely correlates with the figures for hermetic parts [Texas Instruments 1992].

Condra and Pecht [1991] tested mature custom bipolar integrated circuits in commercial plastic and hermetic ceramic (military parts) dual in-line packages (DIPs) on twelve circuit-card assemblies. They subjected the packages to 1000 temperature cycles from -55 to +85°C to compare the functional reliability of the two types of packages. No differences were observed in any of the twenty-six measured parametric values. They then added these parts to untested groups of fifty of the same devices in another set of circuit-card assemblies, with an

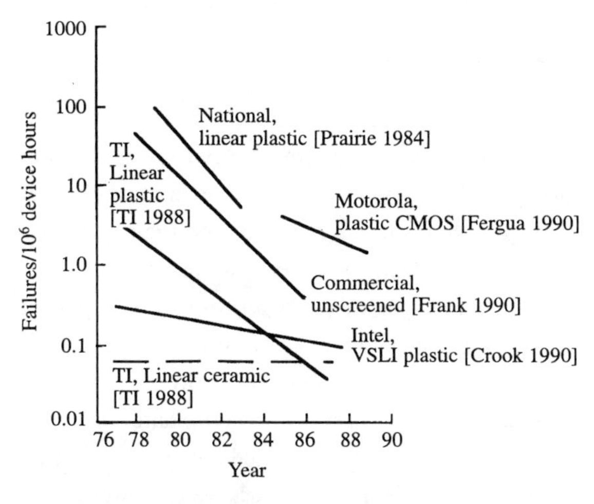


**Figure 1.6** *Microcircuit reliability improvement trends [Condra 1991]*

**Table 1.11 Ratio of ceramic to plastic package cost for three packaging styles, with die costs neglected [Dicken 1989]**

| Lead count | Cost of ceramic package/ plastic package | | |
|---|---|---|---|
| | Dual in-line package | Chip carrier | Pin-grid array |
| 8 | 4.0 | | |
| 16 | 6.7 | | |
| 18 | 6.3 | 14.3 | |
| 20 | 6.0 | 17.1 | |
| 24 | 8.3 | 18.8 | |
| 28 | 7.5 | 17.0 | |
| 40 | 6.9 | 13.5 | |
| 48 | | 13.5 | |
| 68 | | 13.3 | 1.6 |
| 84 | | 11.0 | 1.9 |
| 120 | | | 2.5 |
| 168 | | | 3.5 |
| 208 | | | 2.6 |
| 256 | | | 2.5 |
| Geometric mean for row | 6.40 | 14.6 | 2.40 |

older, discrete version of the card as a control. These were subjected to 1000 hr. of 85°C/85%RH, with 28 V of intermittent bias (30 min on, 30 min off). Previously thermally cycled parts (both ceramic and plastic) could only be tested up to 650 hr. before failure occurred elsewhere on the cards. Among previously untested packages, neither PEMs nor ceramic packages failed. Conservative lifetime estimates for both types of packages in avionic applications were well over thirteen years, even under combined testing.

A study by Rockwell International, Collins Group [Grigg 1986], compared plastic dual in-line packages and plastic surface mounted devices to ceramic dual in-line packages (CERDIPs) during extended temperature cycling of -40°C to +85°C for 3532 hr. (4-hr. cycles). The parts were mounted on circuit boards, biased with 5 V during testing, and subjected to 98% relative humidity. These

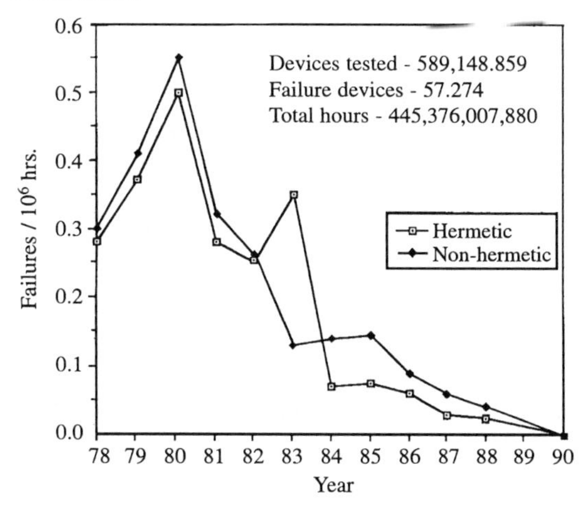


**Figure 1.7** *Comparative failure rate as a function of year [Priore and Farrell 1992]*

conditions were designed to simulate a worst-case avionics environment. The failure rate for the plastic parts was as low as 0.0016% per 1,000 hrs., while the failure rate for the hermetic parts was 0.0061% per 1000 hr, both at 60% confidence. The large difference in failure rates, shown in Table 1.12, is due to the loss of hermeticity of glass seals. The failure mechanism for both plastic and hermetic parts was metallization damage due to moisture. The sample size of the various lots was 2920 plastic parts and 1200 ceramic parts from five different vendors (Texas Instruments, Motorola, National, Signetics, and Fairchild); there were 174.6 million equivalent field device-hours.

Lidback [1987], of Motorola's Semiconductor Product Sector, studied 133,747 PEMs and 46,473 hermetic packages (obtained from Motorola's Semiconductor Product Sector) that underwent 1000 temperature cycles from -65 to + 150°C. The failure rate was 0.083% for the PEMs, and 0.099% for the hermetic

**Table 1.12 Thermomechanical reliability**

| Study | Package | Number failed/ number tested | Failure rate | Comments |
|---|---|---|---|---|
| Grigg (Rockwell Collins) | plastic | 1/2920 | 0.0016%/1K hr | 60% confidence |
| | hermetic | 2/1200 | 0.0061%/1K hr | 60% confidence |
| Villalobos (Motorola TED) | plastic | 23/9177 | 0.44%/1K cycles | % defective |
| | hermetic | 7/1844 | 0.38%/1K cycles | % defective |
| Lidback (Motorola SPS) | plastic | 11/133,747 | 0.083% | % defective |
| | hermetic | 46/46,473 | 0.099% | % defective |

packages. Villalobos [1989] ran 1000 cycles at -65 to +150°C using a smaller sample size of 9240 plastic packages and 1848 ceramic packages (also obtained from Motorola). The ceramic failure rate was 0.38%, while the plastic failure rate was 0.44%. Both of these studies indicated similar failure mechanisms for plastic and hermetic packages under worst-case conditions.

Moisture tests are not considered applicable to hermetic packages, so a comparison with plastic is not usually made. However, highly accelerated tests, including autoclave storage tests run at 103.3 Pa, 121°C, for 24 hrs, showed 0.022% failures; temperature-humidity bias tests run at 85°C/85% relative humidity for 1000 hr showed 0.024% failures in a sample size of 18,255 [Lidback 1987]. In terms of historical improvement, Kurtz [1990] reported that temperature, humidity, bias testing (85°C/85%RH) in 1974 produced 25% cumulative failures after 1000 hr of exposure, compared with 0.1% in 1990.

Probably the best endorsement for PEMs is from automotive manufacturers, which consume plastic-integrated circuits at a rate of over 2.7 million a day [Watson 1991], using some of the commercial industries' most stringent requirements. Motorola's Automotive Industrial Electronics Group (AIEG) buys plastic-encapsulated semiconductors from suppliers who have qualified their product to AIEG's internal qualification standards, which are designed to simulate worst-case conditions in automobiles. For example, automotive qualification includes sample temperature cycling for 1000 cycles, thermal shock (liquid-to-liquid) for 500 cycles, 85°C and 85%RH testing for 1000 hr, life testing for 1000 hr, high-temperature reverse bias for 1000 hr, intermittent operational life testing for 20,000 cycles, and autoclave (live steam) testing for 96 hr. The number of rejects allowed for all these tests is zero. AIEG indicates that most vendors pass these tests without problems, indicating a broad, industry-wide ability to meet or exceed harsh automotive standards [Straub 1990].

In general, PEMs lend themselves well to automatic assembly techniques, which eliminate manual handling and operator error, resulting in high yields and low assembly costs. On the other hand, even in high-volume lines for hermetic packages, automated-assembly pick-and-place machines have been known to crack hermetic seals or chip ceramic packages.

### 1.2.5 Availability

Plastic-encapsulated microcircuits are much more readily available than hermetic devices, mainly because market forces (cost and volume) encourage most designs to be developed first as plastic-encapsulated. In 1993, over 97% of all integrated circuits were packaged in plastic. It is estimated by suppliers that, at any given time, 30% more part functions are available in plastic than in ceramic [Condra and Pecht 1991].

Plastic devices are assembled and packaged on continuous production lines, as opposed to the on-demand production of hermetic parts. Thus, acquisition lead times for plastic packages are significantly shorter. The problems associated with restarting a hermetic line are absent in continuous plastic production lines.

Hermetic packages are developed only when there are perceived high-performance requirements and sufficient market interest. Thus, some parts are simply not available from major manufacturers in hermetic form. Furthermore, the U.S. military and government, the major purchasers of hermetic parts, have become a relatively small portion of the total electronics market (~5% in 1990), though they accounted for nearly 80% of the total market in the 1960s. With package technology moving to surface mount, development of ceramic packages has lagged further in the microelectronic market, making adaptation of plastic-packaged integrated circuits to government and military applications more critical. With global competition, industrial research in materials and manufacturing processes will continue to focus on PEMs.

## 1.3 SUMMARY

Advantages of plastic packages over their ceramic counterparts include smaller form factors, lighter weight, good performance, lower cost, and parts availability. The reliability of plastic packages has also increased substantially since the early days of plastics, with improvements in encapsulants, die passivation, metallization technology, and assembly automation. Plastic-encapsulated microcircuits will continue to account for the vast share of the integrated circuits market in coming years, but hermetic packages, with their special characteristics of ceramics, will continue to have their unique niche applications.

## REFERENCES

Brizoux, M. et al. Plastic-integrated Circuits for Military Equipment Cost Reduction Challenge and Feasibility Demonstration. *40th Electronic Components and Technology Conference* (1990) 918-924.

Condra, L., and Pecht, M. Options for Commercial Microcircuits in Avionic Products. *Defense Electronics* (July 1991) 43-46.

Cook, J. P., and Servais, G. E. Corrosion Failures in Semiconductor Devices and Electronic Systems. *Proceedings of 2nd Automotive Corrosion Prevention Conference,* Dearborn, Michigan (1983) 187-197.

Dicken, H. Physics of Semiconductor Failures, 3rd edition, DM Data Inc. Scottsdale, AZ (1988).

Electronic Materials Handbook, Packaging, ASM International, 1 (1989) 203-212.

Flood, J. T. Reliability Aspects of Plastic-encapsulated Integrated Circuits. *Proceedings of International Reliability Physics Symposium,* Institution of Electrical and Electronic Engineers (1972).

Grigg, K. C. Plastic Versus Ceramic Integrated Circuits Reliability Study. *Report No. WP86-2020*, Rockwell International, Collins Groups (July 1986).

Intel Corporation, *Packaging, A Handbook*, Intel # 240800 (1993) 4-2 and 4-3.

Kurtz, B. E. Implications of Non-hermetic Chip Packaging for Military Electronics. *1990 International Packaging Conference* (September 1990) 848-857.

Lidback, C. A. Plastic-encapsulated Products vs. Hermetically Sealed Products. *Summary Report, Motorola Inc.*, Government Electronics Group (January 1987).

Motorola ICD, Reliability Engineering Report. *Plastic Package Reliability Report*, Bipolar Integrated Circuits, 1974. I/C Reliability Engineering Dept., Motorola (1974).

Nachnani, M., Nguyen, L., Bayan, J., and Takiar, H. A Low-Cost Multichip (MCM-L) Packaging Solution, *Proceedings of International Electron Manufacturing Technology Symposium* (1993) 464-468.

Priore, M., and Farrell, J. Plastic Microcircuit Packages: A Technology Review. *Rept. # CRTA-PEM*, Reliability Analysis Center, Rome, NY (March 1992).

Sinnadurai, N. Advances in Microelectronics Packaging and Interconnection Technologies-Toward a New Hybrid Microelectronics. *Microelectronics Journal*, 16, 5 (1985).

Straub, R. J. Automotive Electronic Integrated Circuits Reliability. *Proceedings of Custom Integrated Circuit Conference* (1990) 92-94.

Texas Instruments, *Texas Instruments Military Plastic Packaging.* Preliminary Handbook (1992).

Villalobos, L. R. Reliability of Plastic-Integrated Circuits in Military Applications. *ESP Report No. PV 620-0530-1*, Motorola Government Electronics Group, Tactical Electronics Division (January 1989).

Watson, G. F. Plastic-Packaged Integrated Circuits in Military Equipment. *IEEE Spectrum* (February 1991) 46-48.

Westinghouse Report, *Policy Directive on ARSR-4 Microcircuit Review* (August 1991).

# 2

# PLASTIC PACKAGING MATERIALS

This chapter discusses the materials used in plastic-encapsulated microcircuits (PEM), their properties and characteristics, their interactions with other materials used in PEM construction, and their use in various applications. The chapter is organized in the same sequence as the materials are used in PEM manufacture. Table 2.1 presents the plastic packaging materials flow for PEM manufacture and the associated properties needed for a high-quality, reliable product.

## 2.1 DIE PASSIVATION

Die passivation is part of the device manufacturing process, rather than of packaging. The purpose of die passivation is to seal the active circuit elements from ambient moisture, ionic contamination, mechanical damage due to handling, and in some cases radiation and electrostatic discharge. Thus, because the die passivation protects the semiconductor circuits and is in direct contact with the

**Table 2.1 Plastic packaging materials flow and materials selection criteria**

| Sequence of material use in PEMs | Material selection criteria |
|---|---|
| Die | • die passivation integrity<br>• compressive residual stresses on the passivation<br>• freedom from edge cracks and spalls<br>• surface finish of die back |
| Leadframe | • coefficient of thermal expansion<br>• thermal conductivity<br>• mechanical strength<br>• adhesion to encapsulant and die-attach material<br>• oxidation and corrosion resistance<br>• stampability, formability, etchability<br>• solderability<br>• design configuration (grooves, anchors, etc.) |
| Leadframe finish | • lead conductivity<br>• adhesion to encapsulant and die attach<br>• leadframe corrosion/dendritic growth<br>• leadframe oxidation resistance<br>• lead solderability<br>• availability of defect-free application process<br>• lead platability |

**Table 2.1 (cont.)**

| Sequence of material use in PEMs | Material selection criteria |
|---|---|
| Die attach | • thermal conductivity<br>• voiding<br>• thermomechanical stress on mounted die<br>• shear strength of die attach interfaces<br>• high-temperature stress compliance<br>• outgassing<br>• material ionic purity<br>• moisture solubility |
| Interconnection | • ultimate tensile strength<br>• flexural strength<br>• bond pull strength<br>• maximum die pad bonding temperature<br>• resistance to intermetallic formation<br>• surface contamination |
| Encapsulating compound | • moldability, cure speed, melt viscosity, flash and bleed, and resistance to voids<br>• hot hardness<br>• mold strain resistance<br>• hydrolyzable ionic impurity level<br>• molding defects<br>• moisture resistance<br>• adhesion to all package elements<br>• curing stress<br>• coefficient of thermal expansion match with leadframe<br>• mechanical strength<br>• glass transition temperature<br>• thermal conductivity<br>• thermal stability |

package molding compounds, the materials, processes, dimensions (especially thickness), and microstructure of the die passivant strongly influence the reliability of the PEM.

About 10,000 Å of 2% to 5% phosphorus-doped, low-temperature (~325°C), plasma-enhanced, chemically vapor-deposited (PECVD or CVD) silicon dioxide is widely used as the die passivation layer. The thickness is a trade-off between the moisture permeation rate and the potential for stress-induced passivation cracking. Although phosphorus is an excellent getter of mobile ionic contaminants in silicon dioxide insulating layers, its concentration must be kept low to prevent the formation of aluminum-corroding phosphoric acid, which can be created and leached out with permeated moisture.

Other passivation integrity concerns include porosity, stress cracks, and nonconformal deposition-induced thickness nonuniformity. Pinholes in chip passivation, a major cause of aluminum interconnect corrosion prior to the late 1980s, have been brought down to negligible levels. Finally, the passivation deposition temperature must be kept well below 400°C to reduce thermal-stress-related failures (hillocks, voiding, thermomigration, and grain growth) of the aluminum metallization.

A more robust chip passivant than silicon dioxide is silicon nitride. It is denser, and therefore a better diffusion barrier to both ionic surface contaminants and moisture. Its disadvantages include susceptibility to stress cracking and proton charging, leading to device surface depletion or accumulation. Both of these characteristics can be controlled by low-temperature chemical-vapor deposition. A thin silicon nitride overcoat on the phosphosilicate glass has also been used as a multilayer passivation. Spin-on glass planarized samples have exhibited more than three times reduction in metallization corrosion-induced EPROM array failure [Gaeta and Wu 1989]. However, with any type of passivation, the aluminum bond pads are opened after passivation for chip-to-package interconnections.

Inorganic coatings can also be deposited directly on the die surface as a designed-in reliability feature. One approach involves a planarizing underlayer of silicon dioxide (~8000 Å), covered by a PECVD layer of silicon carbide (~3000 Å) [Snow and Chandra 1990, Chandra 1991]. The latter is a relatively flexible and impermeable amorphous film. This die-coating process will enhance the reliability of the device under adverse environmental conditions, but the disadvantage of higher cost has discouraged widespread support for this approach.

## 2.2 LEADFRAME

A leadframe consists of a die mounting paddle and leadfingers. The die paddle primarily serves to mechanically support the die during package manufacture. The leadfingers connect the die to the circuitry external to the package. One end of each leadfinger is typically connected to a bond pad on the die by wirebonds or tape automated bonds. The other end of each leadfinger is the lead, which is mechanically and electrically connected to a substrate or circuit board.

The leadframe is constructed from sheet metal by stamping or etching, often followed by a finish such as plating (see Chapter 3 for manufacturing details).

### 2.2.1 Leadframe Materials and Their Properties

The leadframe geometry and material composition will affect the ease of the manufacture, quality, and performance of the package over time. Suitability of a material for use in leadframes depends on cost and the ability of the material to meet the demands generated during leadframe fabrication, package assembly, circuit board mounting, and device performance.

Table 2.2 outlines some leadframe materials, their compositions, and their properties. Common materials for plastic package leadframes include copper alloys and Alloy 42. Critical properties include adherence to the molding compound, a coefficient of thermal expansion as close as possible to both the die and the molding compound, high strength and formability, and high electrical and thermal conductivity.

The leadframe material must adhere to the encapsulant; deadhesion between the leadframe and the encapsulant establishes pathways for direct moisture and contaminant entry into the package. The epoxy molding compound chemically adheres to Alloy 42 better than to copper alloys, probably because Alloy 42 does not form a passivating oxide film as easily as copper alloys do. However, deadhesion is also a function of thermomechanical stresses induced during thermal cycles at the leadframe encapsulant interface by mismatches in the coefficients of thermal expansion (CTEs) of the leadframe material and the encapsulant. For example, common copper alloy leadframe materials have CTEs in the range of 17 to 18 ppm/°C, Alloy 42 has a CTE of about 4.5 ppm/°C, and those of most epoxy encapsulants are from 16 to 20 ppm/°C. Thus, copper alloy leadframes reduce thermomechanical stress delamination at the leadframe encapsulant interface better than Alloy 42 leadframes.

The leadframe's coefficient of thermal expansion should also match with that of the die to reduce thermomechanical stresses at the die leadframe interface.

**Table 2.2 Properties of common leadframe materials [Minges et al. 1989, Zarlingo and Scott 1981, Ptacek and Schuder 1985]**

| Alloy group | Designation and typical composition | Electrical conductivity (% of international annealed copper standard) | Thermal properties | | Solderability (scale of 1 to 4) with rosin flux |
|---|---|---|---|---|---|
| | | | Conductivity at 20°C (W/m°C) | Coefficient of thermal expansion ($10^{-6}$/°C) | |
| Cu-Zr | C-151 (0.1% Zr) | 90 to 95 | 380 | 17.6 | 1 |
| Cu-Fe | C-194 (2.35% Fe; 0.03% P; 0.12% Zn) | 60 to 65 | 260 | 17.4 | 1 to 2 |
| | C-195 (1.5% Fe; 0.8% Co; 0.6% Sn; 0.1% P) | 50 | 200 | - | 1 to 2 |
| Cu-Mg | C-155 (Cu+Ag 99.8%; Mg 0.11%; P 0.06%) | 86 | 344 | - | 1 |
| Fe-Ni | ASTM F-30 or Alloy 42 (58Fe/42Ni) | 2.5 | 12 | 4.0 to 4.7 | 3 to 4 |

The CTE of common silicon dies is 2.3 to 2.6 ppm/°C, which is closer to that of Alloy 42. The stresses in the die due to the thermal expansion mismatch can, however, be minimized by the use of a compliant organic die attach, which transfers only part of the total strain to the die.

After the molding operation, the leadframe is trimmed and the leads are formed into such shapes as DIP leads, gull-wing leads, or J-leads. Forming requires that the lead material be formable; however, formability has to be traded off against strength (typically characterized by the material's yield strength), which prevents damage to the leads during handling and component assembly on a substrate or circuit board. In through-hole mounting, the leads should possess enough strength so that stresses in the leads due to small misalignments during automatic lead insertion do not cause them to buckle.

Typically, alloys (rather than pure metals) are used for leadframes, because the dispersion of the alloying metals in the base metal increases the yield strength. The leadframe material is further strengthened by cold working the material, typically in a cold rolling mill. For a given material, the change in strength of the material depends on the percentage reduction during rolling — that is, the percentage reduction in thickness of the sheet metal with respect to the original thickness. This is sometimes also referred to as the temper of the material; the higher the temper, the greater the cold working and the strength. Tempers are correlated to different tensile strength ranges that result from various percentage reduction ranges. For example, half-hard ZHC copper has a tensile strength of 296.5 to 351.6 MPa (43 to 51 ksi), corresponding to a rolling reduction of 20%, while extrahard ZHC copper has an ultimate tensile strength in the range of 406.8 to 448.2 MPa (59 to 65 ksi), corresponding to a percentage reduction of approximately 60% [Zarlingo and Scott 1981].

Cold working the metal by rolling increases the strength but decreases the ductility of the material, which lessens the ability of the material to undergo further deformation without cracking. This is not favorable for the lead-forming operation, which involves bending leads over small radii (also a cold-working process). Thus, strength and formability go hand in hand. Figure 2.1 plots the minimum radius for right-angle bends for Olin C-194 and C-195 copper alloy leadframe material of 0.25 mm (0.010 in.) gauge for various tempers [Zarlingo and Scott 1981]. Note that a bend with a 0.38-mm (0.015-in.) radius can be made with alloy tempers of yield strengths of 480 MPa (70 ksi) or below. Figure 2.1 describes bends with axes perpendicular to the direction of rolling; using any other bend axis could result in some cracking at the bend.

While Alloy 42 has the maximum yield strength and superior strength and formability, it compares poorly with copper alloys in terms of thermal and

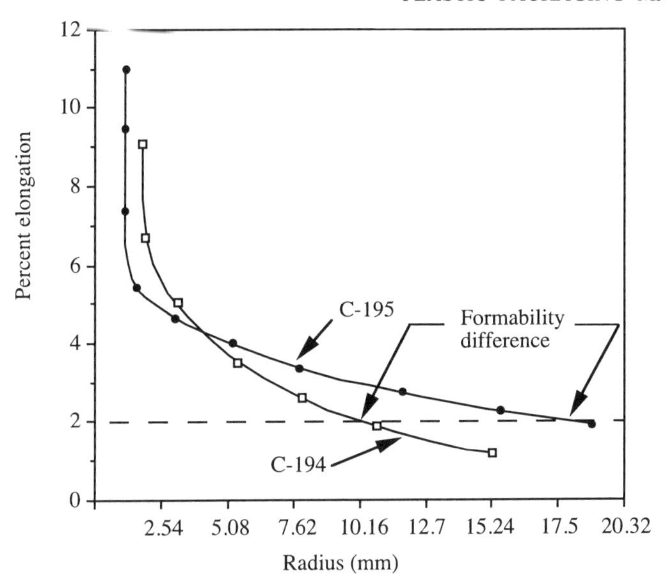


**Figure 2.1** *Plots of the minimum radius for right-angle bends for Olin C-194 copper alloy leadframe material of 0.25 mm gauge for various tempers [Zarlingo and Scott 1981]*

electrical conductivity (Table 2.2). Alloy 42, with its high nickel content, is also significantly more expensive than copper alloys. Some copper alloys — for example, C-195 — provide yield strength comparable to Alloy 42, and thus meet strength requirements while still providing the other advantages of copper alloys.

Formability along both the axes is an important factor for leadframes, such as quad-flatpacks, that support leads on all four sides. Unlike copper alloys, Alloy 42 bends almost equally well along the axes perpendicular to and parallel to the direction of rolling [Shumay 1987]; that is, the minimum bend radius for copper alloys has to be larger when the bend axis is parallel to the direction of rolling than when it is perpendicular.

Cold working during the leadframe rolling process creates elongated grains, making slipping between grains difficult and increasing yield strength. However, because the leads will be subjected to several high-temperature processes during package formation and assembly, recrystallization of the grains (annealing) will occur, accompanied by softening, that is, reduction in yield strength and increase in ductility. Furthermore, for a given alloy at a given elevated temperature, the greater the cold working (i.e., the higher the temper), the greater the tendency is to soften. If loss of strength is a concern, the leadframe should not be exposed to temperatures in excess of the annealing temperature for the alloy.

The annealing temperature is a function of alloy composition. Zarlingo and Scott [1981] exposed different leadframe alloys to elevated temperatures in the range of 50°C to 600°C for one hour to study the resulting reduction in yield strength. He found that pure copper had the least resistance to softening and that annealing occurs below 200°C. The addition of alloying elements increased the annealing temperature. For example, alloys that use iron and cobalt with copper do not soften significantly below 400°C. The maximum resistance to softening is achieved by the addition of nickel—in, for example, C-725, which alloys nickel and tin with copper.

For most applications, the highest temperature experienced by the leadframe corresponds to the temperature of encapsulation and is typically limited to 200°C. Zarlingo and Scott [1981] have presented the change in yield strength of leadframe materials as a function of the exposure time at 350°C. Figure 2.2 illustrates that the yield strengths of both Alloy 42 and C-195 do not decrease appreciably for exposure times of up to 100 min. Copper Alloys C-151, C-155, and C-194 showed approximately 25% yield strength, for exposure periods of 100 min. Commercially pure copper or C-110, which is 99.9% copper, softens rapidly, making it unsuitable for leadframe applications. Thus, appropriate leadframe material selection depends on the specific application.

The leadframe serves to electrically connect the die to the conductors on the printed circuit board, so leadframe materials should have low electrical resistivities. In this respect, copper alloys are significantly superior to Alloy 42. The electrical conductivity of Alloy 42 is only 2.5% of the international annealed copper standard, compared to 50 to 95% for copper alloys.

The leadframe can significantly affect heat transfer from the package. For all common leadframe copper alloys, the thermal conductivity is an order of magnitude better than Alloy 42. The thermal conductivity of Alloy 42 is only 12 W/m°C, compared to 344 W/m°C for C-155, 260 for C-194, and 200 for C-195. Thus, copper alloys are preferable with respect to both thermal and electrical requirements.

Leads can be bent due to misalignment during automatic insertion, especially with through-hole mounted packages. If the leads are straightened, they must be able to withstand any lead bend fatigue. Table 2.3 [Zarlingo and Scott 1981] presents the lead bend fatigue of common leadframe materials. The lead bend fatigue is affected by the ratio of lead width to thickness. The data presented in Table 2.3 were developed from samples with the same lead width-to-thickness ratio: 0.25 mm (10 mils) thick and 3.48 mm (137 mils) wide. Bends were made both parallel to the direction of rolling (longitudinal) and perpendicular to it (transverse).

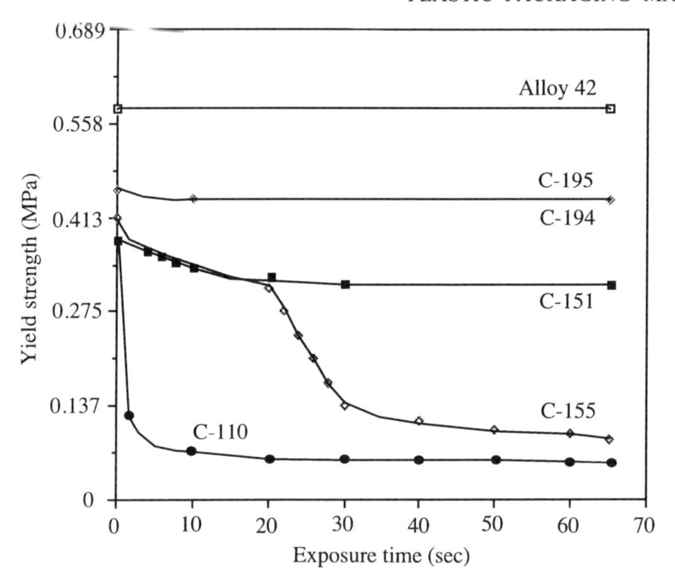


**Figure 2.2** *Yield strength of both Alloy 42 and C-195 not decreasing for exposure times of up to 100 min [Zarlingo and Scott 1981]*

Some companies require that the leads withstand three cycles of bending through 90° followed by straightening.

### 2.2.2 Leadframe Finish Materials

Leadframe finishing aims to minimize corrosion and enhance solderability. The properties to be considered in selecting the leadframe finishing material(s) include platability of the lead base metal, solderability, corrosion resistance, and

**Table 2.3 Comparison of lead bend fatigue of leadframe materials for 0.25-mm (0.010-in.) gauge and 3.48-mm (0.137-in) wide strips [Zarlingo and Scott 1981]**

| Leadframe material | Tensile strength (MPa) | Number of 90° bend cycles to failure | |
|---|---|---|---|
| | | Longitudinal | Transverse |
| C-151 | 420 | 15 to 20 | 14 to 16 |
| C-155 | 420 | 15 to 20 | 14 to 16 |
| C-194 | 420 | 15 to 20 | 14 to 16 |
| C-195 | 517 | 17 | 16 |
| Alloy 42 | 641 | 111 | 102 |

ability to withstand lead-forming stresses without cracking. To effectively prevent corrosion of the base leadframe material, any plating should be non-porous, adhere well to the base metal, and be inherently less susceptible to corrosion than the base metal.

Lead-plating material options are relatively independent of the lead base material for the common leadframe materials. Lead plating can involve one or two layers of plating. When a two-layer plating process is used, an undercoat is plated directly on the lead base material, and a primary finish is applied over the undercoat. The undercoat acts as a barrier layer, preventing diffusion of the lead base material into and through the primary lead finish. Such diffusion can result in the formation of intermetallics that degrade plating adhesion strength and solderability. The diffusion process can expose the lead base material to the external environment and cause corrosion. The undercoat also provides a solderable surface. The primary finish prevents the oxidation of the undercoat, preserving its solderability. The quality of the plating depends substantially on leadframe surface roughness and cleanliness. For more information on solderability and testing for solderability, see Chapter 3 and Wolverton [1991], Tleel [1987], and Davy [1990].

Solderability, the ability of a material to form a strong bond with solder, requires that the plating material be wetted by and form strong intermetallics with solder. Tin and tin-lead (about 60 to 63% tin solder) are usually adequate for preserving solderability for storage times of less than two years. Pure tin is not recommended because of its tendency toward whisker growth. Tin, however, enables higher temperature operation because the melting point of tin is 232°C, compared to 180°C for solder [Intel 1993]. Whisker growth may be prevented by using a tin-plating thickness in excess of 10 μm (400 microinches), hot-dipped tin, or electroplated tin followed by reflow. For more information on whisker growth, see NASA [1992] and Britton [1974].

Solder finishes cannot withstand typical molding temperatures [Foppen 1993], so the finish should be applied after encapsulation. To reduce contamination, it is preferable to apply the finish by hot dipping rather than by electroless or electrolytic plating, because of the corrosive chemicals in plating solutions. Some manufacturers use plated leadframes prior to encapsulation, such as a palladium plating over a nickel undercoat for fine-pitch applications.

Although the leadframe is not generally preplated over its entire surface prior to encapsulation, spot plating on the leadframe with silver to enhance the bondability of wirebonds is common [Slay 1993, Abbot et al. 1991]. However, plated silver has a tendency toward dendrite formation. Furthermore, the use of solder coating restricts the choice of the coating process to hot dipping, unless the ionic contamination introduced as a result of plating solutions is acceptable. Controlling the coating thickness in hot dipping is rather demanding. With fine-pitch surface-mount leads, solder bridging may occur between adjacent leads. Hot-dipped plating also tends to be thin on the lead corners and thicker at the center, which can result in lead noncoplanarity (deviation of a lead from the lead seating plane).

To circumvent the problems associated with silver and solder coatings, palladium plating can be used over nickel. Palladium is more inert than gold at about a quarter of the price [Koford et al. 1981]. Palladium plating with a nickel undercoat can also withstand molding process temperatures and can be applied before die assembly on the leadframe. Palladium provides a suitable surface for wirebonding, eliminating the need for silver spot plating, with its associated spot-plating tooling cost, and the risk of silver migration [Abbot et al. 1991]. When components are soldered to the printed circuit board, the palladium plating of the lead dissolves in the solder and the bond is actually formed with the underlying nickel. Whereas the thermal impedance and reliability of gold wirebonds is comparable to copper-plated copper leadframes, palladium plating increases the adhesion strength of the encapsulant because of its higher free surface energy compared with copper [Abbot et al. 1991]. From the assembly viewpoint, preplating leadframes with a noble metal like palladium offers cost savings, since it eliminates the extra step of either solder dipping or solder coating. Although preliminary reliability data seems promising, some European board assemblers do not use Palladium-plated leadframes due to solderability concerns. Indeed, although the adhesion strength of palladium coated frames is typically excellent, the finish is poor compared with the solder equivalent. Poor aesthetics breeds customer uneasiness.

## 2.3 DIE-ATTACH MATERIALS

Die-attach bonding is used to mechanically attach the die to the leadframe. Other functions of die-attach materials include heat transfer from the dies to the leadframe and, if needed, die backside electrical contact. Material property considerations in choosing die-attach materials include the shear strength of both die-attach interfaces, void density on application, impurity content, volume resistivity, thermal conductivity, manufacturability, and cost-effectiveness.

Early assemblies of plastic-encapsulated devices incorporated Alloy 42 leadframes with gold plating on the die-attach paddle. The backside gold-metallized die was attached with gold-tin eutectic preform at ~300°C. Die cracking became frequent with this process, particularly for large dies.

Consequently, most plastic-encapsulated microelectronics are now assembled with either polymer or solder die attachment methods which are less expensive than high-gold-content solders, are more easily automated for high throughput, and enable the use of silver-plated copper leadframes.

Solventless conductive epoxies are the materials in widest commercial use, comprising about 80% of the die attach market [Kearny 1988]. They are normally filled with 70 to 80% silver, which imparts electrical conductivity and enhances thermal conductivity. There may also be coupling agents on the metallic particles to promote wetting and electrical conductivity. Formulations are available with or without an alumina filler, depending on whether electrical isolation of the die from the paddle is desired. These die-attach materials usually require a separate cure at 150 to 180°C for about an hour, although newer materials that cure in 5 to 10 min have also been developed.

Characteristics of some of the commercially available polymer die-attach adhesives are presented in Table 2.4. Although epoxy die attach has been the workhorse for many years, cyanate ester-based adhesives have found increased popularity for many applications [Nguyen 1994]. Table 2.5 presents some of the unique Johnson Matthey products.

The typical conductive epoxy die adhesive has a coefficient of thermal expansion around 55 ppm/°C and a thermal conductivity around 0.008 W/cm°C. Low-stress die-attach materials have lower coefficients of thermal expansion and a low modulus of elasticity when cured, offering a greater degree of stress relaxation [Okabe et al. 1988]. Lap shear strengths of cured die-attach materials, as measured by ASTM D-1002, should be (and generally are) greater than 6.9 MPa (1 ksi). Die-shear tests using actual devices commonly demonstrate shear strengths between 35 and 69 MPa (5 to 10 ksi). However, if nonwetting occurs, the package will fail a die-shear test. Although X-ray radiography can be used to monitor interface voids, a die-shear test is often the only monitor of bond integrity; bondline thickness of ~25 μm is considered optimal. If the device operates above the glass transition temperature of the polymer, there will be a loss of bond shear strength, but lower die stress will result from the decreasing modulus of elasticity.

The thermal resistance of a silver-filled epoxy adhesive layer is 1 to 2°C/W higher than that of a solder die attachment layer. If the total thermal resistance of a plastic package, from junction to ambient is considered (30 to 100°C/W), this additional resistance is insignificant. These polymer adhesives typically exhibit volume resistivities in the range of $1 \times 10^{-4}$ to $4 \times 10^{-4}$ Ω-cm, and can be modeled using percolation theory [Opila and Sinclair 1985]. For devices requiring

**Table 2.4 Represents true commercial polymer die attach adhesives [Robock and Nguyen 1989]**

| Supplier | Type No. | Resin type | Filler type | Components | Electrical conductive | Maximum cure temperature | Lap shear strength (kg/mm$^2$) | Remarks |
|---|---|---|---|---|---|---|---|---|
| Epotek | H20E | epoxy | silver | 2 | yes | 175 | 1.06 | two-component epoxy adhesive with high electrical and thermal conductivity |
| | H35-175M | epoxy | silver | 1 | yes | 180 | 1.27 | similar to H20E, except single component, no mixing required |
| | H44 | epoxy | gold | 1 | yes | 150 | 2.11 | single-component, gold-filled epoxy adhesive |
| | H81E | epoxy | gold | 2 | yes | 150 | 0.91 | two-component, gold-filled epoxy adhesive |
| | P-1011 | polyimide | silver | 1 | yes | 150 | 0.5 | low cure-temperature polyimide adhesive |
| Amicon | C-840-4 | epoxy | silver | 1 | yes | 200 | 1.6 | low ionic impurity epoxy adhesive can be pre-cured as B-stage on leadframes |
| | C-940-4 | polyimide | silver | 1 | yes | 270 | 1.45 | low ionic impurity polyimide adhesive; excellent heat resistance to 400°C |

**Table 2.4 (cont.)**

| Supplier | Type no. | Resin type | Filler type | Components | Electrical conductive | Maximum cure temperature | Lap shear strength (kg/mm$^2$) | Remarks |
|---|---|---|---|---|---|---|---|---|
| | CT-4042 | epoxy | silver | 2 | yes | 150 | 1.6 | low ionic two-component epoxy adhesives |
| | C-990 | epoxy | silver | 1 | yes | 275 | 1.55 | very rapid cure epoxy adhesive; cures on wirebond heater block |
| | ME-868 | epoxy | oxide | 1 | no | 180 | 1.6 | high thermal conductivity, electrical insulation epoxy adhesion |
| Ablestick | 71-1 | polyimide | silver | 1 | yes | 275 | 0.7 | general-purpose polyimide direct-attach adhesive |
| | 84-1LM1 | epoxy | silver | 1 | yes | 150 | 4.58 | low temperature cure conductive epoxy adhesive |
| | 84-1A | epoxy | silver | 1 | yes | 200 | 4.23 | rapid-cure version of 84-1LM1 |
| | 84-1LM1S1 | epoxy | silver | 1 | yes | 200 | 2.11 | low-viscosity, low-ionic version 84-1LM1 |
| | 941-3 | epoxy | silver | 1 | yes | 175 | 2.11 | 13-stageable epoxy chip adhesive for preprinting leadframes |

**Table 2.5 Johnson Mathey Products [Nguyen 1994]. Each product is a single component attach, with silver as the filler and electrical conduction capability**

| Product | Resin type | Cure temperature (°C) | Remarks |
|---|---|---|---|
| JM 4613 | silver/glass | 440 | silver/glass adhesive |
| JM 6100 | silver/glass | 380 | silver/glass adhesive |
| JM 7000 | cyanate ester | 150 to 300 | rapid-cure heat resistance to 380°C |
| JM 2000 | cyanate ester | 120 | low temperature, very rapid cure |
| JM 2500 | cyanate ester | 150 | low-stress material |

low electrical resistance for the path through the backside, this could be a major limitation in the use of polymer adhesives.

Polyimides are also used for die attach such as on copper leadframes in high-temperature applications. However, because polyimides are cured at higher temperatures than epoxies, and consequently possess high glass transition temperatures (180 to 275°C), high die stresses on cooling can arise, making the die attach more susceptible to cracking. Polyimides have very low ionic contaminants (< 10 ppm), but their die-attach formulations include up to 30% solvents that need to be driven out before cure to avoid void formation.

Another class of die-attach materials is silver-filled glass-resin adhesive. This material becomes wholly inorganic after firing, eliminating many disadvantages of organic materials, while maintaining a form that is still convenient to apply. High-temperature bake cycles of over 400°C for several minutes are required to fire these resins.

Solid-film polymer die attach materials are supplied as precast films of either thermoset (epoxy or polyimide) or thermoplastic formulations (acrylic, polyester, cured polyimide, or B-stage epoxy) or a blend of the two. The absence of die-attach voids in 100% solid films is a particular advantage for large dies, with an added guarantee of uniform thickness. They are either conducting or insulating, and exhibit 7 to 35 MPa die-shear strengths and 0.7 to 7.0 W/m-°C thermal conductivities. They contain low concentrations of ionic impurities and can be applied to the back of the wafer before dicing or put on heated leadframes and cured (the usual way).

Motorola has been shipping production volumes of high-power integrated circuit devices die-bonded with Sn-Sb solder for some time [Owens 1994]. The solder is 92Sn/8Sb, with a liquidus of 246°C and a solidus of 236°C. The principal advantages of this material are low processing temperatures (which reduce residual stresses when the part is cooled to room temperature), low stress, and better resistance to fatigue failure than 95Pb/5Sn solder. Motorola has also shipped high volumes of "J-alloy," a solder composed of 65Pb/25Sn/10Sb with a liquidus of 236°C. The J-alloy solder is generally not being used in new "large die" applications, due to its high stress effect on the die [Owens 1994]. For more information on J-alloy solder, see Winchell et al. [1970].

Solder die attach is used for thermal management of high-power devices in nonoxidizing ambients without flux. To promote solder wetting, die-attach solders are usually applied as preforms, typically with a thickness greater than 50 μm between the metallized leadframe and the back of the die. Some of the common die-attach preform solders are listed in Table 2.6. The liquid temperature and coefficient of thermal expansion of the solder, the major considerations in its choice, ensure minimal die shift and deadhesion during the subsequent process steps of molding and soldering to the circuit board. The most common solders, 95Pb/5Sn and 65Sn/25Ag/10Sb, are used with Ti-Ni-Ag metallizations on both the leadframe and the backside of the die. However, these soft solders are susceptible to fatigue failure under thermal cycling.

With solder die attachment materials, there is usually a requirement to match the solder to the metallization layer on the back of device in order to create an intermetallic bond between the die attach solder and the backside metallization. Most backside metallizations consist of a barrier layer of chrome, vanadium, titanium or some other metal followed by a nickel layer, and finally a silver layer. The silver quickly dissolves into the solder and helps wet the back surface of the die. The nickel dissolves much more slowly, allowing the creation of an intermetallic layer that chemically bonds the die to the solder. Owens [1994] found that several solders, such as 92Sn/8Sb and 95Pb/5Sn, bond well to uncontamined bare copper, eliminating the need to plate the die attachment area.

## 2.4 WIREBOND MATERIALS

Of all the interconnection technologies used in microelectronics packaging, wirebonding is overwhelmingly dominant. Typically, the higher cost flip-chip and high-density interconnection technologies are not used in PEM manufacture, and

**Table 2.6 Die-attach preform solders**

| Composition | Liquid temperature (°C) | Solidus temperature (°C) |
|---|---|---|
| 80Au/20Sn | 280 | 280 |
| 97.5Pb/1.5Ag/1Sn | 309 | 309 |
| 95Pb/5Sn | 314 | 310 |
| 88Au/12Ge | 356 | 356 |
| 98Au/2Si | 800 | 370 |
| 92Sn/8Sb | 246 | 236 |

both TAB and flip-TAB can be plastic-encapsulated only by potting or glob-topping (radial spreading), due to the fragility of leadframes. A comparison of various interconnect technologies is given in Pecht [1994], and wirebonding trends are reviewed in Chenetal [1993].

Table 2.7 lists the properties of various bonding wire materials. Of all the materials in wirebonding, gold is the most common; gold-wire ball bonding is the established interconnection technology between pads on the integrated circuit chip and external leadframe fingers in plastic-encapsulated microcircuits. Major efforts have been made to reduce or eliminate the gold wire for both economic and reliability reasons. Several studies have been conducted on the use of ultrafine wires of copper, silver, and aluminum for thermocompression and thermosonic ball bonding in plastic packages. The three major techniques for wirebonding—thermocompression, ultrasonic, and thermosonic—are discussed in Chapter 3.

### 2.4.1 Gold Wire

Gold is typically used for thermocompression and thermosonic bonding because of its high oxidation resistance. Because of its low work hardening characteristic, gold has a low tendency to crater (overbonding of the wire to the bond pad, which results in cracks in the underlying chip layers bond liftoff, accompanying silicon craters in the chip underneath).

Pure gold is very soft, so small amounts of impurities, such as 5 to 10 ppm by weight of beryllium or 30 to 100 ppm by weight of copper, are applied to make the gold wire workable. Beryllium-doped gold wire is stronger than copper-

**Table 2.7 Properties of common wire materials at 27°C**

| Property | Aluminum (Al) | Copper (Cu) | Gold (Au) |
|---|---|---|---|
| Specific heat (W-sec/kg°C) | 900 | 385 | 129 |
| Thermal conductivity (W/m°C) | 237 | 403 | 319 |
| Specific gravity (kg/m$^3$) | $2.7 \times 10^{-3}$ | $8.96 \times 10^{-3}$ | $1.93 \times 10^{-2}$ |
| Melting point (°C) | 660 | 1083 | 1064 |
| Electrical resistivity (Ω-m) | $2.65 \times 10^{-8}$ | $1.73 \times 10^{-8}$ | $2.25 \times 10^{-8}$ |
| Temperature coefficient of electrical resistivity (Ω-m/°C) | $4.3 \times 10^{-11}$ | $6.8 \times 10^{-11}$ | $4 \times 10^{-11}$ |
| Elastic modulus (Pa) | $3.45 \times 10^{10}$ | $1.32 \times 10^{10}$ | $7.72 \times 10^{10}$ |
| Yield strength (Pa) | $1.03 \times 10^{7}$ | $6.89 \times 10^{7}$ | $1.72 \times 10^{8}$ |
| Ultimate tensile strength (Pa) | | | |
| wire | $4.48 \times 10^{7}$ | $2.21 \times 10^{8}$ | $2.07 \times 10^{8}$ |
| bond pads | $3.93 \times 10^{8}$ | $3.45 \times 10^{8}$ | $4.83 \times 10^{8}$ |
| Coefficient of thermal expansion (1/°C) | $46.4 \times 10^{-6}$ | $16.1 \times 10^{-6}$ | $14.2 \times 10^{-6}$ |
| Poisson's ratio | 0.346 | 0.339 | 0.291 |
| Hardness (Brinell) | 17 | 37 | 18.5 |
| Percentage elongation (%) | | | |
| wire | 50 | 51 | 4 |
| bond pads | - | - | 2 to 3 |

doped gold wire by about 10 to 20%, under most conditions. The increased strength of the beryllium-doped wire is advantageous for automated thermosonic bonding, when high stresses are generated due to high-speed capillary movement.

Gold wirebonded to a gold metallization is very reliable, because the bond is not subject to interface corrosion, intermetallic formation, or other bond-degrading conditions. Even a poorly welded gold-gold bond will increase in strength with time and temperature [Jellison 1975].

Gold bonded to silver metallization is very reliable for long times at temperatures higher than 150°C [James 1977]. This bond system does not form intermetallic compounds or exhibit interface corrosion. However, contaminants such as sulfur reduce the bondability of silver to gold. Thermosonic bonding with temperatures as high as 250°C may be needed to provide high bondability of silver [Harman 1989].

Gold wires bonded to aluminum leadframes can lead to the formation of the brittle intermetallic phases $Au_5Al_2$, $Au_2Al$, $AuAl_2$, $AuAl$, and $Au_4Al$. Intermetallic formation is accelerated at temperatures higher than 175°C. Excessive intermetallic formation is often accompanied by Kirkendall voiding and an increase in bond resistance. Gold-aluminum intermetallics are more susceptible to flexure damage than pure aluminum wires.

Gold wires bonded to copper leadframes can lead to the formation of the ductile intermetallic phases $Cu_3Au$, $AuCu$, and $Au_3Cu$ [Hall 1975]. These phases lead to a decrease in bond strength at higher temperatures (200 to 325°C) because of void formation, and the acceleration of intermetallic compounds. However, Pitt and Needes [1982] studied gold thermosonic bonds to thick-film copper, and found little strength degradation at 150°C for up to 3000 hr, and no failures at 250°C in over 3000 hr. Cleanliness of the bonding surface is extremely important to ensure good bondability in copper-gold systems [Lang 1988, Fister 1982].

### 2.4.2 Aluminum Wire

Pure aluminum is typically too soft to be drawn into a fine wire. Thus, aluminum is often alloyed with 1% silicon or 1% magnesium. The equilibrium solid-state solubility of silicon in aluminum at 20°C is about 0.02% by weight. Only at temperatures higher than about 500°C does silicon attain 1% solubility in a solid solution at equilibrium. Thus, 1% silicon exceeds the solubility of silicon in aluminum at room temperature by a factor of fifty. At ordinary room temperatures, there is a tendency for silicon to precipitate, forming a silicon second phase that can lead to hardening and the development of a potential fatigue crack nucleation site.

Aluminum alloyed with 1% magnesium can be drawn into a fine wire that exhibits a breaking strength similar to that of aluminum-1% silicon. The resistance of aluminum-1% magnesium wire to fatigue failure and to degradation of ultimate strength after exposure to elevated temperatures is superior to that of aluminum-1% silicon wire. These advantages of aluminum-1% magnesium wire over its silicon substitute occur because the equilibrium solid solubility of magnesium in aluminum is about 2% by weight; thus, at 0.5 to 1% magnesium concentration, there is no tendency toward second-phase segregation, as there is with aluminum-1% silicon.

Aluminum ultrafine wires have been used in plastic-encapsulated microcircuits because the material is considerably cheaper than gold, and the aluminum wire to aluminum bond pad interface avoids the formation of brittle intermetallics. However, aluminum wirebonding has been typically restricted to wedge-wedge bonding. Wedge-wedge bonding is time-consuming compared to thermocompression or thermosonic bonding, and fatigue cracks can occur in the heel region of the contact zone and propagate to failure during operational life. Thermocompression or thermosonic ball bonding avoids these disadvantages, and efforts have been made to extend these technologies to aluminum bonding. Ball formation for aluminum is difficult, because even an oxygen content of 1000 ppm results in formation of wrinkles on the ball surface, defects in ball shape, and asymmetry with respect to the axis of the wire [Hirota et al. 1985, Bishoff et al. 1984, Onuki et al. 1984].

Factors affecting aluminum wirebond reliability include aluminum wire composition, ball quality, and bonding conditions (such as an inert argon atmosphere, oxide thickness on the ball, and ultrasonic power during thermosonic bonding) [Onuki et al. 1984]. Onuki compared the aluminum balls formed under various bonding conditions by examining the eccentricity, constriction, and sphericity of the balls. The definitions of the terms are given in Figure 2.3. The oxide thickness on the aluminum ball surface is a critical factor controlling ball shape. Since the aluminum ball shape approaches the shape obtained with gold wires as the oxide film becomes thinner. High electric power and short ball formation period in an argon atmosphere prevent oxidation of the aluminum surface. Factors affecting aluminum bonding strength include ball hardness and deformation, strain caused by ultrasonic power during thermosonic bonding, the length of time over which the power was applied, and the bonding force. The aluminum ball bond breaking force (quantified by shear test using a knife edge) increases with an increase in ball hardness at constant strain. The aluminum bonded area increases in ball hardness and strain.

Bischoff et al. [1984] evaluated the characteristics of various wire compositions, including 25-μm-diameter aluminum-1% silicon, aluminum-1% magnesium, aluminum-4% copper, and aluminum-1% magnesium-1% silicon wires bonded to various substrate metallizations, including aluminum, silver, gold, and copper. They found that the rupture strength of the aluminum wire during bond pull testing at 150°C was low due to the formation of a weakened zone caused by grain coarsening above the ball; the coarse-grained structure develops in the part of the wire exposed to elevated temperature during the ball flame-off process. The intercrystalline fracture typical of a coarse-grained structure was observed in the aluminum-1% silicon wire.

Onuki et al. [1984] evaluated the ball hardness of various aluminum wire compositions and found the hardnesses to vary as follows: Al-Mg ≈ Au > Al-Ni > Al-Si. Hardness was defined as the inverse of the height difference in the ball before and after pressing by a capillary at a load of 90 g.

Onuki et al. [1984] also evaluated the reliability of a 50-μm-diameter Al-Mg wire ball bonded in a resin-encapsulated power transistor to 4-μm-thick aluminum electrodes. For a ball strain defined as 2ln(ball bond diameter/ undeformed ball diameter) of 1.0, the wirebonds showed no degradation after 200 to 400 cycles of -55 to 150°C, 20,000 power cycles of ambient temperature to 110°C, and a pressure cooker test at 121°C, 2 atm for 100 to 125 hr.

Aluminum wires bonded to a nickel coating (used as a substitute for gold) are more reliable than aluminum-silver or aluminum-gold bonds [Harman 1989]. Aluminum wires with a diameter greater than 75 μm (3 mil) have been bonded to nickel plating in power devices [Harman 1989]. The aluminum-nickel system is well suited to high-temperature applications and is less prone to Kirkendall voiding and galvanic corrosion than gold-aluminum bonds.

Aluminum wirebonded to aluminum metallization is reliable because it is not prone to intermetallic formation and corrosion. This type of bond welds best ultrasonically, even though a thermocompression bond can be produced by high deformation.

Aluminum wirebonded to a silver-plated leadframe is not common in PEMs. The silver-aluminum phase diagram consists of many intermetallic phases. Silver-aluminum bonds are seldom used because of their tendency to degrade interdiffusion and to oxidize in the presence of humidity. The aluminum-chlorine corrosion mechanism is responsible for bond degradation, as evidenced by the formation of aluminum hydroxide in the zeta phase of the silver-aluminum intermetallic. Kirkendall voids can also occur in this metal system, but typically only at temperatures higher than 175°C [James 1977, Jellison 1975]. See Hermansky [1972], Shukla and Singh [1982], Jellison [1975], Kamijo [1985], and James [1977] for more informations on aluminum-silver bonding.

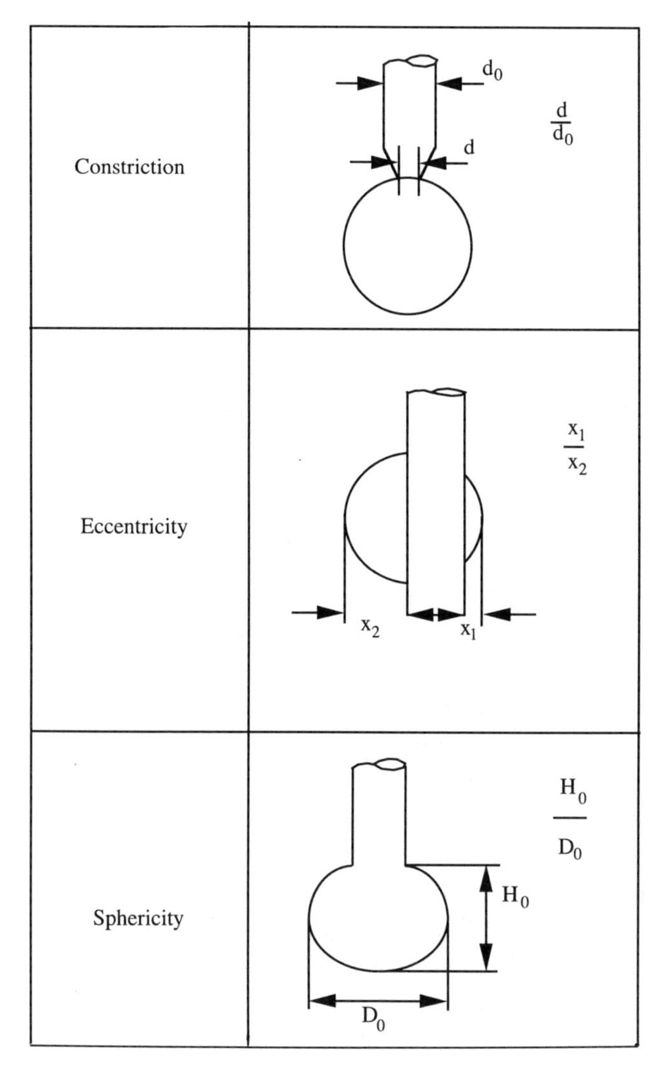


**Figure 2.3** *Definitions for the eccentricity, constriction, and sphericity of the balls [Onuki et al. 1984]*

### 2.4.3 Silver Wire

Molten silver dissolves 2% oxygen under 1 atmosphere pressure. Solid silver also has a small but finite oxygen solubility. The intermetallic growth rate in a silver - aluminum system is one-fourth that of a gold-aluminum system under the same temperature conditions. Figure 2.4 shows the intermetallic layer thickness of silver-aluminum and gold-aluminum bonds for 1 sec bonding time. Due to

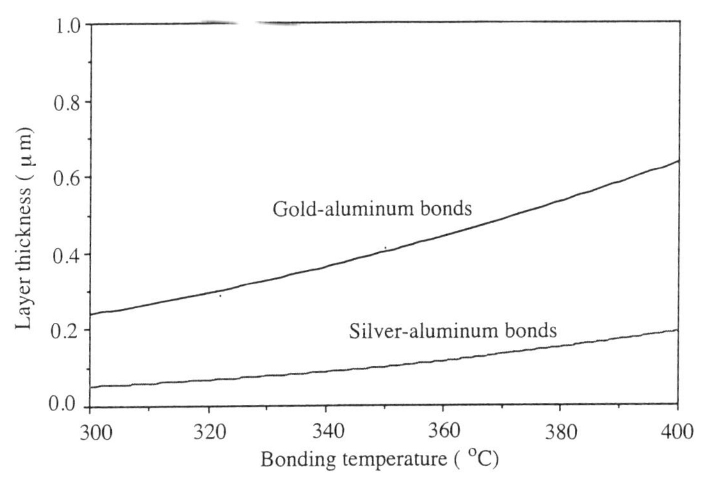


**Figure 2.4** *The intermetallic layer thickness of silver-aluminum and gold-aluminum bonds for 1 sec bonding time [Kamijo and Igarashi 1985]*

constitutional supercooling, silver wire demonstrates cellular solidification when flame-off is conducted in air. Cellular solidification is the formation of a regular corrugated grain structure in which the corrugations are roughly parallel to the direction of crystal growth. Typically, constitutional supercooling occurs in the presence of an advancing solid-liquid interface. This phenomenon is based on the concept that the liquid in contact with an advancing solid-liquid interface will in general have a different composition than that of bulk liquid. The liquidus temperature of the liquid in contact with the solid-liquid interface is lower than the liquidus temperature of the bulk liquid. When the bulk liquid is at a temperature above the solid-liquid interface temperature but lower than the bulk liquid liquidus temperature, the liquid is said to be constitutionally supercooled [Chalmers 1964]. Constitutional supercooling in silver is a result of oxygen dissolved in the silver ball and a rapid cooling rate. Cellularly solidified silver bonded to aluminum typically undergoes severe degradation due to corrosion, even under moderate humidity conditions. Preventing cellular solidification of silver often involves reducing the oxygen concentration of the ball by using a protective gas atmosphere. The critical oxygen concentration for no cellular solidification is 100 ppm [Kamijo and Igarashi 1985].

Silver has a shear modulus approximately 20% larger than gold. The shear modulus of the wire material affects the ball hardness. The silver ball is thus harder than the gold ball and has a higher probability for cratering under identical bonding conditions [Kamijo and Igarashi 1985]. Figures 2.5 shows the ball strain versus mean strength relation for silver and gold wires. The mean shear strength of silver bonds is about 10% less than that for gold wire after humidity testing at 85°C/100%RH for 100 hr. [Kamijo and Igarashi 1985]. Unlike gold wire silver wire shows shear strength degradation after humidity tests. Cross-sections of an as-bonded silver-aluminum system revealed that the silver-aluminum reaction was very slow, and the aluminum metallization remained almost unchanged without the diffusion reaction. Therefore, silver-aluminum bonds often need annealing at 300°C in a nitrogen atmosphere for 60 min for adequate diffusion and intermetallic layer formation [Kamijo and Igarashi 1985].

### 2.4.4 Copper Wire

Copper wires are used primarily because of their economy and their resistance to sweep (movement of the wire in a plane perpendicular to its length) during plastic encapsulation [Kurtz and Cousens 1984, Atsumi et al. 1986, Levine and Sheaffer 1986, Onuki et al. 1987]. Copper is harder than gold, and thus greater attention is needed during the bonding operation to prevent cratering of the chip. The harder copper wire tends to push the softer bond-pad metallization aside during bonding, so bonding systems using copper wire requires harder metallizations [Riches and Stockham 1987, Hirota et al. 1985]. Copper wire also requires an inert atmosphere for bonding because of its tendency to oxidize readily. Olsen and James [1984] observed that for aluminum wirebonded to oxygen-free, high-conductivity copper, thermal aging at 150°C in ambient did not affect the bond strength, while in a vacuum, the bond strength decreased.

Copper wire is susceptible to oxidation during ball formation and to chip cratering during bonding. The tendency of copper to oxidize and crater are interrelated [Kurtz and Cousens 1984]. As a result of being formed in air (or an inert atmosphere), the copper ball has a heavy oxide layer. Subsequent bonding thus requires substantial force and power to break down the oxide skin on the ball this often results in cratering. Several studies have been conducted to evaluate the effect such factors as oxygen concentration, purity, moisture, wire hardness, and bonding atmosphere on the bondability degradation of copper wire.

The fatigue strength of copper wires varies inversely with oxygen concentration. High-purity copper with 10 ppm oxygen has twice the fatigue life of copper wire with 400 ppm of oxygen. Panousis [1978] used precleaned copper leads to make thermocompression bonds to titanium-palladium-gold films on alumina substrates, and found surface oxidation to be the degradation mechanism for pre-weld shelf-life degradation. Pfaelzer and Frisch [1967] showed that the strength of copper wires bonded in a 10-torr vacuum using ultrasonic bonding was higher than that of wires bonded in air. Johnson [1967] found that absorbed

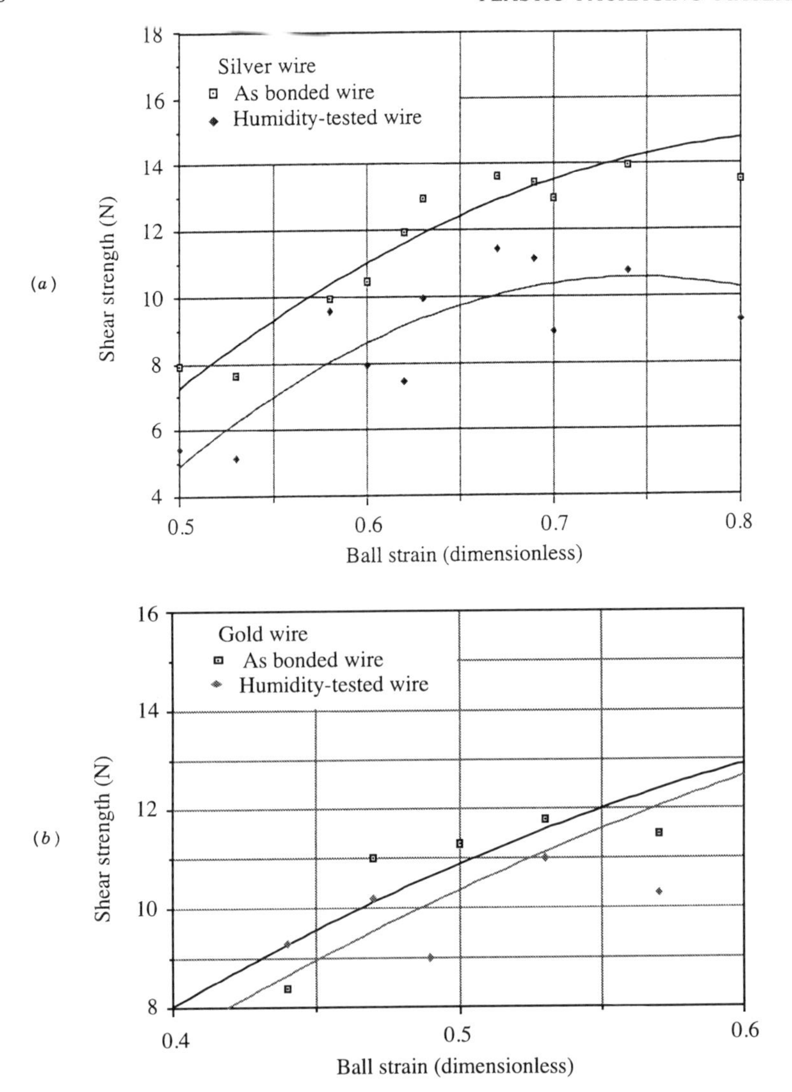


**Figure 2.5** *The ball strain versus mean strength relation for (a) silver and (b) gold wires [Kamijo and Igarashi 1985]*

water on copper surfaces constituted a highly mobile film, that is difficult to remove from the interface and inhibits the intimate contact of the interfaces during ultrasonic welding. Harman and Wilson [1989] found that even a monolayer of water vapor serves as a barrier for prevention of an ultrasonic bond. Kurtz and Cousens [1984] showed that an increase in the exposure time of copper wire to the atmosphere increases the percentage of carbon and oxygen on the ball. Surface contamination causes 30% degradation in copper-copper ultrasonic bond strength after 24 hr. of exposure to 85.3% relative humidity.

Kurtz and Cousens [1984], while evaluating the effects of Vicker's hardness on the wirebond pull strength of copper wire, compared two wires with hardnesses of 64.99 and 77.26, with ultimate tensile strengths of 605 and 325 gf, respectively. The softer copper wire had a lower pull strength, and demonstrated greater deviation and interfacial failure at the second weld. The harder wire showed better weldability, with the wire failing at the heel of the first bond. Figure 2.6 shows the Vicker's hardness measurement method for copper balls. The Knoop hardness of the aluminum pad affects the wirebond characteristics. In conventional gold wirebonding, the Knoop hardness of the pads is in the neighborhood of 35, while for copper, Hirota et al. [1985] found that stable bonding could be achieved under lower ultrasonic vibration condition at a Knoop hardness of 45. Figure 2.7 shows the effect of the aluminum pad's Knoop hardness on copper wirebonding.

Mori et al. [1988] produced a new copper bonding wire for automated bonders by adding dopants to high-purity 6N copper, which has less ball hardness and lower deformation resistance. The loop shape stability of 6N copper wire is, however, lower than that of gold wire because of coarser grain size. The doped

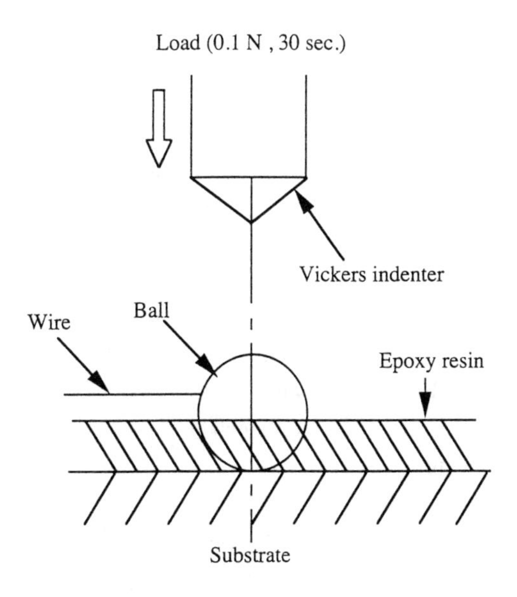


**Figure 2.6** *Vicker's hardness measurement method for copper ball [Mori et al. 1988]*

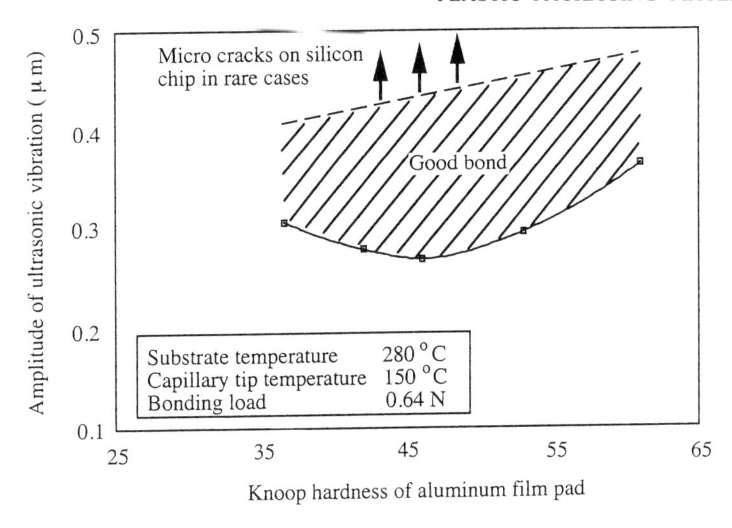


**Figure 2.7** *Effect of aluminum pad's Knoop hardness on copper wirebonding [Hirota et al. 1985]*

6N copper wire has a higher breaking strength and percentage elongation with respect to high-purity copper bonding wire. However, the breaking strength of both the doped wire and the high-purity copper wire decreases, and elongation increases, with an increase in annealing temperature for temperatures higher than 200°C. The annealing temperature at which the elongation increases is higher for doped copper wire than for high-purity copper wire [Mori et al. 1988].

Figure 2.8 shows the breaking strength and elongation after annealing of doped and high-purity copper wire versus gold wire. While evaluating the relative pull strengths of copper and gold wire, Mori et al. [1988] found that copper wires had higher pull strength values (≈ 9.6 to 10.7 gf) than gold (≈ 6.7 gf). The dominant failure mode for copper wire was failure at the heel of the second bond, and for gold wire was failure at the ball neck. Similar data were also obtained by Kurtz and Cousens [1984], who showed that copper wire has a pull strength in the range of 6 to 15 gf and a shear strength in the range of 30 to 50 gf. The bond resistance of copper wires (25 μm-diameter) on aluminum pads was in the neighborhood of 6.3 mΩ, compared with 5.2 mΩ for gold wires of the same diameter. Kurtz and Cousens [1984] showed that copper balls bonded to aluminum pads passed 1000 hr. of operating life at 125°C, 2,000 hrs. bias humidity at 85°C and 85% relative humidity, 100 temperature cycles, 45 thermal shocks, and 480 hr. of steam pressure at 121°C.

Intermetallic growth in copper-aluminum bonds is slower than in gold-aluminum bonds and does not result in Kirkendall voiding, but does produce a low shear strength around 150 to 200°C, due to the growth of a brittle $CuAl_2$

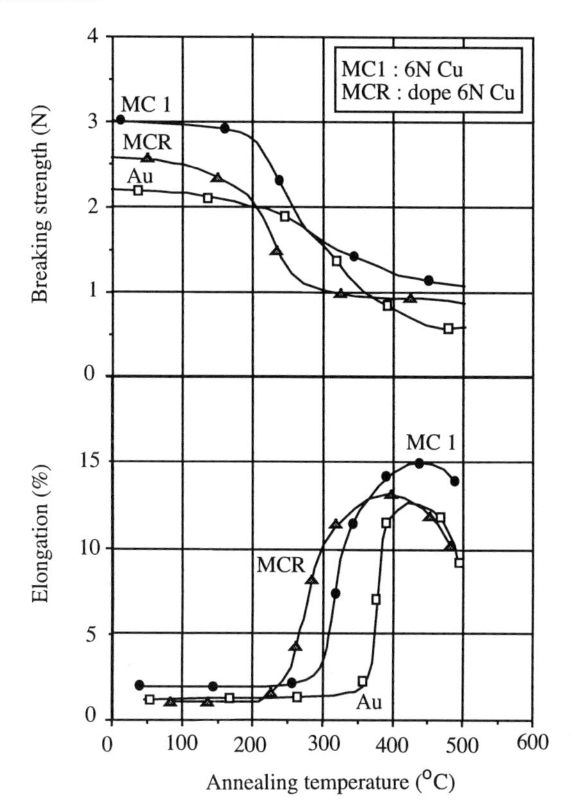


**Figure 2.8** *The breaking strength and elongation after annealing of doped and high-purity copper wire versus gold wire [Mori et al. 1988]*

phase [Atsumi et al. 1986, Onuki et al. 1987]. Copper-aluminum bonds are stronger than gold-aluminum bonds in the presence of brominated flame retardants. In particular, Thomas et al. [1977] observed that copper-aluminum bonds were strong after 1245 hr. at 200°C, but gold-aluminum failed after 700 hr. However, aluminum metallizations containing copper-aluminum intermetallics can corrode readily in the presence of chlorine contamination and water [Nesheim 1984, Thomas et al. 1977, Totta 1977, Zahavi et al. 1984]. Figure 2.9 shows the kinetics of the formation of intermetallic phases in gold-aluminum, copper-aluminum, palladium-aluminum, and nickel-aluminum.

### 2.4.5 Low-profile Wirebond Interconnects

Plastic package designs such as TSOP or TQFP that require low-profile wirebonds have resulted in modifications of wirebonders and wire materials to adjust to the stringent requirements of the new designs [Kew et al. 1991, Levine

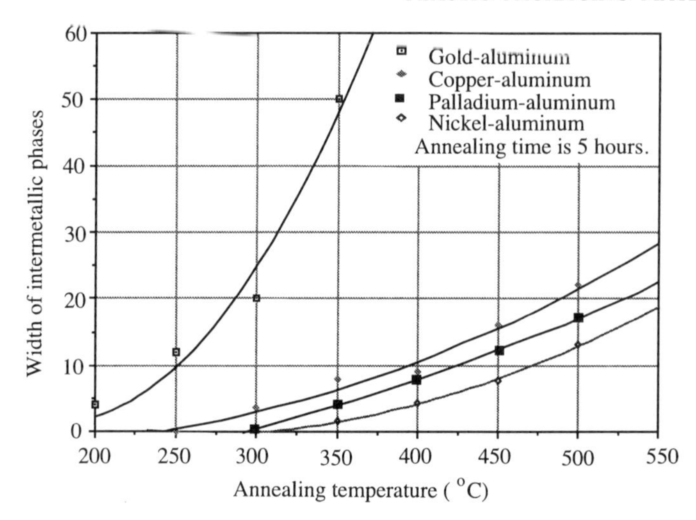


**Figure 2.9** *The kinetics of formation of intermetallic phases in various diffusion couples, including gold-aluminum, copper-aluminum, palladium-aluminum, and nickel-aluminum [Bischoff et al. 1984]*

and Sheaffer 1993]. New loop shapes have been accommodated by specialized wirebond tool motions and materials that can produce low-loop wires with higher impurity doping concentrations (e.g., silver, palladium, platinum in gold wire). Typically, low- profile wirebonding loop heights range from 125 to 175 μm (5 to 7 mils), wire lengths range from 0.7 to 2.0 mm (30 to 80 mils).

Low loop-profile wirebond interconnects differ from conventional loops in that they have three segments: a straight vertical wire above the ball, a horizontal wire length spanning the die to the edge, and a sloping segment to the second bond. The grain size and wire material properties determine the length of the first wire segment or the loop height. During ball bonding, the ball is formed by an arc from the electronic flame-off, which melts the wire tip. As melting progresses, additional wire material is pulled into the sphere due to surface tension and gravity. The thermal excursion of the wire, from ambient temperature to below the melting point then cooling back to ambient, causes the recrystallization of the wire material.

The recrystallization temperature is a material property that defines the minimum temperature at which spontaneous recrystallization or grain structure change occurs. The recrystallized structure consists of new equiaxed grains, which are much larger than the original grains formed by the cold-working process. The stress required to cause deformation for the two grain structures is significantly different. The recrystallized wire length, also called the heat-affected zone (HAZ) can be reduced if the recrystallization temperature is

**Table 2.8 Dependence of dopant species on recrystallization temperature of gold wire (the degree of cold work is fixed at 99%) [Levine and Sheaffer 1993]**

| Purity of gold | 99.9999% | 99.995% | | | |
|---|---|---|---|---|---|
| Dopant species | | Ag, Pd, Pt | Mg, Ni, Si | Co, Cu, Fe, Ga, Ge, In | Al, Be, Ca, Pb, Sn, Ti |
| Recrystallization temperature (°C) | 150 | 150 | 150 to 200 | 200 to 300 | over 300 |

increased. The recrystallized grains have a lower yield strength than the fine-grained unrecrystallized grain structure. Since the HAZ bends at lower stress levels, than the unrecrystallized grain structure the bend radius forms in the HAZ. The radius of bend is determined by the length of the HAZ. The loop height is thus determined by the wire composition. Some of the common compositions used to address low loop heights are given in Table 2.8.

While increased impurity concentration results in a decreased HAZ length and a lower loop height, some additional defect mechanisms can arise: an increase in the microhardness of the wire material, which results in increased probability of cratering, failure to weld, and embrittlement of the intermetallic layer; and shrinkage pores in the molten ball because of increased dopant levels. During solidification, which occurs from the outer skin of the ball inwards to the center, impurities are pushed ahead of the solidification front, enriching the impurity levels at the surface of the central pore. During bonding, the enriched surface of the central pore forms a seam and is pushed towards the bond pad. The seam is a reliability concern in thermal cycling environments.

### 2.4.6 Fine-pitch Wirebond Interconnects

Traditional gold wire used in wirebonding is not used for fine-pitch bonding for several reasons: the wire breaks during handling; there is insufficient pull strength after bonding; manufacturing is difficult with long wire; and tailing occurs easily on the second side. Yamashita et. al [1991] experimented with various gold-based alloys, and found the bond strength varied with diameter and annealing conditions. The minimum value of wire strength was selected to be greater than the maximum second bond strength (including peel resistance). Wires with a high breaking strength (> 3g) and low elongation (< 0.5%) and wires with a low breaking strength (< 2g) and high elongation (> 2%) are more prone to tailing.

Typically, the wires are manufactured by annealing after wire drawing.

Compared to conventional gold wires, the gold-alloy ultrafine wire has a unique annealing curve. In conventional materials during annealing, the tensile strength and hardness of the wire decrease, while elongation increases with a rise in annealing temperature due to the phenomena of recovery, recrystallization, and grain growth. In ultrafine wires, the tensile strength and hardness decrease with annealing temperature, just as with conventional wires. However, the elongation tends to decrease after a maximum annealing temperature is reached. The decrease in elongation occurs when the grain size becomes comparable to the wire diameter (i.e., a bamboo-like structure). Therefore, a great deal of attention must be paid to the optimum annealing temperature for ultrafine pitch wirebonding.

## 2.5 TAPE AUTOMATED BONDS

Tape automated bonding (TAB) is a bonding technique that involves the use of a prefabricated interconnection pattern, one circuit pattern wide and several hundred long, on an organic film carrier. The initial TAB process involves bonding chips to the leadframe metallization pattern on the tape using thermocompression inner-lead bonding. Subsequent processing of testing, encapsulation, screening, and so forth can then be carried out automatically in filmstrip form. Next, the individual pretested and qualified parts are excised from the tape and attached to a board, substrate, or package by outer-lead bonding.

The three basic tape structures are single-, double-, and triple-level tapes. The former is commonly made of 0.07 to 0.08 mm (2.6 to 3 mil) thick etched copper metal, so all leads are shorted until excised from the tape. Double level tape includes a dielectric polymer film on which the 0.02 to 0.04 mm (0.6 to 1.6 mil) thick patterned, leadframe rests. Typically, a thin sputtered chromium-copper adhesion layer (1 μm) is used as a plating base on the polymer for the thicker leadframe metallization. Triple-level tape is very similar to laminated printed circuit board, and uses a copper-foil layer glued on prepunched polymer film. Following the lamination process, the adhesive is cured and the lead pattern is photolithographically defined. The copper leadframes for all tape types are usually gold, tin, or tin-lead plated to ensure better inner-and outer-lead bonding. During chip and package attachment, the plated leads at the inner-and outer-lead bonding sites are reflowed.

Bumped chip or bumped tape technology, with or without flip-chip and bonding pads over the entire chip surface, is needed for TAB bonding. The development of proper bump metallurgy to be used with bonding pads has been the subject of much research. While the tape bump is usually only gold or tin-plated copper, the chip bump layers typically consist of an adhesion layer (Cr, or Ti), a barrier layer (Cu, Pt, Pd, Ni) and a final thick-bump metal layer (Au, Au/Cu, Pb-Sn/Cu). An alternative for tape bumping, called transfer bumping, uses gold bumps plated onto a temporary substrate and transferred to the ends of the inner-leads of the tape by a thermocompression bonding process, yielding copper tape with a gold bump bonded to the inner leadfinger [Hatada et al. 1987].

## 2.6 MOLDING COMPOUNDS

A PEM encapsulant is generally an electrically insulating plastic material formulation that protects an electronic device and die-leadframe assembly from the adverse effects of handling, storage, and operation.

Encapsulant techniques include molding, potting, glob-topping, and conformal coating. Molding compounds are discussed in this section. The key material properties of molding compounds are discussed in Section 2.7. The materials associated with the other techniques are presented in Section 2.8.

In general, the majority of chip packages use epoxies and molding processes. Novolac-based epoxies are supplied as molding bricks (preforms) that are ready for use in transfer-molding machines to produce single-in-line packages, dual-in-line packages, plastic-leaded chip carriers, quad flatpacks, and various types of small-outline integrated circuit packages. Pin-grid array carriers — carriers with cans, for example — frequently use multicomponent liquid epoxies or preforms.

The polymeric encapsulating resin, modified by the additives, must possess adequate mechanical strength, adhesion to package components, manufacturing and environmental chemical resistance, electrical resistance, matched coefficient of thermal expansion, and high thermal and moisture resistance in the use temperature range. Thermosetting encapsulation compounds based on epoxy resins or, in some niche applications, organosilicone polymers, are widely used to encase electronic devices. Polyurethanes, polyamides, and polyesters are used to encase modules and hybrids intended for use under low temperature and high humidity conditions. Modified polymides as encapsulants have the advantages of thermal and moisture stability, low coefficient of thermal expansion, and high material purity. Thermoplastics are rarely used for PEMs, because they require unacceptably high temperature and pressure processing conditions, are low purity materials, and result in moisture induced stresses. The molding compound is a proprietary multicomponent mixture of an encapsulating resin with various types of additives. The principal active and passive (inert) components in a molding compound include curing agents or hardeners, accelerators, inert fillers, coupling agents, flame retardants, stress-relief additives, coloring agents, and mold-release agents. Table 2.9 lists these components and their functions in epoxy molding compounds [May 1989]. Postmolding of devices with molding compounds is generally done with thermosetting polymers, like epoxies. The transfer-molding process technology used with thermosetting polymers requires a low initial viscosity of the heated molding compound so it can flow over the leadframe and wirebonds with minimum possible deformation. The process also demands a high modulus and high thermal distortion temperature in the final cured state to withstand subsequent mechanical manufacturing operations and soldering processes. Conversely, thermoplastics with too high a melt viscosity and too low a softening temperature are inappropriate for plastic encapsulation of microelectronic devices. High melt viscosity causes low PEM yield (Chapter 3) and low encapsulant softening temperature inhibits manufacture of PEM populated printed wiring boards (Chapter 4).

To overcome package failures induced by the high coefficient of thermal expansion of the encapsulant, low-stress epoxy molding compounds were developed in the early 1980s. Large VLSI-memory dies and small-outline packaging demands also needed to be accommodated. This low-stress molding compound formulation was achieved by maximizing the filler loading to bring the cured-encapsulant coefficient of thermal expansion down.

Toughness was improved by devising a dispersed-domain morphology of elastomer modifiers. Plastic package processing yield was increased by the introduction of very low-viscosity molding compounds (from 1500 to 2000 poise to 300 poise at 100 $sec^{-1}$) [Blyler et al. 1986]. The productivity increase was largely due to the reduction of cure time by about two-thirds.

***Epoxies.*** The earliest materials used for plastic packaging of microelectronic devices were silicones, phenolics and bisphenol A epoxies because of their excellent molding characteristics. Silicones were used as molding compounds because of their high-temperature performance and purity. Poor adhesion of silicons with the device and metallization resulted in device failure in salt spray tests and flux penetrants during soldering. Although epoxy materials offer much better adhesion than silicones, the common epoxies (bisphenol A or BPA) have a low glass transition temperature in the range of 100 to 120°C, and high concentrations of corrosion agents such as chlorine. In the formulation of epoxy-molding compounds, novolac epoxies displaced BPA materials because of their higher epoxide functionality and improvement in heat-distortion temperatures.

**Table 2.9 Components of epoxy molding compounds used in electronic packaging**

| Component | Concentration (wt% of resin) | Major function | Typical agents |
|---|---|---|---|
| Epoxy resin | matrix | binder | cresol-novolac |
| Curing agents (hardeners) | up to 60 | linear/cross polymerization | amines, phenols and acid anhydrides |
| Accelerators | very low (< 1) | enhance rate of polymerization | amines, imidazoles, organophosphines, ureas, Lewis acids and their organic salts |
| Inert fillers | 68 to 80 | lower coefficient of thermal expansion, higher TC[a] (w/$Al_2O_3$), higher E,[b], reduce resin bleed, reduce shrinkage, reduce residual stress | ground fused silica (widely used), alumina |
| Flame retardants | ~10 | retard flammability | brominated epoxies, antimony trioxide |
| Mold-release agents | trace | aid in release of package from mold | silicones, hydrocarbon waxes, fluorocarbons, inorganic salts of organic acids |
| Adhesion parameter | trace | enhance adhesion with IC components | silanes, titanates |
| Coloring agents | ~0.5 | reduce photonic activity; reduce device visibility | carbon black |
| Stress-relief additives | up to 25 | inhibit crack propagation, reduce crack initiation, lower coefficient of thermal expansion | silicones, acrylonitrile-butadiene rubbers, polybutyl acrylate |

[a]Thermal conductivity [b]E modulus

ICs molded with phenolics experienced early corrosion failures caused by the generation of ammonia during postmold curing, the presence of high levels of sodium and chloride ions and high moisture absorption. Molding compounds have moved steadily toward all epoxy systems as the high-temperature novolac materials were improved and the hydrolyzable impurity levels were driven below 20 ppm.

All epoxy resins contain compounds from the epoxide, ethoxylene, or oxirane group, in which an oxygen atom is bonded to two adjacent (end) bonded carbon atoms. Because the resins are cross-linkable (thermosetting), each resin molecule must have two or more epoxy groups. Most commonly, the epoxide group is attached linearly to a chain of $-CH_2-$ and is called a glycidyl group. This group is attached to the rest of the resin molecule by an oxygen, nitrogen, or carboxyl linkage forming glycidyl ether, amine, or ester. The resin is cured by the reaction of the epoxide group with compounds containing a plurality of reactive hydrogen atoms (curing agents) like primary, secondary, or tertiary amines; carboxylic acid; mercaptan; and phenol; as such, they are coreactants. Catalytic cures are affected by Lewis acids, Lewis bases, and metal salts, resulting in the linear homopolymerization of the epoxide group forming polyether linkages. The viscosity of the base resin for the encapsulant grade materials is kept low by limiting the average degree of polymerization of the novolac epoxy to values around 5. The different commercial formulations differ primarily in the distribution of the degree of polymerization, with an average molecular weight of about 900. Longer and shorter chain lengths provide looser or tighter crosslink structures, with varying ductility and glass transition temperatures.

In electrical and electronic applications, three types of epoxy resins are commonly used: the diglycidyl ethers of bisphenol A (DGEBA) or bisphenol F (DGEBF), the phenolic and cresol novolacs, and the cycloaliphatic epoxides. Liquid DGEBAs synthesized from petrochemical derivatives are most common. They are readily adaptable for electrical and electronic device encapsulation. DGEBF is less viscous than DGEBA. The epoxy novolacs, essentially synthesized in the same way as DGEBA, are primarily solids. Because of their relatively superior elevated-temperature performance, they are widely used as molding compounds. The cycloaliphatic epoxides or peracid epoxides, usually cured with dicarboxylic acid anhydrides, offer excellent electrical properties and resistance to environment exposure.

A new class of popcorn resistant (during reflow soldering in printed wiring assembly) ultralow-stress epoxy encapsulating resins has been synthesized. Ultralow-stress epoxy molding compounds exhibit almost no shrinkage while curing. The absence of stress is achieved by including chemicals that inhibit cross-links between polymer chains in the plastic. Without cross-links, the chains are less likely to pull toward one another as the plastic cures. These new compounds result in PEMs that routinely survive solder immersion shock tests without failure. Table 2.10 gives the mechanical and electrical properties of selected epoxy resins.

Table 2.11 summarizes PEM encapsulating properties of commonly used polymeric materials. Except for epoxy resins, all others presented have only niche applications in PEM manufacture.

### 2.6.1 Fillers and Coupling Agents

Fillers are employed in epoxy resins for a variety of reasons. They are used to modify the properties and characteristics of epoxies. The principal functions that filler is used for are to control the viscosity, reducing the shrinkage and coefficient of thermal expansion, effecting cost reduction, and coloring the epoxy resins.

ASTM D-883 defines filler as "a relatively inert material added to a plastic to modify its strength, permanence, working properties, or other qualities or to lower costs." Epoxy resins cannot be used alone for encapsulants due to their high CTE and low thermal conductivity. Inert inorganic fillers are added to the molding compound to lower the coefficient of thermal expansion, increase thermal conductivity, raise the elastic modulus, prevent resin bleed at the molding tool parting line, and reduce encapsulant shrinkage during cure (and thus, reduce residual thermomechanical stress). Microstructurally, particle shape, size, and distribution in the chosen filler dictate the rheology of the molten epoxy molding compound. The advantages and disadvantages of the use of the epoxy are given in Table 2.12. Common fillers and the properties they affect are given in Table 2.13.

Historically, the filler with the optimum combination of required properties, has been crystalline silica or alpha quartz. A typical crystalline silica filled molding compound, loaded to 73% by weight, offers a coefficient of thermal expansion of about 32 ppm/°C and a thermal conductivity of around 15 kW/m°C. Crystalline silica was replaced by ground fused silica, which provided lower density and viscosity. A formulation with similar moldability as one with 73% crystalline silica, requires 68% fused silica and produces a coefficient of thermal expansion of around 24 ppm/°C and thermal conductivity about 16 kW/m°C.

Addition of particulate fillers generally reduce strength characteristics such as tensile strength and flexural strength. Fillers do not usually provide any

**Table 2.10 Mechanical and electrical properties of selected epoxy resins**

| Curing agent type | Anhydride | Aliphatic amine | Aromatic amine | Catalytic | High-temperature anhydride | Epoxy novolac anhydride | Dianhydride |
|---|---|---|---|---|---|---|---|
| Typical curing agent | HHPA[a] | DETA[b] | MPDA[c] | $BF_3$-MEA[d] | NMA[e] | NMA | PMDA[f] |
| Parts | 78 | 12 | 14 | 3 | 90 | 87.5 | 55 |
| Resin | DGEBA[g] | DGEBA | DGEBA | DGEBA | DGEBA | Novolac | DGEBA |
| Catalyst, type/phr | BDMA[h]/1 | - | - | - | BDMA/1 | BDMA/1.5 | - |
| Cure cycle (h/°C) | 4/150 | 24/25 + 2/200 | 2/80 + 2/200 | 2/120 + 2/200 | 4/150 + 3/200 | 2/90 + 4/165 + 16/200 | 4/150 + 14/200 |
| Heat deflection temperature, (°C) | 130 | 125 | 150 | 174 | 170 | 195 | 280 |
| Tensile strength (MPa)<br>at 25°C<br>at 100°C | <br>72<br>37 | <br>75<br>32 | <br>85<br>45 | <br>43<br>29 | <br>75<br>46 | <br>66<br>- | <br>22<br>14 |
| Tensile modulus (GPa)<br>at 25°C<br>at 100°C | <br>2.80<br>2.10 | <br>2.87<br>1.80 | <br>3.30<br>2.20 | <br>2.70<br>1.90 | <br>3.40<br>1.40 | <br>2.94<br>- | <br>2.70<br>- |

**Table 2.10 (cont.)**

| Curing agent type | Anhydride | Aliphatic amine | Aromatic amine | Catalytic | High-temperature anhydride | Epoxy novolac anhydride | Dianhydride |
|---|---|---|---|---|---|---|---|
| Elongation (%) | | | | | | | |
| at 25°C | 5.6 | 6.3 | 5.1 | 3.0 | 2.7 | 3.2 | - |
| at 100°C | 11.1 | 9.0 | 7.2 | 9.1 | 7.2 | - | - |
| Flexural strength (MPa) at 25°C | 126 | 103 | 131 | 112 | 112 | 147 | 59 |
| Flexural modulus (GPa) at 25°C | 3.22 | 2.48 | 2.80 | - | 4.80 | 3.69 | 3.61 |
| Compressive strength (MPa) at 25°C | 111 | 224 | 234 | - | 116 | 159 | 254 |
| Compressive modulus (GPa) at 25°C | 5.09 | 1.86 | 2.13 | - | 0.73 | 2.22 | 2.41 |
| Volume resistivity (ohm cm) | | | | | | | |
| at 25°C | $4 \times 10^{14}$ | $2 \times 10^{16}$ | $1 \times 10^{16}$ | - | $2 \times 10^{14}$ | $10^{16}$ | - |
| at 100°C | $2 \times 10^{14}$ | $5 \times 10^{12}$ | - | - | $3 \times 10^{12}$ | - | - |
| at 150°C | $1 \times 10^{10}$ | - | - | - | $3 \times 10^{11}$ | - | - |
| at 200°C | $5 \times 10^{7}$ | - | - | - | $1 \times 10^{9}$ | - | $7 \times 10^{11}$ |

**Table 2.10 (cont.)**

| Curing agent type | Anhydride | Aliphatic amine | Aromatic amine | Catalytic | High-temperature anhydride | Epoxy novolac anhydride | Dianhydride |
|---|---|---|---|---|---|---|---|
| Dielectric constant, 60 Hz | | | | | | | |
| at 25°C | 3.3 | 3.49 | 4.60 | 3.53 | 3.36 | 3.43 | 3.57 |
| at 100°C | 3.3 | 4.55 | 4.65 | 3.51 | 3.39 | - | 3.70 |
| at 150°C | - | 5.8 | 5.40 | 3.69 | 3.72 | - | 3.732 |
| Dielectric constant, 1 KHz | | | | | | | |
| at 25°C | 3.2 | 3.26 | 4.50 | 3.56 | 3.26 | 3.45 | 3.52 |
| at 100°C | 3.4 | 4.65 | 4.60 | 3.37 | 3.01 | - | 3.65 |
| at 150°C | 4.0 | 4.83 | 4.90 | 3.59 | 3.29 | - | 3.70 |
| Dielectric constant, 1 MHz | | | | | | | |
| at 25°C | 3.2 | 3.33 | 3.85 | 3.20 | 2.99 | 3.20 | 3.34 |
| at 100°C | 3.4 | 4.36 | 4.30 | 3.25 | 3.90 | - | 3.52 |
| at 150°C | 3.6 | 4.23 | 4.60 | 3.30 | 3.29 | - | 3.61 |
| Dissipation factor, 60 Hz | | | | | | | |
| at 25°C | 0.005 | 0.005 | 0.008 | 0.008 | 0.008 | 0.0066 | 0.007 |
| at 100°C | 0.003 | 0.002 | 0.008 | 0.008 | 0.009 | - | 0.005 |
| at 150°C | 0.003 | 0.003 | - | - | 0.009 | - | 0.008 |

**Table 2.10 (cont.)**

| Curing agent type | Anhydride | Aliphatic amine | Aromatic amine | Catalytic | High-temperature anhydride | Epoxy novolac anhydride | Dianhydride |
|---|---|---|---|---|---|---|---|
| Dissipation factor 1 KHz | | | | | | | |
| at 25°C | 0.007 | 0.006 | 0.017 | 0.009 | 0.006 | 0.0058 | 0.008 |
| at 100°C | 0.004 | 0.056 | 0.006 | 0.011 | 0.003 | - | 0.004 |
| at 150°C | 0.07 | 0.048 | 0.035 | 0.046 | 0.039 | - | 0.004 |
| Dissipation factor, 1 MHz | | | | | | | |
| at 25°C | 0.013 | 0.034 | 0.038 | 0.024 | 0.021 | 0.016 | 0.022 |
| at 100°C | 0.015 | 0.048 | 0.02 | 0.015 | 0.013 | - | 0.019 |
| at 150°C | 0.02 | 0.033 | 0.015 | 0.012 | 0.013 | - | 0.013 |

[a] hexahydropthalic anhydride; [b] diethylenetriamine; [c] metaphenylenediamine; [d] boron trifluoride monoethylene; [e] nadic methyl anhydride; [f] pyromellitic dianhydride; [g] diglycidyl ether of bisphenol A; [h] benzyldimethylamine

**Table 2.11 Common polymeric encapsulants for microelectronic packages**

| Polymeric material | Properties | Advantages and disadvantages |
|---|---|---|
| Epoxies | • good chemical and mechanical protection<br>• low moisture absorption<br>• suitable for all thermosetting processing methods<br>• excellent wetting characteristics<br>• ability to cure at atmospheric pressure<br>• excellent adhesion to a wide variety of substrates under many environmental conditions<br>• thermal stability up to 200°C | • high stress<br>• moisture sensitivity<br>• short shelf life (can be extended under low temperature storage conditions) |
| Silicones | • low stresses<br>• excellent electrical properties<br>• good chemical resistance<br>• low water absorption<br>• thermal stability up to 315°C<br>• good UV resistance | • low tensile tear strength<br>• high cost<br>• attacked by halogenated solvents<br>• poor adhesion<br>• long cure time |

**Table 2.11 (cont.)**

| Polymeric material | Properties | Advantages and disadvantages |
|---|---|---|
| Polyimides | • good mechanical properties<br>• solvent and chemical resistant<br>• excellent barriers<br>• thermal stability from -190 to 600°C | • high cure temperature<br>• dark color (some)<br>• attacked by alkalis<br>• high cost<br>• require surface priming and/or coupling agents to improve adhesion properties<br>• low moisture resistance<br>• low dielectric constant<br>• lower thermal stability<br>• provides improved stress relief |
| Phenolics | • high strength<br>• good moldability and dimensional stability<br>• good adhesion<br>• high resistivity<br>• thermal stability up to 260°C<br>• low cost | • high shrinkage<br>• poor electrical properties<br>• high cure temperature<br>• dark color<br>• high ionic concentration |
| Polyurethanes | • good mechanical properties (toughness, flexibility, resistance to abrasion)<br>• low viscosity<br>• low moisture absorption<br>• ambient curing possible<br>• thermal stability up to about 135°C<br>• low cost | • poor thermal stability<br>• poor weatherability<br>• flammable<br>• dark colors<br>• high ionic concentration<br>• inhomogeneities due to poor mixing of 2-part system |

**Table 2.12 Advantages and disadvantages of using fillers [Ellis 1993]**

| Advantages | Disadvantages |
|---|---|
| Reduced formulation cost | Increased weight |
| Reduced shrinkage | Increased viscosity |
| Improved toughness | Machining difficulties |
| Improved abrasion resistance | Increased dielectric constant |
| Reduced water absorption | |
| Increased heat deflection temperature | |
| Decreased exotherm | |
| Increased thermal conductivity | |
| Reduced thermal expansion coefficient | |

**Table 2.13 Common fillers and their property modification [Ellis 1993]**

| Filler | Property |
|---|---|
| Alumina | Abrasion resistance, electrical resistivity, dimensional stability, toughness, thermal conductivity |
| Aluminum trioxide | Flame retardation |
| Beryllium oxide | Thermal conductivity |
| Calcium Silicate | Tensile strength, flexural strength |
| Copper | Electrical conductivity, thermal conductivity, tensile strength. |
| Silica | Abrasion resistance, electrical properties, dimensional stability, thermal conductivity, moisture resistance |
| Silver | Electrical conductivity, thermal conductivity |

significant enhancement of glass transition temperature, or other measures of heat distortion temperatures. Fillers also allow the possibility of modifying various thermal characteristics including the thermal conductivity, and thermal expansion coefficient. Thermal conductivity can be increased by a factor of about five by the addition of fillers such as alumina and copper. Generally, an increase in filler concentration increases thermal conductivity. Figure 2.10 shows data on effect of filler content on thermal conductivity. Thermal conductivity can also be enhanced by the addition of other filler materials such as aluminum nitride, silicon carbide, magnesium oxide and silicon nitride [Rosler 1989]. However, Proctor and Solc [1991] have shown the futility of using filler materials for thermal conductivity more than 100 times of that of the base resin, and estimated

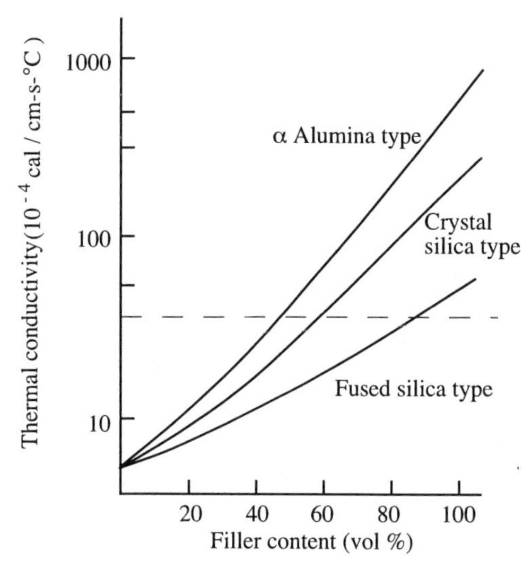


**Figure 2.10** *Relationship of filler content and thermal conductivity of the melting compound [Rosler 1989]*

that a practical limit of thermal conductivity improvement is about 12 times that of the thermoset resin. Maximization of filler loading can only be achieved by using multimodal size and shape distribution of filler particles within the polymer matrix. The influence of varying the thermal conductivity of the epoxy molding compound on the thermal performance of the IC package has also been evaluated [Chen et al. 1992]. In this case, by varying the filler types (e.g., fused and crystalline silica, aluminum oxide, aluminum and boron nitride, silicon carbide, and diamond), the particle sizes, and the filler distribution, the thermal conductivity of the composites was estimated. The resulting effectiveness in dissipating heat away from the package was computed for three different surface mount packages, namely, SOIC 8-lead, 16-lead wide, and 24-lead wide. The results indicated that an enhanced mold compound can cause a decrease in thermal resistance of the package equivalent to what can be achieved with a molded-in heat spreader. However, when thermally enhanced leadframes are used, the modified compounds are less effective.

In low-stress epoxy resins, spherical silica particles are usually blended with crushed silica to further lower the coefficient of thermal expansion, by increasing filler loading with a nonlinear increase of viscosity. However, an increase in the melt viscosity of the molding compound can increase void density and increase the difficulty of achieving a uniform encapsulant flow over large areas. Variables such as particle size, particle size distribution, particle surface chemistry, and particulate volume fraction have been found to be the most important variables necessary for optimum property enhancement. Figure 2.11 shows an SEM of the particle size distribution of a typical spherical fused silica additive used in the found fused silica filler stock. The effect of the filler figure ratio (angular: spherical) as measured by the molding compound spiral flow length for two volume-percentages filler loadings is shown in Figure 2.12. However as Figure 2.13 shows, moisture ingress susceptibility of the molding compound as measured by the "popcorn" effect (see Chapter 4) during 215°C, 90-sec. vapor phase soldering increases with higher percentages of spherical silica filler.

Rosler [1989] evaluated various other fillers for optimizing the coefficient of thermal expansion and the thermal conductivity for specific applications. Figure 2.14 presents the effect of lowering the coefficient of thermal expansion of the molding compound as a function of the crystalline silica, $\alpha$-alumina, and fused silica volume percentage.

Most polymers undergo shrinkage during the polymerization and cross-linking process which can be damaging in many applications, including electronics. Incorporation of fillers reduces shrinkage by simple bulk replacement of resin with an inert compound which does not participate in the cross-linking process. Filler addition results in increased viscosity and improved toughness. Investigations have shown that the incorporation of particulate fillers such as silica, glass microspheres, and alumina trihydrate can increase the toughness of various epoxy formulations [Moloney et al. 1983, 1984, 1987, Spanoudakis et al. 1984]. One of the major advantages here is that there is an improvement in the modulus too.

For many electronic and electrical applications, electrically conductive resins are required. Most of the polymeric resins exhibit high levels of electrical resistivity. Electrical conductivity can be improved however, by the judicious use of fillers. One of the examples of improving the electrical conductivity, using fillers is in epoxy where silver is used as a filler. Silver is employed in either flake or powdered form. Sometimes other fillers, such as copper, have been used but with reduced efficiency. The popularity of silver is due to the absence of an oxide layer formation which imparts the electrical insulating characteristics. For this reason metallic fibers such as aluminum are hardly considered for this application.

A filler can be used to its best advantage if the adhesion between the polymer and itself is good. In particular the filler particle-polymer interface will not be stress-bearing and therefore provides a point of mechanical weakness. Coupling agents are used to increase the adhesion between fillers and polymers by linking them with covalent bonds [Brydson 1985]. Commonly used coupling agents include silanes, titanates, aluminum chelates, and zircoaluminates. Interfacial

(a)

(b)

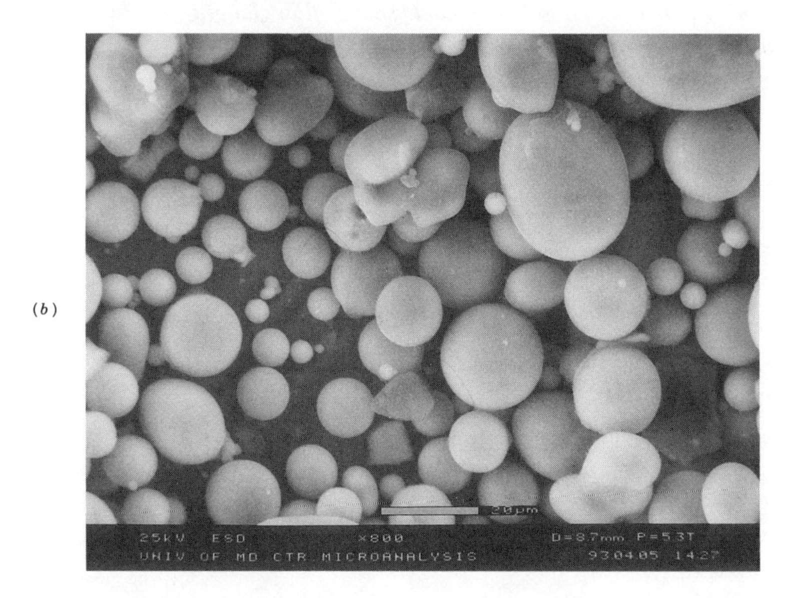


**Figure 2.11** *Micrograph of (a) a typical ground fused silica and (b) a spherical fused silica filler used in mixing with ground angular fused silica in molding compound [Courtesy of Minco Inc. TN]*

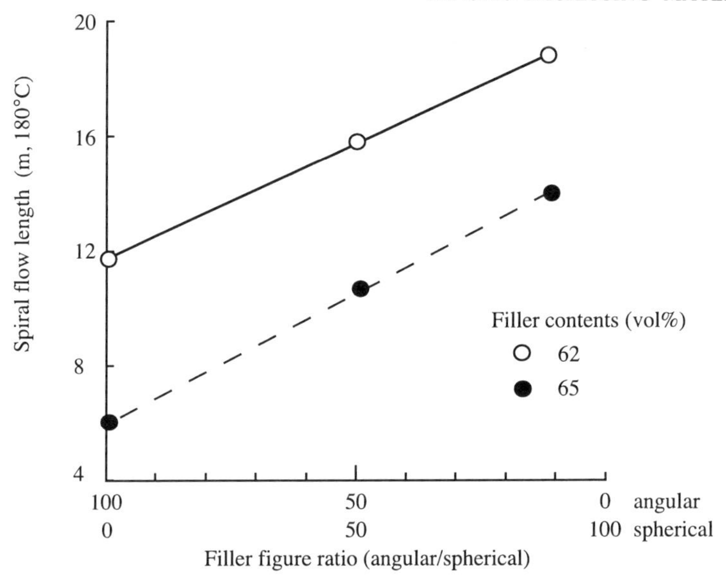


**Figure 2.12** *The effect of the filler figure ratio (angular, spherical) as measured by the molding compound spiral flow length for two volume-percentages filler loadings*

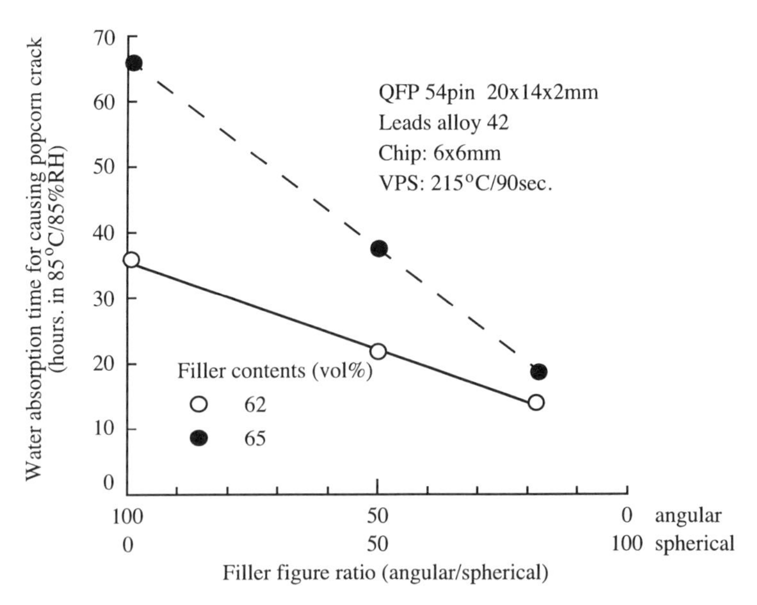


**Figure 2.13** *Moisture ingress susceptibility of the mold compound as measured by the popcorn effect*

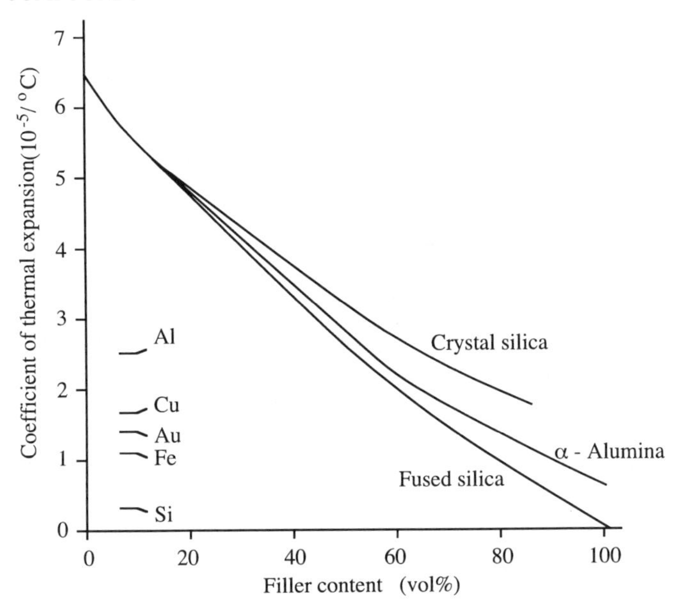


**Figure 2.14** *Effect on lowering the coefficient of thermal expansion of the molding compound as a function of the crystalline silica, α-alumina, and fused silica volume percentage*

adhesion enhances mechanical strength and the heat resistance of the encapsulant. Although, coupling agents also improve processability, some adversely affect releasability from the mold if not correctly administered. This adhesion of fillers to the polymer matrix through coupling agents also extends to the device and leadframe, reducing delamination failure.

The main source of radiation in plastic packages comes from fillers. However, the use of synthetic high-purity silica fillers in recent years has brought down the alpha emission rates (AER) of molding compounds below the detection level of most proportional counters. Ditali and Hasnain [1993] studied AER sources in an electronic part and developed Table 2.14, which shows that outside the major alpha-source contamination of the die itself, the leadframe material is more contaminated (per unit surface area) than the plastic encapsulant, and, ceramic package materials are three to four hundred times more alpha-emitting material contaminated than plastic encapsulants.

An approximate conversion factor of AER to soft-error rate is 0.001 AER ≡ 0.1% SEU per thousand operating hours for the contaminated material in contact with the die active surface and alpha-particle energies in the range of 4-6 MeV. AER of about 0.001 can be obtained from 10 ppb of U-238 (4.2 MeV alpha) or 4 ppb of Th-230 (4.7 MeV alpha).

**Table 2.14 Alpha emission rate (AER) of various processing films, leadframes, and packaging materials [Ditali and Hasnain 1993]**

| Substance (alphas/cm $^2$/hr) | AER |
| --- | --- |
| Bare silicon (Si) | 0.00020 |
| Si + CVD ox(TEOS) | 0.00164 |
| Si + plasma oxide | 0.00188 |
| Si + plasma nitride | 0.00443 |
| Si + tungsten | 0.00308 |
| Si + aluminum | 0.00682 |
| Si + polysilicon | 0.00098 |
| Si + field oxide | < 0.00010 |
| Si + BPSG | < 0.00010 |
| Si + CVD nitride | < 0.00010 |
| Fully processed w/o $WSi_x$ | 0.02400 |
| Fully processed w/$WSi_x$ | 0.04230 |
| Die coat (polyimide) | < 0.00010 |
| Leadframes | |
| 1 Meg DIP | 0.00677 |
| 1 Meg ZIP | 0.00258 |
| 256K DIP | 0.00124 |
| 64K DIP | 0.00109 |
| Packaging material | |
| Plastic | 0.00080 |
| Ceramic DIP (vendor A) | 0.02320 |
| Ceramic LCC (vendor A) | 0.02530 |
| Ceramic DIP (vendor B) | 0.03230 |
| Ceramic DIP (vendor C) | 0.02610 |

### 2.6.2 Curing Agents and Accelerators

The binder system of a plastic encapsulant product consists of an epoxy resin, hardener or curing agent, and an accelerating catalyst system. The conversion of epoxies from the liquid (thermoplastic) state to tough, hard thermoset solids is accomplished by the addition of chemically active compounds known as curing agents. Epoxidized phenol novolac and epoxidized cresol novolac are most often used. The specific composition of the resin and the hardener system, therefore, is optimized for the specific application. For example, incorporation of phenol novolacs in the matrix resin can be used to increase cure speed. In general, the glass transition curing temperature is maximized at an epoxide to hydroxyl ratio of about unity.

Curing agents and accelerators perform the primary function of setting the extent and rate of polymerization of the resin. Consequently, the selection of the agent and cure chemistry are just as important as the choice of resin. Often, the same agent is used for both cure and acceleration. The cross-linking reactions of epoxy resins are affected by aliphatic or aromatic amines, carboxylic or its derivative acid anhydrides, and complex phenols, whereas homopolymerization is affected by Lewis acids or bases and their organic salts. The most widely used curing agents are amines and acid anhydrides. While aliphatic amines react rapidly at room temperature, aromatic amine curing agents impart higher thermal stability and improved chemical resistance. Depending on the curing agent used, the reactions proceed through different chemistries. Because this reaction is virtually quantitative, resin and curing agent are mixed in the ratio of one amino hydrogen to one epoxide. In the case of anhydrides and phenols as curatives, cross-linking occurs through pendant hydroxyls. Table 2.15 lists the curing properties of different classes of curatives used with epoxy resins. Because of their excellent moldability, electrical properties, and heat and humidity resistance [Kinjo et al. 1989], phenol novolac and cresol novolac hardeners have become the dominant curatives for microelectronic packaging.

To promote cure within a reasonable period of time, it is necessary to use accelerators. Accelerators reduce the in-mold cure time and improve productivity by catalytic activity. Typical accelerators include aliphatic poylamines or tertiary amines, phenols, nonyl phenol, resorcinol, or semi-inorganic-derived accelerators such as triphenyl phosphite and toluene-*p*-sulforic acid.

For transfer molding of microelectronic devices epoxy resins are B-staged to produce storage-stable thermosetting epoxy slugs known as prepregs. The B-staged resin is formed by reacting terminal epoxides with aminohydrazides at 70 to 80°C to yield B-staged resins with excellent storage stability.

### 2.6.3 Stress-release Additives

The toughness and stress-relaxation response of the epoxy resins can be enhanced by the addition of flexibilizers and stress-relief agents. Stress-release additives lower the thermomechanical shrinkage stresses that can initiate as well as propagate cracks in the molding compound or in the device passivation layer. In terms of molding-compound property modification, stress-release agents lower the elastic modulus, improve toughness and flexibility, and lower the CTEs.

**Table 2.15 Pure DGEBA epoxy resin curing agents used in molding compounds [May 1989]**

| Curative type | Concentration range (parts/100 parts of resin) | Typical cure | | Typical postcure | | Pot life at 25°C (hr.) | Heat deflection temperature (°C) |
|---|---|---|---|---|---|---|---|
| | | Time | Temperature (°C) | Time (hr.) | Temperature (°C) | | |
| Aliphatic tertiary amines and derivatives (room temperature cure) | 12 to 50 | 4 to 7 days | 25 | 0 to 1 | 150 to 200 | 0.25 to 3 | 55 to 124 |
| Cycloaliphatic amines (moderate temperature cure) | 4 to 29 | 0.5 to 8 hr. | 60 to 150 | 0 to 3 | 150 | 0.25 to 20 | 100 to 160 |
| Aromatic amines[a] (elevated temperature cure) | 14 to 30 | 4 hr. | 80 to 200 | 2 | 150 to 200 | 5 to 8 | 145 to 175 |
| Carboxylic and anhydrides[a] (elevated temperature cure) | 78 to 134 | 2 to 5 hr. | 25 to 150 | 3 to 4 | 150 to 200 | 24 to 120 | 74 to 197 |
| Lewis acids and bases[a] (elevated temperature cure) | 3 | 7 hr. | 120 to 200 | 4 | ~200 | > 250 | 175 |
| Latent curing agents (elevated temperature cure) | 6 | 1 hr. | 175 | 1 | 175 | ∞ | 135 |

[a] Two-step cure

Inert flexibilizers, like phthalic acid esters or chlorinated biphenyls, remain as a separate phase. Reactive flexibilizers remain in the epoxy matrix as a single-phase material, and thus lower the tensile modulus and improve the ductility of the encapsulant. They also reduce the exotherm and in certain circumstances reduce shrinkage too. Flexibilizers can also improve adhesive joint properties such as lap shear, peel strength, impact strength and low temperature crack resistance. Table 2.16 gives the influence of flexibilizers on epoxy resins. Stress-relief agents, on the other hand, remain as a second dispersed phase either during [Manzione et al. 1981] or prior to cure [Nakamura et al. 1986].

In epoxy molding compounds, the major stress-relief agents used are silicones, acrylonitrile-butadiene rubbers, and polybutyl acrylate (PBA). Silicone elastomers, with their high-purity and high-temperature properties, are the most favored stress-relief agents. Silicone elastomers interface-modified with polymethyl methacrylate (PMMA) possess uniform domain sizes (1 to 100 μm), and inhibit passivation layer cracking, aluminum line deformation, and cracking.

Stress-release additives lower the thermomechanical shrinkage stresses that can initiate as well as propagate cracks in the molding compound or in the device passivation layer. In terms of molding-compound property modification, stress-release agents lower the elastic modulus, improve toughness and flexibility, and lower the coefficient of thermal expansion.

Inert flexibilizers, like phthalic acid esters or chlorinated biphenyls, remain as a separate phase. Reactive flexibilizers remain in the epoxy matrix as a single-phase material, and thus lower the tensile modulus and improve the ductility of the encapsulant. Stress-relief agents, on the other hand, remain as a second dispersed phase either during [Manzione et al. 1981] or prior to cure [Nakamura et al. 1986].

In epoxy molding compounds, the major stress-relief agents used are silicones, acrylonitrile-butadiene rubbers, and polybutyl acrylate (PBA). Silicone elastomers, with their high purity and high temperature properties, are the most favored stress-relief agents. Silicone elastomers interface-modified with polymethyl methacrylate (PMMA), possess uniform domain sizes (1 to 100 μm), and inhibit passivation layer cracking, aluminum line deformation, and package cracking.

### 2.6.4 Flame Retardants

Flame retardants in the form of halogens are added to the epoxy resin backbone because epoxy resins are inherently flammable. One of the more important flame retardants is the homologous brominated DGEBA (diglycidyl ethers of bisphenol

**Table 2.16 Influence of flexibilizers on epoxy resins**

| | Difunctional amine | | | Polysulfide | | | Polyamide | |
|---|---|---|---|---|---|---|---|---|
| Flexibilizer | - | 25 | 25 | 50 | 25 | 50 | 43 | 100 |
| Epoxy resin | 100 | 100 | 100 | 100 | 100 | 100 | 100 | 100 |
| Amine hardeners | 20 | 13.2 | 20 | 20 | 20 | 20 | - | - |
| Pot-life (min) | 20 | 69 | 44 | 76 | 13 | 6 | 150 | 150 |
| Viscosity 25°C (cP) | 3700 | 1070 | 870 | 490 | - | - | 21,0000 | 21,0000 |
| Flexural strength (MPa) | 110 | 99 | 122 | - | 105 | - | 73 | 80 |
| Compressive yield stress (MPa) | 103 | 96 | 98 | - | 85 | - | 88 | 73 |
| Impact strength (J/1.27-cm notch) | 0.95 | 1.11 | 1.4 | 10.85 | 0.68 | 2.3 | 0.41 | 0.43 |
| Heat distortion temperature (°C) | 95 | 44 | 40 | < 25 | 53 | 32 | 81 | 49 |

A). When used with normal nonhalogenated resin, these give self-extinguishing properties, flame retardancy being achieved by the bromine liberated at the decomposition temperature. Approximately 13 to 15% (wt) of bromine is required to make an unfilled epoxy flame retardant [May 1989]. Table 2.17 compares the electrical properties of a brominated bisphenol A-based epoxy resin with those of a "standard" epoxy resin. Previously, halogen-containing additives were used (as hardeners and/or fillers), but these tended to lower the heat distortion temperature. Use of halogenated epoxy resins can be used without sacrificing properties.

Antimony trioxide used as a filler is another commonly used flame retardant. However, the cost is higher. Homologous brominated DGEBA (diglycidyl ethers of bisphenol A) epoxy and heterogenous antimony oxides can also be used together [Rosler 1989].

**Table 2.17 Properties of standard and brominated epoxy resins [May et al. 1973]**

| | | |
|---|---|---|
| Formulation (parts by weight):<br>Bisphenol A-epichlorohydrin epoxy resin | 100 | - |
| Brominated bisphenol A resin | - | 100 |
| 4,4' -Methylenedianiline | 26 | 19.3 |
| Gel time, 50 g at 100°C (min) | 15 | 14 |
| Heat distortion temperature (°C) | 150 | 148.8 |
| Flexural strength (MPa) | 125.5 | 131.7 |
| Dielectric constant (10 MHz) | 3.31 | 3.28 |
| Dissipation factor (10 MHz) | 0.029 | 0.030 |
| Volume resistivity (1 min/cm) | $3.1 \times 10^{14}$ | $3 \times 10^{14}$ |

Other ways of incorporating flame retardancy in epoxy resins include using a nonreactive phosphorus-containing diluent, chlorinated waxes, aluminum hydrate, and various phosphorus derivatives. These are less popular methods because they can lead to deterioration of physical properties of the system, and limit the choice of the curing agent and filler.

Although flame retardants are a source of contamination in encapsulant formulations, the use of nonhydrolyzable compounds and ion-getters can control their corrosion-inducing effects. However, studies at Plaskon Electronic Materials [Bates 1993] show improved high-temperature storage performance of Plaskon 3400 epoxy molding compound when flame retardants are eliminated as per Figure 2.15.

### 2.6.5 Mold-release Agents

The excellent adhesion of epoxy resins to all types of surfaces is the major reason for the wide application of release agents in microelectric packaging. But these same properties make release from compounding and molding equipment difficult. Consequently, mold-release agents are needed that do not degrade the epoxy adhesion to the package components. This is achieved by controlling the release-agent activity as a function of temperature.

Mold-release agents are usually in the form of microplates that vary from

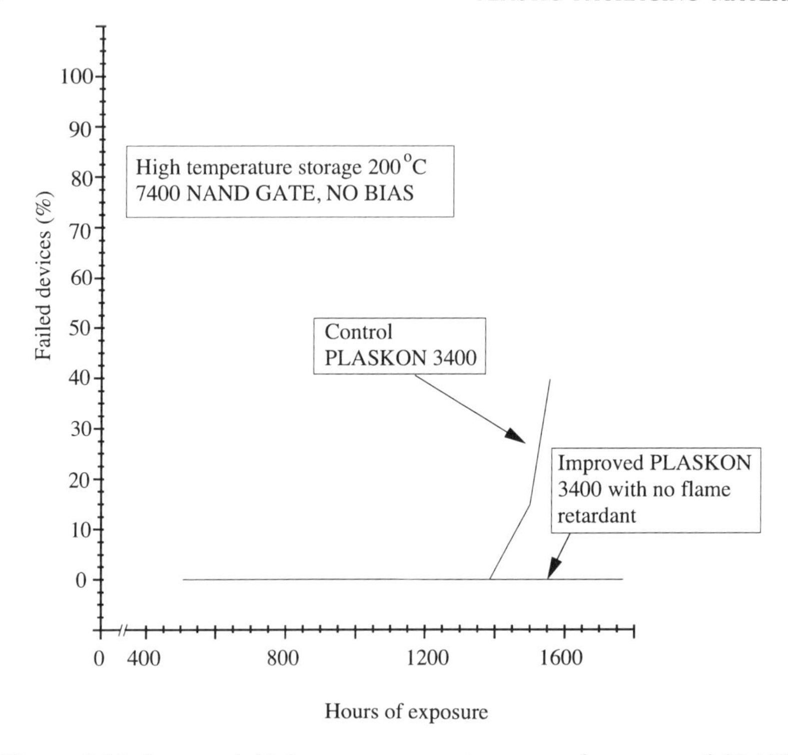


**Figure 2.15** *Improved high temperature storage performance of PLASKON 3400 epoxy molding compound when flame retardants are eliminated [Courtesy of PLASKON Electronic Materials]*

liquids to pasty solids to finely divided powders. In general, the release agent should be insoluble in the resin mixture, should not melt at the curing temperature, and should be applied in a continuous film. The selection of the release agent is determined by the material of construction of the mold and the type of encapsulation selected.

Different release agents in epoxy molding compounds are used for compounding (~100°C) and molding (~175°C) operations. Room-temperature adhesion of epoxy molding compound to leadframe is provided by the higher temperature (~175°C) active mold-release agent solidification and consequent loss of effectiveness. The release agent used for compounding, if not completely degraded, could become active at near 100°C use temperature and lead to some delamination in the package. Mold-release agents for epoxy molding compounds include silicones, hydrocarbon waxes, inorganic salts of organic acids, and fluorocarbons. Of these, hydrocarbon waxes, such as carnauba wax, are the most common for molding compounds used in microelectronic encapsulation.

Silicones and fluorocarbons have poor functional temperature selectivity, while organic acid salts can corrode metallic package elements.

### 2.6.6 Ion-getter Additives

The objective of ion-getter additives is to reduce the conductivity of any accumulated water at the metal-encapsulant interfaces inside the package, and thus to retard any electrolytic corrosive degradation processes. The incorporation of alkali-and halide-ion gettering agents into the epoxy molding compound makes residual $Na^+$, $K^+$, $Cl^-$, and $Br^-$ ions in the epoxy unavailable for dissolution into any diffused, accumulated water in the package. Ion getters in epoxy molding compounds are hydrated metal oxide powders with particles several microns in diameter. These materials react with highly active alkali and halide ions to release $OH^-$ and $H^-$ ions inside the molding compound, and form nearly water-insoluble alkali and halide compounds. In about 5 wt% concentration, ion-getter additives have been reported to reduce aluminum cathodic and chemical corrosion by one-and-one-half to three times [Mizugashira et al. 1987]. The main application of such formulated molding compounds is in packages where the encapsulant thickness is around a few mils, as in ultrathin outline packages.

### 2.6.7 Coloring Agents

Coloring agents are normally used to distinguish different device types in packages, reduce photonic activity of the device, and eliminate device visibility through the normally pale yellow epoxy encapsulant. The colors are produced by the addition of thermally stable organic dyes or pigments. Carbon black is used in most plastic-encapsulated silicon integrated circuits, even though it does add slightly to the electrical conductivity of epoxy, and can reduce moisture resistance. The concentration of carbon block is usually less than 0.5% to avoid problems related to moisture absorption and impurities.

## 2.7 CHARACTERIZATION OF MOLDING COMPOUND PROPERTIES

The molding compound is typically characterized by a set of properties that determine the suitability of the molding compound for a given application and process. From a manufacturing perspective, viscosity and flow characteristics and curing times and temperatures are important factors in determining which encapsulant materials should be used. From a performance viewpoint, key characteristics range from electrical, mechanical, and thermal properties to chemical and humidity resistance, the water solubility of contaminants, and the adhesion to chip/component surfaces. Key properties include the following:

- mold filling characteristics, resin bleed, hot hardness, and mold staining;
- spiral flow length;
- shear viscosity at low shear rates;
- cure time;
- flow resistance;
- glass transition temperature;
- thermomechanical properties;
- crack sensitivity;
- hydrolyzable ionic purity;
- moisture solubility and diffusion rates;
- adhesion strength to leadframe materials;
- moisture sensitivity of viscosity and glass transition temperature;
- alpha particle emission (if applicable);
- popcorn resistance for surface mount technology.

Technical information on the above properties and information on optimum processing conditions for the molding compound is provided by the mature suppliers. Table 2.18 gives such an example characterization for a typical molding compound. The characterization of molding compounds is given in the sub-section below.

### 2.7.1 Shrinkage Stresses

Shrinkage stresses in a molded plastic package result from polymerization and the disparities in the coefficients of thermal expansion among the various materials that are contained within the package and are in intimate contact. The following relationship can be used to calculate the magnitude of the thermal shrinkage stress, $\upsilon$, as the package goes through a thermal excursion from $T_1$ to $T_2$:

**Table 2.18 Typical characteristics of molding compound properties**

|  |  |
|---|---|
| Coefficient of thermal expansion | $16 \times 10^{-6}$ °C$^{-1}$ |
| Glass transition temperature | 155°C |
| Flexural modulus | 13,700 MPa |
| Flexural strength | 147 MPa |
| Thermal stress | 4.4 MPa |
| Thermal conductivity | $16 \times 10^{-4}$ cal/cm-sec°C |
| Volume resistivity | $7 \times 10^{16}$ Ω-cm |
| Hydrolyzable ionics | < 20 ppm |
| Uranium content | 0.4 ppb |
| Spiral flow length | 76 cm (30 in.) |
| Gel time | 23 sec |
| Hot hardness | 85 (Shore D) |

$$\sigma = c_5 \int_{T_1}^{T_2} \frac{[\alpha_p(T) - \alpha_i]}{\dfrac{1}{E_p(T)} - \dfrac{1}{E_i}} dT \qquad (2.1)$$

where $\alpha$ is the temperature-dependent coefficient of thermal expansion, $E$ is the modulus of elasticity, $c_5$ is a design-dependent geometric constant, the subscript $p$ refers to the molding compound, and the subscript $i$ refers to either the semiconductor device or leadframe, depending upon the desired calculation.

Since $1/E_i \ll 1/E_p$, $\alpha_p$ and $E_p$ can be replaced with single values for the glassy region denoted by subscript $g$ (up to the glass transition temperature). Since the encapsulant is compliant above the glass transition temperature, $T_g$, the above integral can be approximated by:

$$\sigma^* = (\alpha_{pg} - \alpha_i) E_{pg} (T_g - T_1) \qquad (2.2)$$

to provide the stress parameter $\sigma^*$.

The stress $\sigma$, or the stress parameter $\sigma^*$, is only a crude approximation of the stress level in the material and does not account for stress concentration points or other geometric and interfacial features that influence delamination, bending or cracking. The thermomechanical property values needed for $\sigma$ or $\sigma^*$ calculation are normally provided by suppliers.

The ASTM F-100 test is designed to determine the thermal stress parameter of Equation 2.2 for the molding compound. ASTM F-100 is a photoelastic experiment where the molding compound is molded around a glass cylinder. The deformation of the glass at the test temperature is determined by counting the fringe patterns produced by polarized light.

### 2.7.2 Coefficient of Thermal Expansion and Glass Transition Temperature

The coefficient of thermal expansion for any material represents a change in dimension per unit change in temperature. The dimension can be either volume, area, or length. However, the rate of expansion varies from material to material and with the temperature. Therefore, the fact that different materials expand differently with the same increase in temperature necessitates that elements attached together have the same or similar coefficients of thermal expansion to avoid the possibility of delamination. The glass transition temperature is an inflection point in the expansion versus temperature curve above which the rate of expansion (and therefore the coefficient of thermal expansion) increases significantly [Murray 1993].

The coefficient of thermal expansion (CTE) and the glass transition temperature are two very important properties measured by thermomechanical analysis (TMA) and reported by most suppliers. The tests are described in ASTM D-696 or SEMI G 13-82 standards. The ASTM D-696 uses the fused quartz dilatometer to measure the CTE. The specimen is placed at the bottom of the outer dilatometer tube with the inner one resting on it. The measuring device, which is firmly attached to the outer tube, is in contact with the top of the inner tube and indicates the variations in length of the specimen with changes in temperature. Temperature changes are brought about by immersing the outer tube in a liquid bath or other controlled temperature environment maintained at the desired temperature. Typically, a graph of expansion versus temperature is plotted. The coefficient of thermal expansion is the slope of the plot and the glass transition temperature is the intersection point between the lower temperature coefficient of thermal expansion ($\alpha_1$) and the higher temperature rubbery region coefficient of thermal expansion ($\alpha_2$). The glass transition temperature separates the glassy temperature region from the rubbery temperature region of an amorphous polymer. The glass transition temperature, being a manifestation of the total viscoelastic response of a polymer material to an applied strain, depends on the rate of strain, the degree of strain, and the heating rate. Since improper molding and postcure conditions can affect both the coefficient of thermal expansion and the glass transition temperature, most PEM manufacturers remeasure these parameters on a predetermined quality control schedule.

### 2.7.3 Mechanical Properties

It is generally recognized that lowering the stress parameter, as per Equation 2.3, leads to improved device reliability and can be accomplished by lowering the modulus, the coefficient of thermal expansion, or the glass transition temperature. The flexural strength and flexural modulus are derived from a standardized ASTM D-790-71, ASTM D-732-85 and reported by suppliers. The ASTM D-790 suggests two test procedures to determine the flexural strength and flexural modulus. The first method suggested is a three-point loading system utilizing center loading on a simply supported beam. This procedure is designed principally for materials that break at comparatively small deflections. In this procedure the bar rests on two supports and is loaded by means of a loading nose midway between the supports. The second procedure involves a four-point loading system, utilizing two load points equally spaced from their adjacent support points with a distance of either one-third or one-half of the support span. This test procedure is designed particularly for large deflections during testing. In either of the cases, the specimen is deflected until rupture occurs in the outer fiber. The flexural strength is equal to the maximum stress in the outer fiber at the moment of break. It is calculated using

$$S = \frac{3P_{rupture}l}{2b_{beam}^{2}d} \tag{2.3}$$

where $S$ is the flexural strength, $P_{rupture}$ is the load at rupture, $l$ is the support span, $b_{beam}$ is the width of the beam, and $d$ is the depth of the beam. Flexural modulus is calculated by drawing a tangent to the steepest initial straight-line portion of the load deflection curve and is given by

$$E_B = \frac{l^3 m}{4b_{beam}d^3} \tag{2.4}$$

where $m$ is the slope of the tangent to initial straight line portion of the load deflection curve, and $E_B$ is the flexural modulus. The tensile modulus, tensile strength and percent elongation are derived from ASTM D-638 and D2990-77 test methods.

Tensile properties of molding compounds, determined according to ASTM D-638, uses "dog-bone" shaped molded or cut specimens with fixed dimensions and held by two grips at the ends. Care is taken to align the long axis of the specimen and the grips with an imaginary line joining the points of attachment of the grips to the machine. They are incrementally loaded to obtain stress-strain data at any desired temperature. A typical curve is shown in Figure 2.16. The tensile strength can be calculated by dividing the maximum load (in N) by the original minimum cross-section area of the specimen (in $m^2$). The percentage elongation is calculated by dividing the extension at break by the original gage length and this ratio is expressed as a percentage. The modulus of elasticity is obtained by calculating the slope of the initial linear portion of the stress-strain curve. If Poisson's ratio for the material is known or separately determined from tensile strain measurements, the shear modulus of the molding compound can be estimated. It is important to note that the stresses encountered in PEMs are actually a complex mixture of tensile and shear stresses.

The evaluation of the cracking potential of the molding compound is particularly important for devices where a relatively small amount of molding compound surrounds a relatively large die (e.g., memories, SOPs, and ultrathin packages). In the absence of any standard procedure for such evaluation, ASTM D-256A and D-256B Izod impact test procedures are commonly followed. The specimen is held as a vertical cantilever beam in test method ASTM D-256A and is broken by a single swing of the pendulum with the line of initial contact being at a fixed distance from the specimen clamp and from the centerline of the notch and on the same face as the notch. A variation of this test is the ASTM D-256B, where the specimen is supported as a horizontal simple beam and is broken by the single swing of the pendulum with the impact line midway between the supports and directly opposite the notch. These are overstress tests for the cracking potential of epoxy molding compounds even under extreme thermomechanical stress conditions and do not test their important region of viscoelasticity. However, they are simulative of trim and form, and handling impact-induced cracking susceptibility.

A fracture test that models the actual strain history of the package for thermomechanically induced failure is the previously mentioned ASTM D-790-71 three-point flexural bending test used to determine flexural modulus. Here a 0.05-mm (2-mil) diameter center-notched rectangular specimen is center strained at a rate that simulates the manufacturing cycle, i.e., 20%/min for liquid-to-liquid thermal shock and 0.1%/min for device on-off operation in air (1°C/min). The area under the stress-strain curve is proportional to the energy to break at the test temperature. The low-temperature data are usually the discriminating factor between molding compounds because the molded body experiences the greatest stress at lower temperatures far removed from the molding temperature.

Since the melt viscosity of a molding compound is shear rate-dependent and a typical mold subjects different shear rates at different points of the molding compound flow channel, the expected shear rates for a specific molding tool need first be calculated under no-slip boundary conditions. The general ranges of experienced shear rates are hundreds of reciprocal seconds in the runner, thousands through the gate, and tens in the cavity. Considerations must also be given to the temperature and time dependence of the shear rate-dependent molten

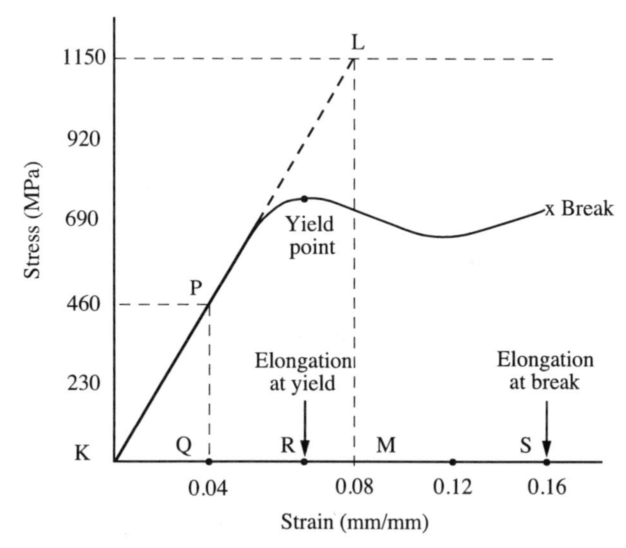


**Figure 2.16** *A typical curve of incrementally loaded specimens from stress-strain data*

molding compound viscosity.

Selection of molding compounds based on shear dependence of viscosity should identify one that has the lowest viscosity at low shear rates and high cavity temperatures for wire sweep and/or paddle shift prone devices and in multicavity molds where complete filling before gelation is a concern [Nguyen 1993]. A material whose viscosity is more temperature insensitive performs better in all suboptimal tool designs. The time dependency of molten molding compound viscosity originates from two opposite phenomena. Although the epoxy curing process will increase the average molecular weight and hence viscosity with time, the decrease of viscosity with increasing molding temperature in the tool will overwhelm it in the early phases. But ultimately, both the molecular weight and the viscosity go to infinity at gelation. Flow-induced stresses, particularly in distant cavities, could thus be very significant at the latter stages of mold filling. Consequently, molds with longer flow lengths and longer flow times need molding compounds with longer gel times at the 150 to 160°C mold filling temperature. This requirement is a trade-off for higher productivity.

### 2.7.4 Spiral Flow Length

The ASTM D-3123 or SEMI G 11-88 test consists of flowing molding compound through a spiral coil of semicircular cross section until the flow ceases. Although it is not a viscometric test, it is a measure of fusion under pressure, melt viscosity, and gelation rate. The spiral flow test is used both to compare different materials and control molding compound quality. However, it cannot resolve the viscous and kinetic contributions to the flow length. Higher viscosity and longer gel time could compensate each other to provide identical flow lengths. The molding tool used in this test is showed in Figure 2.17.

A "ram-follower" device, specified in SEMI G 11-88, is used to separately record ram displacement versus time, and consequently separate the viscosity-induced flow time from the gel time in the total spiral length formation time of different molding compounds. The molding tool used in spiral flow test covers the several hundred per second shear rate range, and thus the test results have no impact on yield and productivity. However, as an indicator of gel time or maximum flow time of thermoset molding compounds, the results of this test can help define mold flow lengths and flow times compatible with the particular molding tool.

### 2.7.5 Molding Compound Rheology from Molding Trial

This is a test for rheological compatibility of a molding compound with a device to be packaged in a molding tool by trial molding operation. Rheological incompatibility can cause wire sweep, die paddle shifting, or incomplete filling of the mold cavity resulting in voids. X-ray analysis of the molded packages and cross-sectioning of the molded bodies through the paddle support are the primary methods of evaluation for these trials. Long wire bond spans (> 2.5 mm) and oversized paddle supports are normally used to create worst case scenarios of wire sweep and paddle shift.

The mold-filling characteristics are controlled by the pressure drop through the gates where the molding compound experiences maximum shear stress. Incomplete fill problems originate from molding compounds that have higher viscosities at these high deformation rates in the gates. This flow closely approximates flow through a sudden contraction or a converging channel, and has both shear and extensional components.

Statistically confident sampling of all stochastic phenomena, such as gate clogging by gel or filler particles, requires a large number of molded packages. The analysis of mold trial results should consider molding compound density and the consequent package sectioning analysis for porosity appraisal. Figure 2.18 shows a typical mold map displaying a variety of runners.

Incomplete filling of the mold (because of low packing pressure) in just a few cavities effectively ruins all the devices loaded because it leads to high porosity

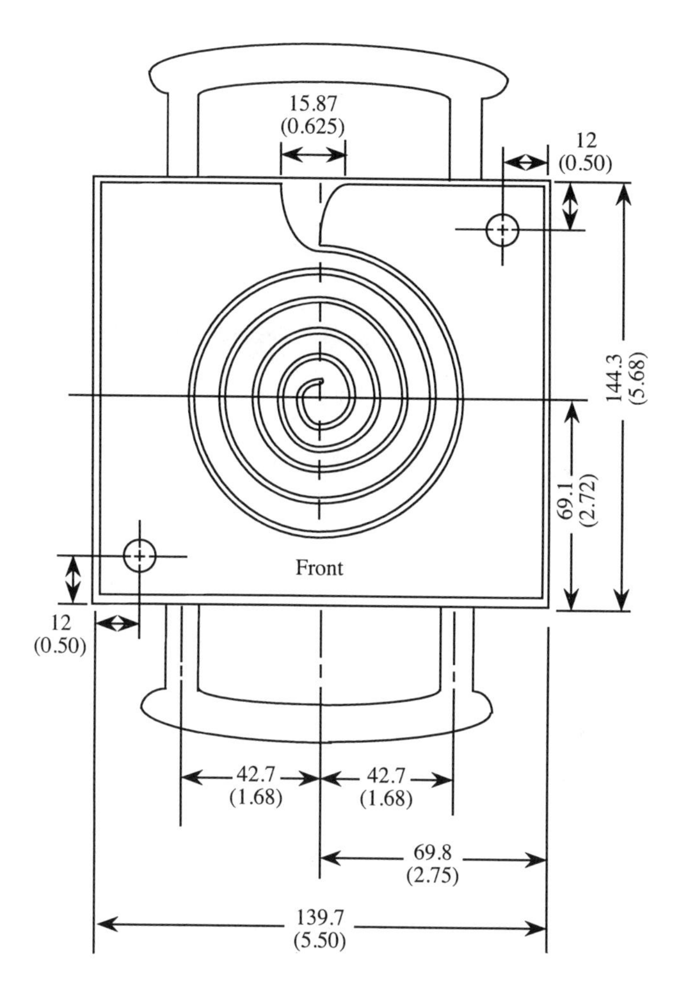


**Figure 2.17** *Molding tool used for the spiral flow length test from ASTM D-3123*

in the encapsulant that would lead to excessive moisture penetration. It is more common in molding tools with a large number of cavities (> 150) and with packages of large volume such as PQFPs and large chip carriers or with package designs with four-sided leads requiring corner gating.

Resin bleed and flash are molding problems where molding compound oozes out of the cavity and onto the leadframe at the parting line of the mold. Whereas flash is caused by the escape of the entire molding compound, resin bleed includes only the strained out resin. Although root causes of these two problems can be traced both to the processing conditions of molding, and mold design and

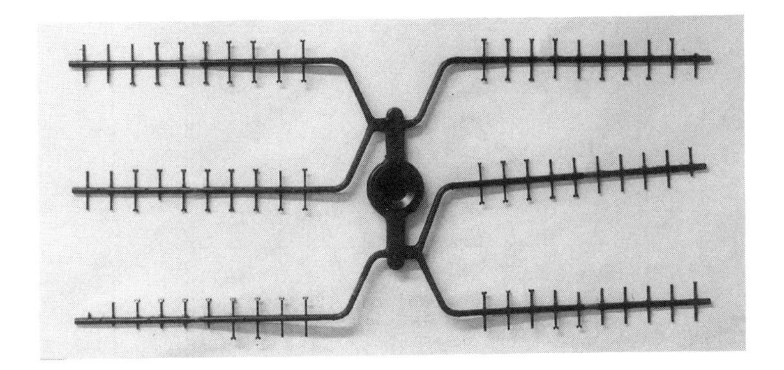

**Figure 2.18** *Molding mapping showing the runners*

mold defects, resin bleed is considered to be more molding compound related. Resin bleed occurs more often with formulations containing low viscosity resin and large filler particles, and with process conditions where excessive packaging pressure is maintained after cavity fills and the use of too low a clamping pressure on the mold halves.

SEMI G 45-88 is a standardized test for assessing a material's potential for resin bleed and flash. It is a transfer molding experiment that measures the flow of molding compound in a shallow channel mold (6 to 75 μm) and simulates flash and bleed in production tools. Propensity of resin bleed and flash from improper molding compound properties is indicated by long spiral flow lengths obtained in that test.

Moisture has a profound effect of decreasing the viscosity of epoxy molding compound and the degree of this effect is a function of the additives and curing agents used in different formulations [Blyler et al. 1986]. Compared to dry conditions, this decrease of molten molding compound viscosity can be 40% or more (at ~0.2 wt% water or higher).

The effect of moisture on the shear thinning behavior, shown in Figure 2.19, indicates that viscosity is simply lowered with little change in shear rate dependence and power-law index. Not withstanding the fact that moisture-induced melt viscosity lowering is beneficial in overcoming flow-stress induced and mold filling problems, excessive moisture content can cause excessive resin bleed and voids. Consequently, moisture uptake sensitivity, determined by the shear rate dependent viscosity test, is a necessary factor in the selection of a molding compound.

### 2.7.6 Characterization of Molding Compound Hardening

Hardening of a molding compound is the progressive thermal polymerization of the matrix resin that leads to gelation in the latter stages of the molding process, an ejectable state of hardness during cure, and complete chemical conversion during postcure. The productivity of plastic package molding depends on the rate of this chemical conversion.

Mold filling can occur in as little as 10 sec at 150 to 160°C (specified by the supplier), and the cure time required before the parts can be ejected from the mold can range from 1 to 4 min. The cure time is about 70% of the molding cycle time. Shorter cure times will generally have shorter flow times into mold cavities before gelation. Multiplunger machines are designed to handle these short flow times and cure times to provide high molding productivity. Most molding tools require molding compounds that flow for 20 to 30 sec and then cure to an ejectable state in less than one additional minute. Standard evaluation tests for these characteristics of molding compounds are discussed below.

### 2.7.7 Gel Time

The gel time of a thermoset molding compound is usually measured with a gel plate. In gel time evaluation with a gel plate, a small amount of the molding compound powder is softened to a thick fluid on a precisely controlled hot plate (usually set at 170°C) and periodically probed to determine gelation. The gel time is a qualitative point in the process where the material can not be smeared into a thin coating. SEMI G 11-88 standard recommends using the spiral flow test as a comparative evaluator. Gel times indicate the productivity of a molding compound. Shorter gel times lead to faster polymerizations rates and shorter times for mold cycle, increasing production.

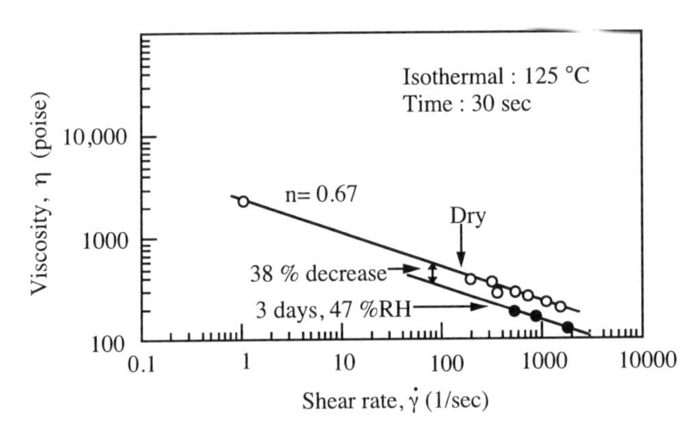


**Figure 2.19** *Effect of moisture on the shear thinning behavior [Blyler et al. 1986]*

### 2.7.8 Adhesion

Poor adhesion of a molding compound to the die, die paddle, and leadframe can lead to such failures as "popcorning" during assembly (see Chapter 4), corrosion, stress concentrations and subsequent thermomechanical failure, package cracking, chip cracking, and chip metallization deformation (See Chapter 5). Consequently, adhesion is one of the most important discriminating properties of a molding compound to be chosen for a particular physical and materials design of a package. The theory and practice of adhesion of integrated circuits molding compounds to package elements have been treated by Kim [1991] and Nishimura et al. [1989].

A common adhesion test for leadframes to the molding compound is to mold plugs of the molding compound on the properly surface-treated leadframe tab, and then determine adhesion strength in a tensile tester. Molding process used in this test must be identical to that used in production. For maximum simulation of manufacturing conditions, a custom designed leadframe is used in a production mold to generate the adhesion test specimens. Determination of adhesion strength to other package materials such as silicon die, polyimide, and silicone overcoat is accomplished by molding the compound on the flat surface of the material and measuring the force required to pull them apart. Needless to say, very careful interpretation of the adhesion test data is required to differentiate between most state-of-the art molding compounds.

### 2.7.9 Polymerization Rate

The novolac epoxy polymerization reaction includes several competing reactions among three or four reactive species. The chain segments that form are complicated and difficult to predict. Thus, thermal analysis methods, which assume that the fraction of the total heat of reaction liberated is proportional to the fraction of complete chemical conversion, are preferred for these types of highly filled opaque systems. Several different empirical forms have been offered to fit conversion data for the epoxy molding compounds. They do not reflect the molecular dynamics of the reaction, but are instead phenomenological in that they assess the engineering behavior of the reaction without a theoretical basis for the reaction mechanism or reaction order. Hale et al. [1987, 1988, 1989] developed one of the most noteworthy forms:

$$\frac{dX}{dt} = (k_{r1} + k_{r2}X^{m_r})(1 - X)^{n_r} \tag{2.5}$$

where the four fitting parameters for the conversion of epoxide groups, $X$, as a function of reaction time are as follows: $m_r$ and $n_r$ are the pseudo-reaction orders, and $k_{r1}$ and $k_{r2}$ are rate constants. For a typical epoxy molding compound $m_r = 3.33$, $n_r = 7.88$, $k_{r1} = \exp(12.672 - 7560/T)$, and $k_{r2} = \exp\ (21.835 - 8659/T)$ [Hale et al. 1989].

Figure 2.20 shows isothermal fractional conversion of epoxide groups with a drop-off in reaction rate near complete conversion. These conversion constants differ from one molding compound to the other and thus form the basis of evaluation of polymerization rate of the compound in question.

Differential scanning calorimetry (DSC) has been used to obtain the degree of polymerization of filled molding compounds [Hale 1988]. Measuring the heat of reaction versus time during an isothermal cure, it expresses the fractional conversion as a function of time as equal to the fractional total liberation of heat:

$$\frac{\Delta H_{t=t_1}}{\Delta H_{total}} = \frac{X}{100} \tag{2.6}$$

An extensive analysis of the polymerization kinetics is generally not required for material selection. The secondary effects of cure kinetics such as gel time, mechanical properties, and glass transition temperature are sufficient to effectively compare different molding compounds.

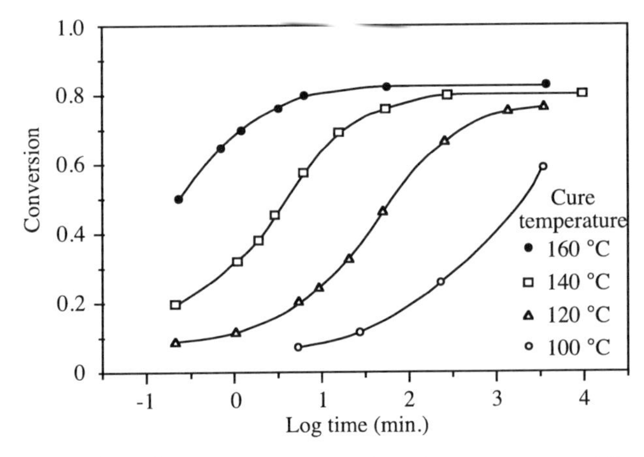


**Figure 2.20** *A plot of conversion versus time for an epoxy molding compound during cure [Hale et al. 1987]*

### 2.7.10 Hardening

Hot hardness or green strength is the stiffness of the material at the end of the cure cycle. Certain degree of this hot hardness is required before the molded strip of parts can be ejected safely from specific molds. Ejection of the strip from the molding tool is also dependent on the characteristics of the mold. These mold characteristics are the draft angle of the vertical surfaces, the surface finish of the tool, and the number and size of ejector pins. Different molding compounds attain this green strength at different points of cure cycle due to either percentage conversion achieved or low modulus above glass transition temperature. It is thus a productivity issue and can either be determined by molding trial or supplied by the vendor. A hot hardness value of about 80 on the Shore D scale within 10 sec of opening the mold is considered acceptable.

### 2.7.11 Postcure

Most epoxy molding compounds require about 4 hr. of postcure at 170 to 175°C for complete cure.

### 2.7.12 Moisture Effects on the Epoxy Hardening Process

Moisture content of molding compounds has a strong impact on the viscosity, void density, polymerization kinetics, and the properties of the cured material. The lowering of glass transition temperature by the use of moist molding compounds indicates that the final inalterable cross-linked network is different under humid conditions. The effect of moisture on the polymerization rate originates from the fact that moisture binding to catalyst sites diminishes its effectiveness. Because of the differences in the formulation of different molding compounds, the effect of moisture content could vary widely and needs to be evaluated from the viewpoints of required encapsulant properties.

### 2.7.13 Thermal Conductivity

Thermal conductivity is an important property of an encapsulant used for high heat dissipating devices or for devices with long duty cycles. Although they are measured with standard thermal conductance testers, the reported values vary widely depending on the instrument and the precision used.

### 2.7.14 Electrical Properties

The state-of-the art VLSI parts require close control of several electrical properties of the molding compound or the cured encapsulant for superior performance. They include dielectric constant and dissipation factor (ASTM D-150), volume resistivity (ASTM D-257), and dielectric strength (ASTM D-149). ASTM D-257 suggests various electrode systems to determine the volume resistivity by measuring the resistance of the material specimen by a measurement of the voltage or current drop under specified conditions, and the specimen and electrode dimensions. The test specimen may be in the form of flat plates, tapes, or tubes. Figure 2.21 shows the application and electrode arrangement for a flat plate specimen. The circular geometry shown in the figure is not necessary though convenient. The actual points of measurements should be uniformly distributed over the area covered by the measuring electrodes. The dimensions of the electrodes, the width of the electrode gap, and the resistance are measured with a suitable device having the required sensitivity and accuracy. The time of electrification is normally 60 sec and the applied voltage is 500 +/- 5 V. The volume resistivity is given by

$$\rho_v = \frac{A_{elec}}{t} R_v \tag{2.7}$$

where $A_{elec}$ is the effective area of the measuring electrode, $R_v$ is the measured volume resistance, and $t$ is the average thickness of the specimen.

ASTM D-149 requires that alternating voltage at a commercial power frequency, normally 60 Hz, be applied to a test specimen. The voltage is increased from zero, or from a level well below the breakdown voltage, until dielectric failure of the test specimen occurs. The test voltage is applied using

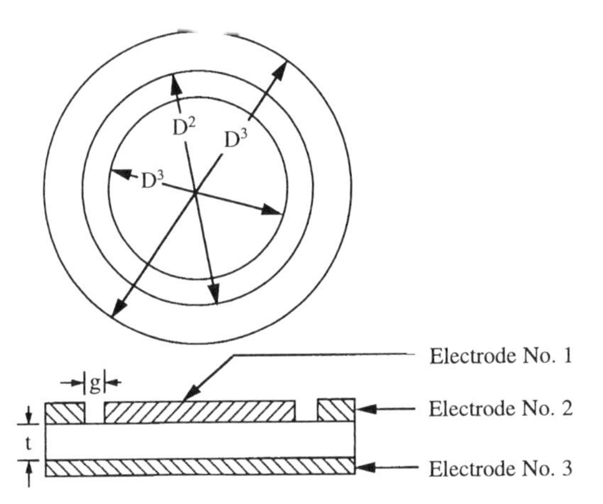


**Figure 2.21** *Electrode arrangement for measurement of volume resistance for a flat specimen [ASTM D-257 1994]*

simple test electrodes on opposite faces of specimens. The specimens may be molded, cast, or cut from a flat sheet or plate. Methods of applying voltage include a short-time test, a step-by-step test, and a slow rate-of-rise test. The second and third methods usually give conservative results.

Epoxy composites in dry environments and at room temperatures have similar electrical properties. Deterioration of some materials may occur after being stored in a moist environment at high temperatures.

### 2.7.15 Chemical Properties

Contamination level of a molding compound ultimately determines the long-term reliability of the parts fabricated with it and used under rugged conditions. SEMI G 29 standard procedure is used to determine water-soluble ionic level in epoxy molding compounds. A water extract is first tested for electrical conductivity and then quantitatively analyzed by column chromatography. Separate determination of hydrolyzable halides (from the resin, flame retardants, and other impure additives) is particularly crucial in assuring long-term reliability of PEMs. Long term (48 hrs), high pressure, and sometimes hot water (up to 100°C) extraction from the molding compound, and subsequent elemental analysis is needed for such evaluation. Modern molding compound formulations contain as little as 10 ppm of these corrosion-inducing ionics. Atomic absorption spectroscopy and X-ray fluorescence techniques are used to determine the content of other undesirable contaminants such as sodium, potassium, tin, and iron. Encapsulants used for memory devices, where single-event upsets from alpha-emitting impurities in the filler silica must be minimized, require determination of uranium and thorium content in the molding compound.

### 2.7.16 Flammability and Oxygen Index

Encapsulation compounds and plastic-encapsulated parts must conform to Underwriters Laboratory flammability ratings (UL 94 V-0, UL 94 V-1, or UL 94 V-2). Molding compounds are evaluated for flammability by UL 94 vertical burn and ASTM D-2836 oxygen index tests. In UL 94 test a 127-mm x 12.7-mm (5-in. x 1/2-in.) cured epoxy test bar of a predetermined thickness is ignited multiple times in a gas flame, and the burn time per ignition, total burn time for 10 ignitions (5 specimens), and extent of burning are recorded for proper UL rating. In the oxygen index test, the minimum volume fraction of oxygen in an oxygen-hydrogen mixture that will sustain burning of a 0.6 x 0.3 x 8-cm molded bar of epoxy molding compound is specified.

### 2.7.17 Moisture Ingress

All plastic packages prefer low moisture diffusivity for high reliability and minimum soldering heat damage. At room temperature, epoxy molding compounds absorb water to approximately 0.5% by weight on long-term exposure to 100% relative humidity. Moisture absorption rate to saturation is usually determined by weighing molded dry parts exposed to a given relative humidity and temperature for some extended periods. Desorption rates of moisture saturated parts are similarly determined by drying at an elevated temperature. The selection of an epoxy molding compound from moisture ingress evaluation is a very recent phenomenon, and is currently quite subjective even for a specific part and its application environment.

## 2.8 OTHER ENCAPSULANTS

By far, epoxy molding compounds make up the largest share of polymers used in the plastic-encapsulated microcircuit industry. A smaller portion of the encapsulant market concerns niche applications involving potting compounds, glob top materials, and conformal coatings.

Potting compounds are actually the forebears of molding compounds. They were formulated to encapsulate transformers and similar parts to prevent arcing. Applications have migrated since the 1950s to hybrid modules and discrete devices. Low ionic purity and low stress are desirable but not critical requirements. Similarly, glob-top resins were originally used to protect devices mounted in hybrid modules and stress-sensitive dies in molding compounds. As consumer packaging trends migrate toward chip-on-board applications, more development efforts are being devoted to glob-topping techniques and materials. The conformal coating is another extension of the glob-top method, except that thinner layering can be achieved.

### 2.8.1 Potting Compounds

A potting compound is usually a liquid two-component system containing a resin and a hardener or accelerator. Potting compounds can either be mixed and dispensed for encapsulation at the point of use or acquired as a premixed, frozen formulation. Because of a low viscosity requirement for room temperature application, potting compounds contain relatively low filler loading level and have high coefficient of thermal expansion. The requisite properties of potting compounds include low viscosity and long pot-life; exothermic and fast cure at low temperatures, good adhesion to all package components, resistant to filler settling, high purity, good thermal and moisture stability, low thermomechanical stress, and good electrical insulating properties.

### 2.8.2 Glob-top Encapsulants

A glob-top encapsulant is a highly viscous paste and is used to encapsulate a chip-on-board or a hybrid on a substrate. Glob-top encapsulants are thermoset epoxy or silicone resin formulations, filled to a medium level with inorganic fillers, and are traditionally meant for low-end disposable electronic products, although some glob-top encapsulants are supplied as a single-component frozen mix in a dispenser. After cure, epoxy-based glob-tops yield a rigid matrix with moderate glass transition temperature, and silicone or polyurethane-based glob-tops provide a flexible compliant mass. During application and cure, glob-tops must possess balanced viscosity, thixotropy, and bleed resistance properties. Low viscosity is particularly required for encapsulant flow into thin gaps in flip-chip and flip-TAB encapsulations. Other desirable glob-top characteristics include fast cure without voids, low level of hydrolyzable ionics, adequate mechanical strength, low stress, good moisture resistance, good adhesion to all encapsulated surfaces, appropriate thermal conductivity, and good insulating properties.

***Silicones.*** Silicone encapsulants are especially suited for protecting semiconductor devices. They have excellent electrical properties, elasticity, low moisture uptake, low ionic purity, and high thermal stability, can withstand continuous operation at temperatures of up to 250°C and short exposures at up to 375°C, and are very cost-effective. Silicones resist thermal and oxidative deterioration, electrical arcing and discharges, ozone, and ultraviolet radiation. Silicones are generally fire-resistant and have excellent values of limiting oxygen index. The relatively low surface tension of silicones (~24 dynes/cm) promotes water and moisture resistance. A continuous moisture film does not form at the silicon device/paddle/leadframe interface, to inhibit conductive pathways. Moreover, silicones do not contain mobile ions and, of themselves, do not readily absorb water. Since they are elastomers or elastoplastic resins, they can withstand high thermal and mechanical shock. The principal disadvantage of silicones is their poor adhesion to most substrates and the consequent migration of corrosion-inducing contaminants along these interfaces.

Silicone materials are ideally suited for use in demanding automotive environments, where vibration damping, thermal stability from -50 to +250°C, and electrical insulation is important. The four different applications of silicones for a single hybrid module are the following [Cook and Servais 1983]:

- a high-strength, flexible, electronic-grade adhesive for the assembly of hybrid modules;
- a thermally conductive, flexible, electronic-grade adhesive to attach the hybrid substrate to the heat sink;
- a soft, totally stress-free gel encapsulant to protect the entire assembly from humidity and other pollution;
- a high-purity coating or gel for localized and selective chip protection.

Unlike organic polymers, silicones are characterized by Si-O-Si linkages, instead of C-C linkages. Silicones are generally produced through a condensation reaction of hydroxysilanes or silanols. A pure silicone material would have only silicon and oxygen in the main chain with pendant hydrogen, but the commercial materials used in microelectronics packaging are often organosilicone materials with hydrocarbon groups pendant to the Si-O backbone. Cross-linked silicone materials are obtained through hydrocarbon- or oxygen-containing bridges to other Si-O chains. They are available in a wide variety of forms, including room-temperature vulcanizing (RTV), clear or colored resins or gels, rigid resins, and transfer-molding compounds. In spite of the purity of the silicones, some catalysts used in silicone compounds can cause corrosion in aluminum metallization.

Although silicones are low-stress, high-purity materials, the generally poor adhesion between silicones and epoxy molding compounds, the low tear strength, and the very high coefficients of thermal expansion of silicones are all major drawbacks. Though the water solubility of silicones is low, the diffusivity of water in silicones is high, resulting in a greater steady-state water flux in silicones than in epoxy materials.

Flexible, primerless silicone encapsulants are available in forms ranging from an elastomer (hardness 65, shore A) to a soft dielectric gel. Silicone dielectric gels, SILGARD 527, are polymers with a low level of cross-linking based on polydimethyl siloxanes. They exhibit good pressure-sensitive adhesion to most substrates, and are therefore extremely good at protecting electronic circuits against corrosion and metal migration. Silicone gels form a cushioning gel-like mass that utilize the stress relief and self-healing properties of a liquid, thereby preventing stress damage to delicate components, even when they are subjected to heat aging. Special silicone dielectric gels retain their softness, adhesion, and electrical properties after more than 1000 cycles from -50 to +200°C.

### 2.8.3 Conformal Coating Materials

Conformal coating of integrated circuit devices are applied on top of the primary passivation layer to provide mechanical protection to the underlying circuitry and to provide environmental protection. The primary environmental protection is from moisture, water vapor condensation, corrosive chemical vapor, salt spray, and dust.

Most conformal coatings start as liquid resin formulations. They may be one- or two-part components, with the latter requiring mixing. Being liquid initially they conform to the topography of the printed wiring board or chip carrier and components, hence the term conformal coating. The consideration for coating materials parameters needed for manufacturing are number of material components, pot life of each component for inventory matching, viscosity and flow characteristics of the formulated coating material, amount of solvents in the liquid material stock (for disposal and precure treatments); and curing conditions (time and temperature). The considerations for cured conformal coating material performance in use include resistance to humidity and specific chemical environment, ease of removal and replacement of failed components, usable range of temperature, dielectric constant and loss factor, and dielectric strength. Mechanical and thermal property considerations include resistance to abrasion, flexibility, modulus of elasticity, thermal conductivity, and the coefficient of thermal expansion. The dielectric constant of the conformal coating material affects signal propagation speed on the circuit board, whereas the loss factor determines signal attenuation. The modulus of elasticity and the coefficient of thermal expansion of the cured coating control compliancy and stress on components at all use temperatures.

As shown in the Table 2.19, polyurethanes, acrylics, epoxies, and silicones are the resins most frequently used for conformal coatings. Polyimides, polyamide-imides, and silicones have been developed for electronic products that can tolerate high cure temperatures and that need protection at elevated temperatures. They are most frequently used either directly on the chip as passivating layers (5 to 15 μm), or on the chip carrier (25 to 200 μm). Silicones, with their frequency-independent low dielectric constants and loss factors, are particularly suitable for microwave circuit packaging. Polyurethanes, fluoropolymers, silicones, acrylates, and epoxies with buffer coatings have good moisture resistance and are most commonly used for components and printed wiring boards. Parylenes (polyxylene derivatives), deposited via vapor-phase polymerization on surface-mount technology printed wiring boards, provide excellent coverage and, when used in the soluble form, can be easily removed. More on the advantages and disadvantages of the use of different conformal coatings can be found in the article by Waryold [1982].

***Polyimides.*** Polyimides represent a general class of materials obtained from the reaction of aromatic dianhydrides and diamines, which, when baked to 250°C to 450°C, undergo a condensation imidization reaction to form linear polyimide polymers. The synthesis of polyimide is usually a two-step process. Polyamic acid is first synthesized by reacting an aromatic diamine with a tetracarboxylic acid anhydride. Polyimide is produced by curing the polyamic acid at temperatures between 200 and 400°C. The most common polyimide is synthesized from pyromellitic dianhydride (PMDA) and oxydianiline (ODA). There are many aromatic diamines and aromatic dianhydrides that can be reacted to provide high-stability polyimide structures. When copolymerized with other network modifiers (such as amideimide, epoxy imides, polyester imides, imide phenolics, and bismaleimide), they can be either thermosetting or thermoplastic, whereas pure linearly polymerized polyimides are always thermoplastics.

In plastic packaging, fully aromatic polyimides are generally used for passivations (both primary and secondary), stress buffers, alpha-particle barriers, and moisture barriers. Table 2.20 gives the properties of various commercially available fully aromatic polyimides. Although the chemical structures of the various polyimides are different they usually possess similar mechanical, thermal, and electrical properties. Generally, when cured, they possess excellent chemical

**Table 2.19 Common conformal coating materials**

| Coating material | Properties | Comments |
|---|---|---|
| Polyurethane resin | • good mechanical properties: toughness, abrasion resistance and flexibility<br>• good adhesion on components and boards<br>• high chemical resistance<br>• low moisture absorption<br>• low viscosity<br>• low cost | • most widely used<br>• stable only up to 135°C<br>• flammable<br>• difficult to rework<br>• unsuitable for high frequency circuits |
| Acrylic resin | • excellent moisture resistance<br>• good dielectric properties<br>• poor chemical resistance<br>• poor mechanical abrasion resistance<br>• easy application<br>• fast drying | • easy to rework<br>• suitable for automated, high-volume production<br>• ultraviolet curable modification available (urethane acrylic) |
| Epoxy resin | • excellent chemical resistance<br>• excellent mechanical properties<br>• excellent wetting for application<br>• good adhesion to all components on a board<br>• acceptable moisture barrier | • needs compliant buffer coating for low stress<br>• rework is very difficult<br>• short shelf life (one component) |
| Silicone resin | • excellent electrical properties: dielectric constant and loss factor<br>• mechanically tough and flexible<br>• high temperature resistant<br>• lower moisture absorption than epoxy<br>• low stress after cure | • high coefficient of thermal expansion<br>• low resistance to hydrocarbons<br>• poor adhesion<br>• high cost |
| Parylene (Paraxylylene) | • penetrates hard to reach places (applied as vapor)<br>• excellent chemical, mechanical, electrical and adhesion properties<br>• excellent moisture resistance<br>• excellent control in manufacture | • masking of noncoated areas difficult<br>• difficult to rework<br>• needs expensive equipment to apply<br>• low thermal stability |

**Table 2.20 Properties of some commercially available fully aromatic polyimides [Craig 1991]**

| Property | Type I (PMDA) | Type III (BTDA) | Type III (BTDA photosensitive) | Type V (BPDA) |
|---|---|---|---|---|
| Tensile strength (MPa) | 103 | 133 | 133 | 343 |
| Elongation (%) | 40 | 15 | 15 | 25 |
| Elastic modulus (GPa) | 1.4 | 2.4 | - | 8.3 |
| Density (g/cm$^3$) | 1.42 | 1.41 | 1.61 | - |
| Moisture absorption (%) | 2 to 3 | 2 to 3 | 1 to 3 | 0.5 |
| Decomposition temperature (°C) | 580 | 560 | 550 | 620 |
| Glass transition temperature (°C) | >400 | >320 | >320 | >400 |
| Coefficient of thermal expansion ($10^{-6}$/°C) | 20 | 40 | 40 | 3 |
| Thermal conductivity (W/m.K) | 0.155 | 0.147 | 0.147 | - |
| Specific heat (J/g.K) | 18 | 18 | 18 | - |
| Dielectric constant at 1 kHz, 50% RH | 3.5 | 3.3 | 3.3 | 2.9 |
| Dissipation factor at 1 kHz | 0.002 | 0.002 | 0.002 | 0.002 |
| Volume resistivity (Ω x $10^{16}$) | >1 | >1 | >1 | >1 |
| Surface resistivity (Ω x $10^{15}$) | >1 | >1 | >1 | >1 |

PMDA, pyromellitic anhydride; BTDA, 4-4′-benzophenone dicarboxylic dianhydride; BPDA, biphenyl dianhydride resistance to most normal process solvents. However, they are attacked by alkalies and organic acids, and absorb moisture to ~3% of their weight. Their good electrical properties and low outgassing make them popular for use under the extreme conditions of space environments and cryogenic to high temperatures (~400°C). Since all polyimides have > 10% elongation and > 100 MPa tensile strength, they easily survive processing and thermal cycling-induced mechanical stresses. As a class, they do not degrade rapidly until temperatures are well over 500°C and develop very low encapsulation-induced stresses (always tensile). They have a glass transition temperature of about 320°C and a coefficient of thermal expansion (CTE) $\simeq$ 40 ppm/°C. Low-coefficient of thermal expansion polyimides (9.6 ppm/°C) have been developed by Sumimoto Bakelite Co. with low dielectric constants (3 to 3.5 at 1 kHz), low dissipation factor (0.002 at 1 kHz), and high breakdown strength (300 kV/mm). Most fully aromatic polyimides require the help of an adhesion promoter, which can be applied as a primer or incorporated into the polymeric solution. Although polyimide silicones have good adhesion to most surfaces, excellent adhesion can be obtained only by using coupling agents, such as an amino-silane and aluminum chelates or aluminum alcoholate. Polyimides have fairly good adhesion to epoxy molding compounds.

Polyimides, in general, have the ability to absorb alpha ray particles from materials. An alpha-ray particle with an energy of 7 MeV is completely absorbed by a 30 to 40-μm-thick film. Improved device designs and purity of molding compounds have made it possible to use a 10-μm-thick polyimide film as a protective layer [Wilson et al. 1990].

***Silicones.*** The starting materials for the preparations of polyorganosiloxanes are usually monomeric organosilicon compounds simultaneously containing organic groups and silicon-functional groups. The term silicone was used due to the fact that the structure of silicone $R_2SiO$ (R being the organic radical) is similar to the ketone ($R_2CO$) but unlike the C=O double bond, the Si=O double bond is unstable, at least at low temperatures. The behavior of the two functional groups is different. They are excellent choices as conformal coating materials when thermal cycle stresses on sensitive components need to be minimized. Silicones have excellent thermo-oxidative stability as well as stable electrical and physical properties over a broad temperature range. They have excellent resistance to oxidation. Silicones generally have a greater coefficient of thermal expansion, but they have a lower elastic modulus which can offset the greater expansion. Silicones are also useful in applications requiring ozone resistance. Silicones are inherently pure, making them compatible for applications requiring extreme precautions against corrosion. Silicones are suitable for high frequency applications due to their low dielectric constant and lower dissipation factor. Silicones have only moderate tensile strength, tear strength and resistance to abrasion compared to polyurethanes and epoxies. Silicones have only fair to poor chemical and solvent resistance. Silicones are effective in providing protection from moisture and humidity, particularly at elevated temperature. Silicone conformal coatings can be solvent-borne resin dispersion consisting of a solvent dispersion of high molecular weight resin or room temperature vulcanized silicone. Table 2.21 gives the typical property ranges of silicone conformal coating.

***Phenolics.*** Phenolic resins are the reaction product of phenol and formaldehyde. They can be molded by compression, transfer, or injection. They are usually selected for their low cost, high temperature resistance, and high mechanical strength. Poor in arc resistance, they are also attacked by strong acids and alkalies, and are available only in dark colors.

***Polyurethanes.*** Urethanes, also called isocyanates, are thermosetting resins that can yield rigid or rubber-like solids or foams. When cured, they are tough and possess outstanding abrasion and thermal shock resistance. They also exhibit favorable electrical properties and good adhesion to most surfaces. They can be used from cryogenic temperatures to 130°C for continuous usage, with the ability to endure short exposures to about 150°C. Some polyurethanes are solvent-removable and allow rework.

## 2.9 SUMMARY

This chapter has outlined the basic materials of construction used in plastic-encapsulated microcircuits, following the manufacturing flow. As the electronics industry charts out its roadmap (as an example, the reader is referred to the Semiconductor Industry Association roadmap for the year 2000 [SIA 1993]), functional as well as quality and reliability requirements will be more exacting and critical: tighter wirebond pitch; thinner die; higher operating temperatures; shorter design cycle time; more cost-effective manufacturing; and "greener" environmental standards. Improvements in materials, processes, and manufacturing will have to be accomplished to meet those goals.

**Table 2.21 Typical property ranges of silicone conformal coatings [Kreider 1991]**

| Property | Value |
|---|---|
| Viscosity (Pa.sec) | 0.120 to 70 |
| Shelf life (months) | 6 to 12 |
| Pot life (days) | 0.5 to 30 |
| Useful temperature range (°C) | -55 to 150 |
| Dielectric strength (kV/mm) | 20 to 50 |
| Volume resistivity (Ω-cm) | $5 \times 10^{14}$ to $5 \times 10^{16}$ |
| Dielectric constant at 25°C<br>At 100 Hz<br>At 100 kHz | <br>2.5 to 3.0<br>2.6 to 3.1 |
| Dissipation factor at 25°C<br>At 100 Hz<br>At 100 kHz | <br>0.0016 to 0.0018<br>0.0004 to 0.002 |
| Durometer hardness Shore A | 20 to 70 |
| Tensile strength (MPa) | 1.7 to 6.2 |
| Elongation (%) | 50 to 1200 |

## REFERENCES

Abbot, D.C., Brook, R.M., McLelland, N., and Wiley, J.S. Palladium as a Lead Finish for Surface Mount Integrated Circuit Packages. *IEEE Transactions on Components, Hybrids, and Manufacturing Technology*, 14 (September 1991) 567-572.

Atsumi, K., Ando, T., Kobayashy, M., and Usuda, O. Ball Bonding Technique for Copper Wire. *36th Proceedings of IEEE Electronic Components Conference,* Seattle WA (May 1986) 312-317.

Bates, B. Molding Compounds Technology for Military Applications. *Commercial and Plastic Components in Military Applications Workshop*, Indianapolis, IN (1993).

Bischoff, A., and Aldinger, F. Ball Bonding with Low Cost Ultrafine Wires. *Electronics Component Conference* (1982) 254-261.

Bischoff, A., Aldinger, F., and Heraeus, W. Reliability Criteria of New Low Cost Materials for

Bonding Wires and Substrates. *Proc. 34th Elec. Comp. Conf.*, IEEE (May 1984) 411.

Blyler, L.L., Blair, H.E., Hubbauer, P., Matsuoka, S., Pearson, D. S., Poelzing, G. W., and Progelhof, R.C. A New Approach to Capillary Viscometry of Thermoset Transfer Molding Compounds. *Polym. Eng. Sci.* 26(20) (1986) 1399.

Britton, S.C. Spontaneous Growth of Whiskers on Tin Coatings: 20 Years of Observation. *Transactions of the Institute of Metal Finishing* 52 (1974).

Chalmers, B. Principles of Solidification. J Wiley, New York (1964) 150-159.

Chandra, G. Low Temperature Ceramic Coatings for Environmental Protection of Integrated Circuits. *Proceedings Materials Research Society Symposium,* 103 (1991) 97.

Cook, J.P., and Servais, G.E. Corrosion Failures in Semiconductor Devices and Electronic Systems. Proc. 2nd *Automotive Corrosion Prevention Conference,* Dearborn, Michigan (1983) 187-197.

Davy, J.G. Accelerated Aging for Solderability Testing. *6th National Conference and Workshop on Environmental Stress Screening of Electronic Hardware* (1990) ESSEH, *Proceedings of the Institute of Environmental Sciences* (1990) 49-58.

Ditali, A., and Hasnain, Z. Monitoring Alpha Particle Sources During Wafer Processing. *Semicond. Intl.* (June 1993) 136-140.

Fister, J., Breedis, J., and Winter, J. Gold Leadwire Bonding of Unplated C-194. *20th Proceedings IEEE Electronic Components Conference* San Diego, CA (1982) 249-253.

Foppen, R. Tin-Lead Plating on Encapsulated Semiconductor Leadframes. *Metal Finishing* (January 1993) 27-31.

Gaeta, I.S., and Wu, K.J. Improved EPROM Moisture Performance Using Spin-on-Glass for Passivation Planarization. *27th IRPS* (1989) 122-126.

Hale, A. Epoxies Used in the Encapsulation of Integrated Circuits: Rheology, Glass Transition, and Reactive Processing, *Thesis, University of Minnesota, Department of Chemical Engineering* (1988).

Hale, A., Bair, H.E. and Macosko, C.W. The Variation of Glass Transition as a Function of the Degree of Cure in an Epoxy-Novolac System. *Proceedings of SPE ANTEC* (1987) 1116.

Hale, A., Garcia, M., Macosko, C.W., and Manzione, L.T. Spiral Flow Modelling of a Filled Epoxy-Novolac Molding Compound. *Proceedings of SPE ANTEC* (1989) 796-799.

Hall, P.M., Panousis, N.T., and Manzel, P.R. Strength of Gold Plated Copper Leads on Thin Film Circuit Under Accelerated Aging. *IEEE Transactions on Parts, Hybrids, and Packaging,* PHP-11, 3 (1975) 202-205.

Harman, G.G. Reliability and Yield Problems of Wire Bonding in Microelectronics. *A Technical Monograph of the ISHM* (1989).

Harman, G.G., and Wilson, C.L. Materials Problems Affecting Reliability and Yield of Wire Bonding in VLSI Devices. *Proceedings of the Materials Research Society 1989, Electronic Packaging Materials Science IV, 154, San Diego CA, 1989, quoted from Harman, G.G.* Reliability and Yield Problems of Wire Bonding in Microelectronics. *A Technical Monograph of the ISHM* (1989).

Hatada, K., Fujimoto, H. and Matsunaga, K. New Film Carrier Assembly Technology: Transferred Bump TAB. *IEEE Trans.* CHMT-10, No.3 (1987) 335-340.

Hermansky, V. Degradation of Thin-Film Silver-Aluminum Contacts. Fifth Czech Conference on Electronics and Physics, Czechoslovakia (October 1972) II.C-11.

Hirota, J., Machida, K., Okuda, T., Shimotomai, M., and Kawanaka, R. The Development of Copper Wire Bonding for Plastic-Molded Semiconductor Packages. *35 Proceedings of IEEE Electronic Components Conference*, Washington, DC (May 1985) 16-121.

Iannuzzi, M. Bias Humidity Performance and Failure Mechanisms of Non-Hermetic Aluminum SICs in Environment Contaminated with $Cl_2$. *20th Annual Proceedings of Reliability Physics Symposium,* San Diego, California (March 1982) 16-26.

Intel literature, *Packaging Handbook.* Intel Literature Sales, Intel Order No. 240800, P.O. Box 7641, Mt. Prospect, IL 60056-7641 (1993).

James, K. Reliability Study of Wirebonds to Silver-Plated Surfaces. *IEEE Transactions on Parts Hybrids, and Packaging* PHP-13 (1977) 419-425.

Jellison, J.L. Susceptibility of Microwelds in Hybrid Microcircuits to Corrosion Degradation. *13th Annual Proceedings Reliability Physics Symposium,* Las Vegas, NE (1975) 70-79.

Johnson, K.I., and Keller, D.V. Effect of Contamination of Adhesion of Metallic Couples in Ultra High Vacuum. *Journal of Applied Physics* 38 (March 1967) 1896-1904.

Kamijo, A., and Igarashi, H. Silver Wire Ball Bonding and Its Ball/Pad Interface Characteristics. *Proceedings of 35th Electronic Components Conference* (1985) 91-97.

Kearny, K.M. Trends in Die Bonding Materials. *Semiconductor International* (1988) 84-89.

Kew, T.G. Tow C. K., and Fuaida H. Low Loop Wire Bonding for Thin Small Outline Package (TSOP). *Third International Microelectronics and Systems '93 Conference*, Kuala Lumpur, Malaysia (August 1993) 68-76.

Kim, S. The Role of Plastic Package Adhesion in IC Performance. *Proceedings of the 41st Electronic Components and Technology Conf.* (1991) 750-758.

Kinjo, N., Ogata, M., Nish, K., and Kaneda, A. Epoxy Molding Compounds as Encapsulation Materials for Microelectronic Devices. *Advances in Polymer Science* 88. Springer, Berlin (1989).

Koch, T., Richling, W., Whitlock, J., and Hall, D. A Bond Failure Mechanism. *Proc. IRPS, IEEE* (1986) 55-60.

Koford, S. Lamczyk, R., and Buszkiewicz, B. Palladium Plating: A Replacement for Gold. *IEEE Electronic Components Conference* (1981) 405-414.

Kurtz, J.D., and Cousens, M. Copper Wire Ball Bonding. *Electronics Components Conference* (1984) 1-5.

Lang, A., and Pinamenni, S. Thermosonic Gold Wirebonding to Precious-Metal-Free Copper Leadframes. *38th Proceedings IEEE Electronic Components Conference,* Los Angeles, CA (1988) 546-551.

Lee, H., and Neville, K., *Handbook Of Epoxy Resins*, McGraw-Hill, Inc. New York, (1967).
Levine, L., and Shaeffer, M. Copper ball Bonding. *Semiconductor International* (August 1986) 126-129.

Levine, L., and Sheaffer, M. Wirebonding Strategies to Meet Thin Packaging Requirements: Part I. *Solid State Technology* (March 1993) 63-70.

Manzione, L.T. Plastic Packaging of Microelectronic Devices, Van Nostrand Reinhold, New York (1990).

Manzione, L.T., Gillham, J.K., and McPherson, C.A. Rubber Modified Epoxies, Transitions and Morphology. *J. Appl. Polym. Sci.* 26. (1981) 889.

May, C.A. Epoxy Materials. *Electronic Materials Handbook*, 1-Packaging, ASM Intl. (1989) 825-837.

Mih, W.C. Am. Chem. Soc. Symp. Series No. 242. (1984) 273.

Minges, M.L., et al. *Electronic Materials Handbook*, 1, ASM Intl., Materials Park, OH (1989).

Mizugashira, S., Higuchi, H., and Ajiki, T. Improvement of Moisture Resistance by Ion-Exchange Process. *IRPS IEEE* (1987) 212-215.

Mori, S., Yishida, H., and Uchiyama, N. The Development of New Copper Ball Bonding-Wire. *Proc. 38th Elec. Components Conf., IEEE* (1988) 539-545.

Nakamura, Y., Tabata, H., Suzuki, S., Iko, K., Okubo, M. and Matsumoto, T. *J. Appl. Polym. Sci.* 32 (1986) 4865.

NASA. *Requirements to Preclude the Growth of Tin Whiskers*, information received from Jack Shaw, Manager, NASA Parts, Projects Office, NASA Goddard Space Flight Center, Greenbelt, MD 20771 (May 1992).

Nesheim, J.K. The Effects of Ionic and Organic Contamination on Wire Bond Reliability. *Proceedings of the 1984 International Symposium on Microelectronics* Dallas TX (September 1984) 70-78.

Nguyen, L.T. Reactive Flow Simulation in Transfer Molding of IC Packages. *Proceedings of the 43rd Electronic Components and Technology Conference* (1993) 375-390.

Nguyen, M. John Matthey Electronics, San Diego. Personal communications (1994).

Nishimura, A., Kawai, S., and Murakami, G. Effect of Leadframe Material on Plastic-Encapsulated Integrated Circuits Package Cracking Under Temperature Cycling. *IEEE Trans. on Comp., Hybrids, and Manuf. Tech.* 12 (1989) 639-645.

Okabe, Y., et al. High Reliability Silver Paste for Die Bonding. *Proc. 38th Electron. Comp. Conf.* IEEE (1988) 468.

Olsen, D.R., and James, K.L. Evaluation of the Potential Reliability Effects of Ambient Atmosphere on Aluminum-Copper Bonding in Semiconductor Products. *IEEE Transactions on Components Hybrids, and Manufacturing Technology,* CHMT-7 (1984) 357-362.

Onuki, J., Suwa, M. Iizuka, T., and Okikawa, S. Study of Aluminum Ball Bonding for Semiconductors. *Electronics Components Conference* (1984) 7-12.

Opila, R.L., and Sinclair, J.D. Electrical Reliability of Silver Filled Epoxies for Die Attach. *23rd IRPS*, IEEE (1985) 184.

Owens, N. Motorola Arizona. Personal communications (1994).

Panousis, N.T. Thermocompression Bondability of Bare Copper Leads. *IEEE Transactions on Components, Hybrids, and Manufacturing Technology,* CHMT-1 (1978) 372-276.

Pecht, M.G. *Integrated Circuit, Hybrid and Multichip Module Design Guidelines: A Focus on Reliability*, J Wiley, New York (1994).

Pfaelzer, P.F., and Frisch, J. Ultrasonic Welding of Metals in Vacuum. Final Report on Contract No. W-7405-eng-48, California University, Berkley, California, College of Engineering

(December 1967) Quoted from Ling, J., Albright, C.E. The Influence of Atmospheric Contamination on Copper Ultrasonic Welding. *Electronics Components Conference* (1984) 209-218.

Pitt, V.A., and Needes, C.R.S. Thermosonic Gold Wire Bonding to Copper Conductors. *IEEE Transactions on Components, Hybrids, and Manufacturing Technology* CHMT-5 (1982) 435-440.
Planting, P.J. An Approach for Evaluating Epoxy Adhesives for Use in Hybrid Microelectronics. IEEE Trans. Parts. Hybrids. Packg. 11 (1975) 305.

Proctor, P., and Solc, J. Improved Thermal Conductivity in Microelectronic Encapsulants. *Proc. 41st Electron. Comp. Conf., IEEE* (1991) 835-842.

Ptacek, R.J., and Schuder, D.E. Surface Mount Devices: Design Considerations in Leadframe Material Selection, presented at the *Fifth Annual International Electronics Packaging Society Conference*, Orlando, FL (October 1985).

Riches, S.T., and Stockham, N.R. Ultrasonic Ball/Wedge Bonding of Fine Cu Wire. *Proceedings of 6th European Microelectronic Conference (ISHM),* Bournemouth, UK (June 1987) 27-33.

Robock, P.V., and Nguyen, L.T. Plastic Packaging. *Microelectronics Packaging Handbook*, eds. Tummala, R. R., and Rymaszewski, E. J., Van Nostrand Reinhold, New York (1989) 556-557.

Rosler, R.K. Rigid Epoxies. *Electronic Materials Handbook,* 1- Packaging. ASM Intl. (1989) 810-816.

Semiconductor Industry Association, Semiconductor Technology Workshop Working Group Reports (1993).

Shukla, R., and Singh-Deo, J. Reliability Hazards of Silver Aluminum Substrate Bonds in MOS Devices. *20th Annual Proceedings Reliability Physics Symposium,* San Diego, CA (1982) 122-127.

Shumay, W.C. Copper's Expanding Role in Microelectronics. Advanced Materials and Processes Inc., *Metal Progress* (December 1987) 54-59.

Slay, B. Phone interview. Texas Instruments, P.O. Box 70448, Midland, TX 70711-0448 (April 1993).

Smith, J. D. B. *J. Appl. Polym. Sci.* 26. (1981) 979.

Snow, S.S., and Chandra, G. A Novel Packaging Concept. Surface Protected Electronic Circuits, *GOMAC Digest of Papers* (1990) 529-532.

Thomas, R.E., Winchell, V., James, K., and Scharr, T. Plastic Outgassing Induced Wire Bond Failure. *27th Proceedings of Electronic Components Conference*, Arlington, VA (May 1977) 182-187.

Tleel, R.E. How Pretinning, Testing Help Improve Solderability of Surface Mount Components. *Surface Mount Technology* (October 1987) 31-36.

Totta, P. Thin Films: Interdiffusion and Reactions. *J. Vac. Sci. Technol.* 14, (1977) 26.

Waryold, J. Conformal Coatings for PC Assemblies. *Insul. Circuits* (January 1982) 40.

Wolverton, M. Component Solderability. *Circuits Assembly* (March 1991) 34-42.

Yamashita, T., Kanamori, T., Iguchi, Y., Arao, Y., Shibata, S., Ohno, Y., and Ohzek, Y. Development of Ultra Fine Wire and Fine Pitch Bonding Technology. *IEICE Transactions,* E 74, No. 8 (August 1991).

Zahavi, J., Rotel, M., Huang, H.C., and Totta, P.A. Corrosion Behavior of Al-Cu Alloy Thin Films in Microelectronics. *Proceedings of International Congress on Metallic Corrosion,* Toronto, Canada (June 1984) 311-316.

Zarlingo, S.P., and Scott, J.R. Leadframe Materials for Packaging Semiconductors. *First Annual International Electronic Packaging Society Conference*, Cleveland, OH (November 1981).

# 3

# MANUFACTURING PROCESSES

Plastic package manufacturing involves various steps, depending on the package style. Figure 3.1 presents a typical flow chart of the baseline plastic encapsulating microcircuits (PEM) manufacturing processes, assuming the chip is ready for bonding, and encapsulation will utilize a transfer-molding technology. This chapter generally follows this flow in describing key PEM manufacturing processes, but also overviews some of the less common PEM manufacturing methods. Table 3.1 lists key process operations and associated process controls.

## 3.1 LEADFRAME FABRICATION

The PEM architecture consists of a chip mounted onto and electrically connected to a leadframe, which is then encapsulated. Leadframes are manufactured by either progressive sheet-metal stamping or chemical etching. Leadframe

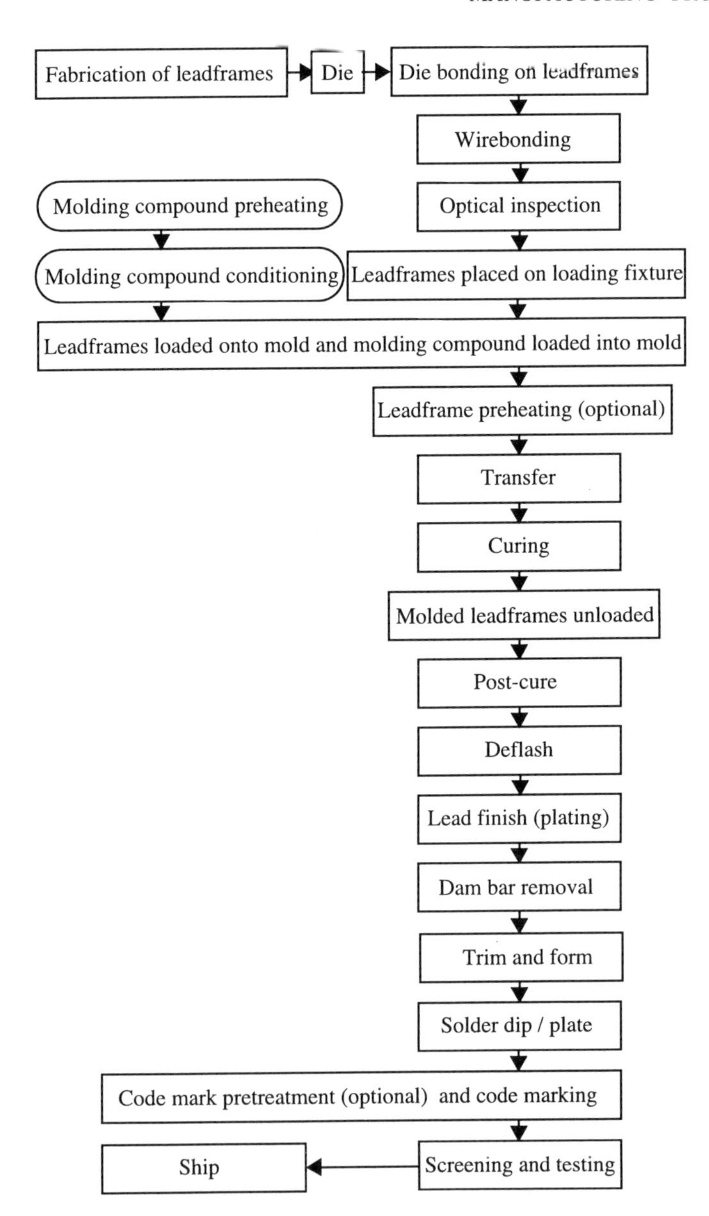


**Figure 3.1** *A typical flow chart of the baseline (PEM) manufacturing process sequence using transfer molding technology*

**Table 3.1 PEM manufacturing operations and process controls**

| Operation | Control |
|---|---|
| Wafer mount | • orientation to ring carrier |
| Water saw | • feed<br>• speed<br>• blade dress and replacement interval<br>• visual check for excessive chipping or passivation damage |
| First optical | • gross chip defects and cracks intruding into active areas<br>• 50% narrowing of metal lines<br>• passivation on bond pads |
| Chip bond | • heater block temperature (eutectic only)<br>• chip push test on specified sample size<br>• twice per shift/machine<br>• cure-oven atmosphere, time, and temperature |
| Wirebond (one pass bond rework allowed) | • capillary pressure<br>• preheater and bonding stage temperature<br>• scrub frequency<br>• wire-pull tests on specified number of wires<br>• twice per shift/machine<br>• fail on break of less than 3.5 grams or any ball lift from pad |
| Second optical | • wire dress<br>• 75% haloing on eutectic bond<br>• snowplowing over chip with polymer bond<br>• any silver epoxy on chip surface |
| Molding | • ram pressure and speed<br>• mold clamping force<br>• temperature across chases, cycle time<br>• pellet internal temperature at preheat |
| Postmold cure | • oven temperature, profile time |
| Deflash | • chemical-bath temperature<br>• immersion time<br>• bath renewal cycle<br>• clearing water temperature/time<br>• mechanical-grit removal<br>• lead damage |
| Backside mark | • correct wafer lot number |
| Lead finish, if electroplate (one pass rework allowed) | • chemical bath assays<br>• time and temperature<br>• clearing bath time and temperature<br>• plating uniformity and thickness |

**Table 3.1 (cont.)**

| Operation | Control |
|---|---|
| Dam bar shear | • shear punch and die maintenance frequency<br>• punch offset<br>• ragged shear edges, splinters |
| Lead trim-and-form | • lead-holding anvil pressure<br>• lead spread (DIPs)<br>• package cracking<br>• coplanarity (surface mount packages)<br>• dimensions |
| Lead finish, if solder dip (one pass rework allowed) | • belt speed, flux type, solder-bath temperature<br>• resistivity of last cleaning zone<br>• coating uniformity, thickness |
| Shorts and opens (electrical test) | • yield |
| Mark | • ink-cure time and temperature<br>• rework allowed<br>• marking durability<br>• laser readability |
| Final quality control gate | • external dimensions<br>• correct part-number marking<br>• electrical<br>• lead finish<br>• solderability<br>• correct shipping container<br>• cosmetic |

stamping is used when low lead-count packages (< 100 leads) are manufactured in sufficient volume (> $10^6$ packages) to justify the tooling cost. Stamping has process control problems for high-quality, high lead-count (> 100 leads), fine-pitch (0.5-mm) packages. Chemical etching is then required.

Stamping is a highly automated, high-speed process, suitable for high production rates that justify the high initial tooling expense. The sheet metal, typically in roll form, is pierced along both edges to create indexing holes that help position the sheet during further processing; the location holes are used to advance the sheet metal strip through the stamping machine. Die-and-punch sets specific to the leadframe geometry are required. Each machine may accommodate from 40 to over 200 die-and-punch sets, and the stamping machines are typically capable of up to a thousand strokes per minute [Minges et al. 1989].

The process is typically accomplished as a series of stampings that progressively approach the final leadframe geometry; the number of steps depends on the geometrical complexity of the leadframe. Figure 3.2 shows the progressive stamping stages for a quad-flatpack (QFP) leadframe.

Following stamping, the leads or leadframes are deburred to remove excess metal that can create electrical and dimensional tolerance problems during assembly and testing of the final product. Burrs can also cause shorts or plating problems. Following deburring, the leads are cleaned and coated to ensure good bonding of further nickel, copper, or tin/lead overplatings to the base metal.

Stamping has some inherent disadvantages. The process can cause lead twisting due to the residual stresses produced in the sheet material by cold rolling. Leads may also develop fine cracks that can propagate, causing an increase in lead resistance or even lead fracture.

Photochemical etching is one of the most widely used methods for manufacturing lead frames. It is a high unit cost process that can be tooled at low costs and time.

Etching lead frames may be done either in a continuous reel-to-reel form or in flat sheets. Both the processes have identical steps except for the equipment used for handling the etching material. Both sides of the lead frame are coated with the material with the help of a photoresist film. Next, the preclean and activation steps are performed followed by lamination. The photoresist is then exposed to the required lead frame pattern with the aid of an ultraviolet source and a precision pattern glass. The areas to be retained as metals are coated with resists and the etched parts are kept free of the resist in the finished pattern. The final steps include running past the material through a series of nozzles spraying etchant, after which the protective photoresist film is stripped from the finished lead frames. The final etched lead frame is then collected on reels or in unit-cut lengths. After the stamping or chemical machining process, the leads are cleaned and spot-plated with silver at the inner lead fingers to facilitate wirebonding. The silver spot plating provides a suitable surface for chip-to-lead interconnections. The plating process is usually conducted on a reel-to-reel basis; thereafter, the leadframes are cut in strips from lengths 150 to 250 mm for further processing [Foppen 1993]. Silver, however, has a tendency to undergo dendritic growth [Krumbein 1987, Dumoulin et al. 1982] and is considered a reliability hazard by some manufacturers. An alternative lead finish process plates with palladium, with a nickel undercoat (See Section 3.7). Tin and tin-lead are unsuitable for spot plating because they cannot withstand encapsulation temperatures.

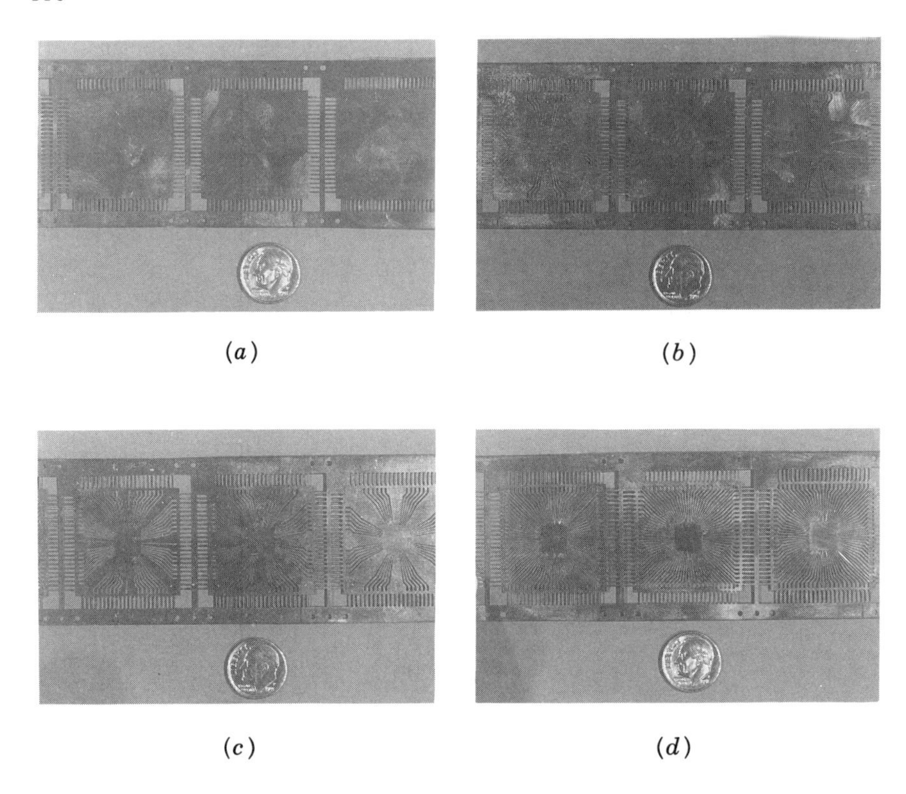

**Figure 3.2** *Progressive stamping stages for a quad-flatpack leadframe*

After encapsulation, the leads are deflashed and the exposed surfaces of the leadframe are plated with tin or a tin-lead alloy to prevent lead corrosion and enhance solderability. The outer leads of the leadframe, are not plated with silver during inner-lead plating because of the narrow tolerances and the softness and smearing characteristics of the plating, especially at encapsulation temperatures. The leads are then trimmed, the excess leadframe metal is cut, and the leads are formed into various shapes, including dual in-line, gull-wing, or J-leads. Details of these processes are presented in Sections 3.6, 3.7, and 3.8.

## 3.2 DIE ATTACHMENT

The integrated circuit diced from the wafer is attached to the center pad of the leadframe, called the die-attach pad or die-paddle, with a soft solder, a gold-silicon eutectic, or, most commonly, a polymer adhesive. The common polymer adhesives are epoxy and polyimide, often filled with silver or alumina to make them more thermally conductive.

The physical process of transferring dies to the die pad requires a collet or a surface pickup tool at the end of the transfer arm that picks up the die and deposits it precisely on the die pad. The collet is typically metal and machined to the exact die size. The surface pickup tool is often of rubber, teflon, or delrin, and may be used over a range of die sizes. It is predominantly used for applying dies to polymer or silver-loaded, glass-resin, die-attachment materials; the collet is used only for high-temperature, eutectic, die-attachment applications. Soft-solder die-attachment may use either collets or surface pickup tools, but latter will require special high-temperature materials (good up to processing temperatures of about 340°C). Otherwise, contamination of the die surface may result from the breakdown of the pickup tool surface, causing package delamination at the die surface during temperature cycling and creating long-term reliability problems. This effect has been analyzed with scanning acoustic microscopy by Olson and Owens [1991].

Gold-silicon eutectic bonding is accomplished by cutting a preform, generally of the 94Au/6Si alloy, placing it on the die-attach pad of the leadframe (Alloy-42), then putting the chip on the preform and heating above the melting temperature of the eutectic (300°C). This forms a rigid joint between the leadframe and the silicon that is not prone to solder fatigue. However, thermal mismatch stresses may be transmitted to the die, sometimes producing horizontal die cracks. Furthermore, gold-silicon eutectic die attachment is a low-throughput manual method that cannot be reliably adapted for high-speed automation [Nguyen et al. 1994].

Polymer die-attach adhesives require curing subsequent to chip placement, usually for about 1 hr at 150°C for epoxies, and longer at higher temperatures for polyimides. The curing schedule is usually determined by differential scanning calorimetry.

Polymer die-attach materials are normally used in low-power PEM parts in plastic dual in-line package (PDIP), plastic-leaded chip carrier (PLCC), small outline integrated circuit (SOIC), and plastic quad flatpack (PQFP) packages. Polymer die-attach materials are widely used, due to their relative cost-effectiveness (compared with solder and eutectic die-attach) and their adhesive compliance under thermal cycling conditions. The major concerns about their use are related to very long-term degradation, especially under high-temperature storage; and interfacial void formation can cause die cracking, and high thermal resistance in the voided area, which can cause parametric shifts due to a local increase in junction temperature. Since polymer adhesives absorb moisture under high-humidity conditions, they have been tentatively correlated with horizontal PEM case cracking during soldering to a substrate or circuit card (see Chapter 4).

The properties of polymeric die-attach materials also dictate manufacturing decisions, such as whether to use a single-component or two-component system. In the latter, the resin or resin-filler is combined with the hardener just prior to use, which results in an excellent shelf life at room temperature. On the other hand, single-component materials avoid problems associated with reproducible mixing and component ratios, but even when stored below 5°C, they are guaranteed for only six months.

The polymer die-attach material can be dispensed with a needle or syringe, or by stamping or screen printing. The viscosity must be selected by the manufacturer for the particular application, with machine adjustments to control the proper thickness and fillet. The fillet provides significant strength for small dies, but must not be allowed near the surface of the die because of the potential for silver migration, when silver is used as a filler. Formation of voids, due to improper dispensing of the adhesive or the release of solvents during cure, is the major concern in polymer die attachment. The effect of voids on the quality and reliability of plastic packages has been extensively studied [van Kessel et al. 1983, Edwards 1987, Mahalingam et al. 1984], and C-mode scanning acoustic microscope has been effectively used for their characterization [Kessler et al. 1987] (see Chapter 6). However, problems caused by improper dispensing have largely been solved by the use of multiple nozzles; void-free results have been demonstrated on chips as large as 20 x 20 mm, using fully automatic process control, made possible through image processing of the dispensing pattern.

The next step in the die bonding process is to place the chip on the adhesive-coated die pad. When large chips are involved, placement errors of less than 25 μm (1 mil) and angular errors of less than 0.3° are maintained by stable pick-and-place units, accurate positioning, and image-processing techniques. Bonding pressures of up to 5 N/cm$^2$ are required to achieve 15-to 30-μm-thick adhesive layers and acceptable visible protrusion of fillets around the edges of the chips. Inaccurate die placement may lead to several manufacturing problems such as high die stress from voiding; epoxy bridging to leadfingers, resulting in inner-lead bonding problems; lead sweep at the wirebond, resulting in low stresses on the bonding wires on one side of the package and high stresses on the opposite side; and low wirebond yield and throughput, due to the extra time needed to search the die position.

A priority of die-bonding manufacturing methods is the reduction of chip stresses caused by the different coefficients of thermal expansion of the chip, molding mass, and die pad. While the stress is mainly bending stress, tensile stress develops at the chip surface, which can cause chip cracking. In some cases, especially when quick-setting adhesives with poor stress relaxation properties are used, stresses can lead to delamination or cracking. Major manufacturers have addressed this by choosing either suitable adhesives and process parameters, or an improved leadframe design (e.g., dividing the die pad into several decoupled subareas).

## 3.3 WIREBONDING

In PEM manufacture, wirebonding is the dominant interconnection technology, although it has changed substantially since it was first introduced. Gold wire is the common material for interconnects in plastic-encapsulated microcircuits. It is bonded using thermocompression and thermosonic bonding because of its easy ball formation and oxidation resistance.

To reduce plastic package costs, studies have been conducted to develop the use of other wire materials, including aluminum, copper, silver, and palladium. While the focus of this section is on gold wirebonding processes, process modifications to extend the bonding processes to copper, aluminum, and palladium are also presented.

### 3.3.1 Thermocompression Bonding

Thermocompression bonds occur when two metal surfaces are brought into intimate contact in a controlled time, temperature, and pressure cycle. During the process (Figure 3.3), the wire and the underlying metallization undergo both plastic deformation and atomic interdiffusion. The plastic deformation at the mating surfaces ensures close contact between the bonding elements, an increase in the bonding area, and the breakdown of the interface film layers. Interdiffusion

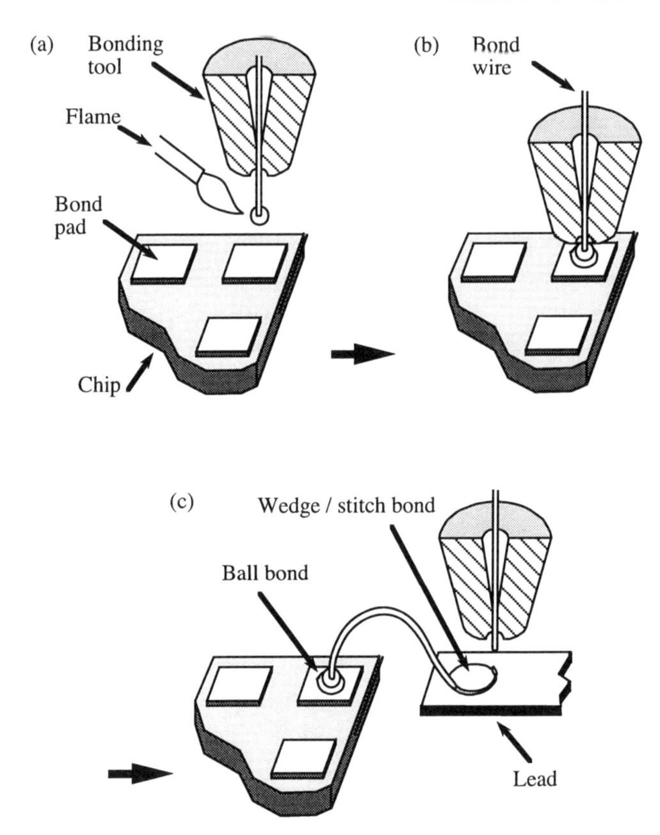


**Figure 3.3** *Thermocompression bonding*

can result in a uniform interface. Surface roughness, oxide-layer formation, and absorbed chemical species or moisture layers can prevent intimate metal-to-metal contact, inhibiting interfacial welding and reducing the inherent strength of the bond.

Typical bonding temperatures during thermocompression range from 300 to 400°C. Heat is generated during the manufacturing process, either by a heated pedestal on which the assembly is placed for bonding, or by a heated capillary feeding the wire. The capillary, made of alumina, tungsten carbide, ruby, or other refractory materials, has a typical wire-feed angle of 90°. Vertical tear separates the wire at the second bond. Typically, bonding time is around 40 msec, which translates to a machine speed of approximately two wires per second [Doane and Franzon 1993, Pecht 1993].

### 3.3.2 Ultrasonic Bonding

Ultrasonic bonding is a low-temperature process for which the source of energy for metal welding is a transducer vibrating the bonding tool parallel to the bonding pad in a frequency range from 20 to 60 kHz. In ultrasonic bonding, the wire is threaded through a hole in the wedge and trailed under the bonding tip. It is secured with a clamp, and the bonding tool is positioned over the first bond. The wedge is lowered and the wire is pressed tightly between the wedge and the bond pad. With the interface under firm compressive stress, a burst of ultrasonic energy is applied; a combination of pressure and vibrational energy breaks down surface oxides and forms a bond at the wirebond pad interface (Figure 3.4).

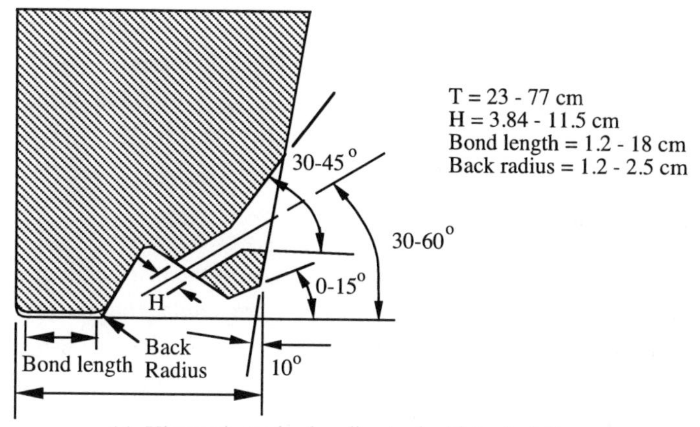


(a) Ultrasonic wedge bonding tool with typical dimensions

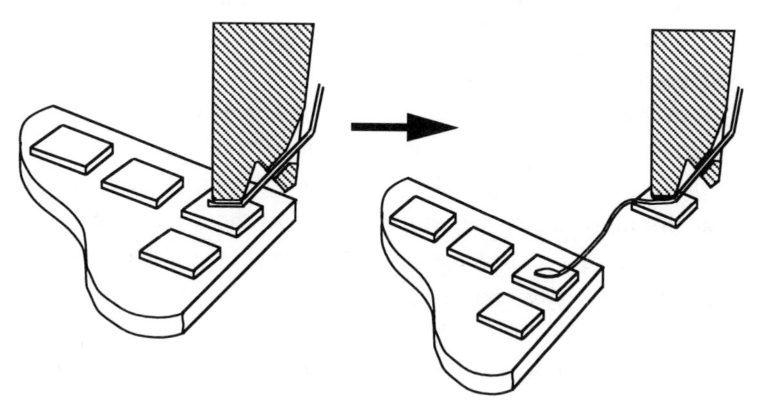

(b) Ultrasonic wedge bonding process

**Figure 3.4** *Ultrasonic wedge bonding: (a) ultrasonic wedge bonding tool with typical dimensions; (b) ultrasonic wedge bonding process*

Normally, the first bond is made to the die and the second to the substrate. This forward bonding is the preferred procedure, because it is less susceptible to edge shorts between the die and the wire. After the second bond, the wire is cut, using either a clamp or table tear. In clamp tear, used for 25-to 50-μm (1-to 2-mil) wires, clamps break the wire while the machine bonding force is maintained on the second bond. Table tear involves keeping the clamps stationary and raising the bonding tool off the second bond to tear the wire. Clamp tear typically offers greater yield, reliability, and speed. The typical wire-feed angle from the capillary is 45 to 90° for table tear, and 30° to 45° for clamp tear.

A negative feature of wedge bonding is the necessity to maintain the directional first-to-second alignment by rotating and aligning the die and substrate with the direction of the wire. Thus, while ultrasonic bonding results in an advantageous small bond pitch, thermosonic bonding, is faster. Typical bonding time is in the range of 20 msec, which translates to a machine speed of four wires per second [Doane and Franzon 1993, Pecht 1993].

### 3.3.3 Thermosonic Bonding

Thermosonic bonding combines ultrasonic energy with the ball-bonding capillary technique of thermocompression. The processes are similar, but in thermosonic bonding, the capillary is not heated and substrate temperatures are maintained between 100 and 150°C. Ultrasonic bursts of energy make the bond. Wire diameters are typically less than 75 μm (3 mils), so the capillary tool can sufficiently deform the wire for easy separation. Typically, at the second bond, the capillary leaves a characteristic circular pattern called a crescent bond. The usual bonding time is in the neighborhood of 20 msec, which translates to a machine speed of ten wires per second for ball bonding and three wires per second for wedge bonding [Doane and Franzon 1993].

While thermosonic bonding as described above is typical of gold wirebonding, process modifications can be made to accommodate the requirements for reliable silver, aluminum, copper, and palladium ball bonding: stable size (≈ 65 μm diameter), high sphericity, axis symmetry with respect to the wire, complete melting, and an oxidation-free ball surface. A silver-wire ball needs a protective gas to prevent oxidation and cellular solidification. Kamijo and Igarashi [1985] devised a torch rod, whereby a torch electrode is placed in a protective gas stream and the stream directed upward. They [1985] were able to obtain a stable discharge using the new torch electrode construction for high speed silver wire ball bonding. Hirota et al. [1985] found that the conditions for stable silver ball bonding to aluminum included a forming temperature of 200°C for 0.15 sec for a 100-Å-thick layer of intermetallic. Both copper and silver can form a stable ball if the oxygen content in the air is as much as 4000 ppm.

Onuki et al. [1984] showed that stable aluminum balls can be produced on 50-μm-diameter wire in an argon-hydrogen atmosphere (to prevent aluminum oxidation) with a discharge voltage of 1000 V, a 5 A discharge current, and 0.38 msec discharge time. The sphericity of aluminum balls increases, while constriction and eccentricity decrease with an increase in discharge times in the range of 0.5 to 30 min. The oxide on the aluminum ball becomes thinner as the discharge time decreases and the discharge current increases. For aluminum, an oxygen content of even 1,000 ppm produces wrinkles on the ball surface, defects in ball shape, and asymmetry with respect to the axis of the wire [Hirota et al. 1985].

Critical parameters affecting the aluminum ball bonding force are ball hardness and ball strain produced by ultrasonic power, bonding time, and bonding load. At a constant strain, the true bonded area between aluminum wire and aluminum bond pads increases with an increase in ball hardness. Typically, during aluminum thermosonic bonding, wire flow caused by the bonding load and ultrasonic power produces mutual surface cleaning between the bonding surfaces by breaking up oxides.

Onuki et al. [1984] evaluated the effect of ultrasonic power during thermosonic bonding on the aluminum ball diameter after deformation and the bond-breaking force of the ball bond (during a shear test using a knife edge attached to the end of a tension gauge). The bond-breaking force and deformed ball diameter increase with increase in ultrasonic power in the range of 0.025 to 0.3 W. Bond-breaking force increases with ball strain (defined as $2\ln(r_{wire}/r_{ball})$, where $r_{wire}$ is the wire radius and $r_{ball}$ is the ball radius) in the range of 0.45 to 1.4 μm for gold, aluminum-magnesium, aluminum-nickel, and aluminum wires. Aluminum-magnesium wire had a bonding strength comparable to gold wire [Onuki et al. 1984].

Copper ball formation requires a reducing atmosphere (flow rate ≈ 3 $cm^3$/sec.) during ball formation to prevent oxidation and a higher work holder temperature (≈ 300°C) than that for gold wire (≈ 225°C). Multiple rectangular shaped electrical pulses are required for copper-ball formation; typically, 600-V discharge pulses are applied for 1.5 msec, then left off for 0.25 msec. After ball formation a 5-msec argon soak allows the ball to cool without oxidizing. Discharge parameters for ball formation include discharge voltage, discharge width, and repetition rate. Low discharge times and current result in unstable or insufficient ball bonding; high times and current result in excessive bonding.

Copper-ball formation with straight (wire is positive) and reverse (wire is negative) electrode polarities affects the ball microstructure [Hirota et al. 1985]. In the case of straight polarity (SP), melting starts at the end of the wire and the volume of fusion increases while the spherical form is maintained. In the case of reverse polarity (RP), the surface skin of the wire is melted predominantly in the initial stage, and a horn-shaped piece of solid metal is left over at the lower end of the ball; the horn disappears gradually as the discharge time increases. The microstructural properties of balls resulting from straight and reverse polarities is different: while SP results in a completely melted ball, RP results in an unmelted wire structure left in the middle of the ball (Figure 3.5). RP is traditionally used with gold wire; it can also be applied to copper ball bonding, though ball stability is not maintained. During discharge, a wire with negative polarity can expand its electrode region, seeking points of higher stability and a lower work function for thermionic emission. Work function is defined as the minimum energy per unit electronic charge required for an electron to break free from the surface of a metal into vacuum. Typically, the oxide along the length of the wire has a lower work function than the wire tip. If the wire has negative polarity, the arc is formed at the top of the wire, and heat is supplied to the surface skin of the wire; if the wire has positive polarity, a stable melting mode is assured, since the arc is formed only at the wire tip that the electrons can reach in the shortest distance (Figure 3.6) [Hirota et al. 1985].

Increasing the capillary temperature reduces the vibration amplitude requirement for sound ball bonding. Mori et al. [1988] compared the effect of ultrasonic power on the shear strength of high-purity 6N copper and doped copper wire versus the shear strength of the bond, and found it increased with an increase in ultrasonic power; however, the acceptable range of ultrasonic powers was wider for doped copper wires. Mori et al. [1988] used the bonding schedule of Table 3.2 for bonding copper wires. Hirota et al. [1985] compared the conditions to form a 100Å thick layer of interdiffusion with copper-aluminum, silver-aluminum, and gold-aluminum bond systems and found that copper-aluminum required the maximum forming time of 17 seconds at 200°C compared to 0.15-0.25 seconds for silver-aluminum and gold-aluminum systems respectively (Table 3.3).

**Table 3.2 Bonding schedule for 25-$\mu$m-diameter copper and gold wire, bonded to 1.3-$\mu$m-thick aluminum-1% silicon metallization on a silicon chip with 0.8-$\mu$m-thick oxide layer underneath the bond pad [Mori et al. 1988]**

| Bonding condition | | First bond | | Second bond | |
|---|---|---|---|---|---|
| | | Copper | Gold | Copper | Gold |
| Force (gf) | | 80 to 180 | 30 to 80 | 80 | 180 |
| Ultrasonic power (mW) | | 10 to 27 | 10 to 25 | 35 | 35 |
| Ultrasonic time (msec) | | 30 | 10 | 20 | 10 |
| Temperature (°C) | | 300 °C | | | |
| Shield gas | Ball | argon + 10 % hydrogen, 1 liter/min | | | |
| | Leadframe | nitrogen + 10 % hydrogen, 6 liter/min | | | |

**Table 3.3 Conditions to form a 100-Å-thick layer of interdiffusion with copper-aluminum, silver-aluminum, and gold-aluminum bond systems [Hirota et al. 1985]**

| | Copper-aluminum | Silver-aluminum | Gold-aluminum |
|---|---|---|---|
| Forming time at 200°C (sec) | 17 | 0.15 | 0.25 |
| Forming temperature in 0.15 sec (°C) | 320 | 200 | 220 |

### 3.3.4 Special Manufacturing Considerations

Manufacturing process modifications to wirebond interconnects have been largely driven by the move towards ultrathin packages (some ≈ 0.4 mm thick) [Iscoff 1994]. The low wirebond loops (4 to 5 mils) used for thin packages are fragile and more taut than traditional loops (8 to 12 mils). This requires that magazine to magazine handling does not incur excessive vertical or horizontal motion. Thin packages have also resulted in bond pad placements which do not always occur on the die's perimeter. These are termed as "in-board" bond pads. While handling the in-board bond pads the bonder must provide special looping algorithms. These wire algorithms allow the wire to be bonded while allowing clearance of the wire to the die edge. Further, small voids in the encapsulant

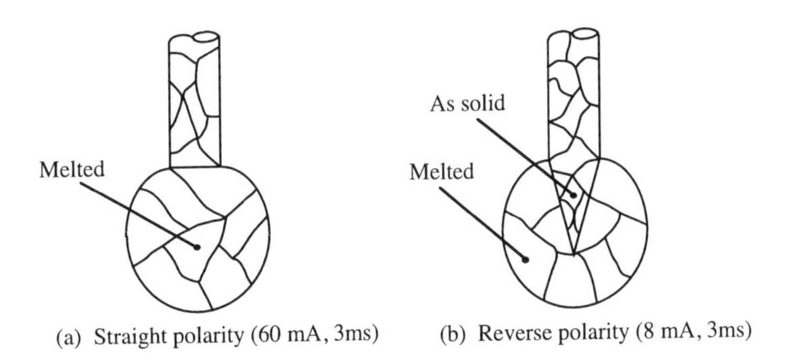


**Figure 3.5** *Comparison of microstructures of each ball [Hirota et al. 1985]*

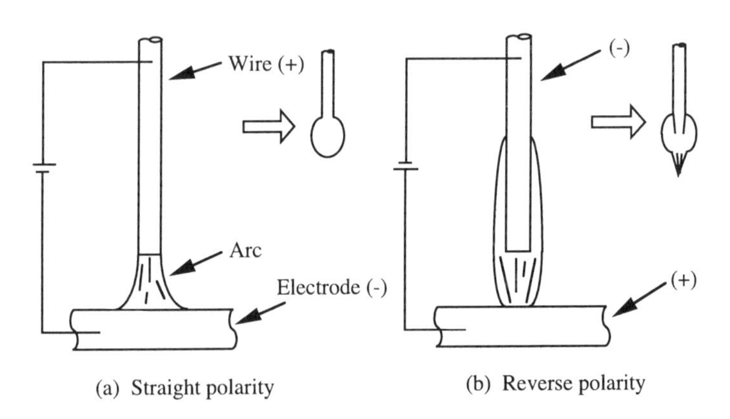


**Figure 3.6** *Schematic diagram of heat input [Hirota et al. 1985]*

material around the connecting gold wires may result in various types of reliability issues, such as moisture or chemical traps and stress points. Some other precautions for ultrathin packages include: low impact for reduced cratering, low bond force such that thinner bond layers can be used without damage, higher ultrasonic frequency (100 - 120 kHz, which assists in higher pull strength and ball shear strength values), minimum process parameter variation, controlled heat up and cool down of the lead frame, and no die movement relative to leads to prevent loop distortion. Some changes in wire bond morphology to enable meet the above requirements, include: nail-head bonding, low-profile wirebonding, and ultra-fine-pitch wirebonding [Iscoff 1994].

***Nail-head bonding.*** Flat ball bonding, or nail-head bonding, was introduced to improve process capability for finer pitches and lower loop heights. A nail-head bond is 0.5 to 0.8 the height of a conventional ball bond; the reduction in ball thickness is achieved by increasing either the applied bonding pressure or the ultrasonic energy. Furthermore, the slew rate — the vertical bonding head speed towards the bond pad — for nail-head bonding is increased by a factor of 1.5 to 2.0 over the normal rate for conventional ball bonding.

The advances that have enabled nail-head bonding are associated with modification in equipment, processes, and materials. The capillary used for conventional ball bonding is modified to include a larger chamfer which facilitates the production of a smaller, rounder ball, accompanied by less flare. Abdullah [1993] developed an equation to help select the chamfer:

$$V_{cham} = V_{fab} - (\pi r_{ch}^2 h_{bn}) - (\pi r_{bb}^2 h_{bb}) \tag{3.1}$$

where $V_{fab}$ is the free-air ball volume, $V_{cham}$ is the chamfer volume, $r_{ch}$ is the capillary hole radius, $h_{bn}$ is the nail-head ball-neck height (typically $h_{bn} \approx 0.15$ free-air ball diameter, $r_{bb}$ is the formed ball-bond radius (typically $\approx 1.25$ of the free air ball diameter), and $h_{bb}$ is the formed ball-bond height (typically $\approx 0.30$ of the free air ball diameter) (Figure 3.7). The capillary hole radius is calculated as:

$$1.55 < \frac{r_{ch}}{r_{wire}} < 1.70 \tag{3.2}$$

where $r_{ch}$ is the capillary hole radius and $r_{wire}$ is the wire radius. The chamfer angle is approximated by:

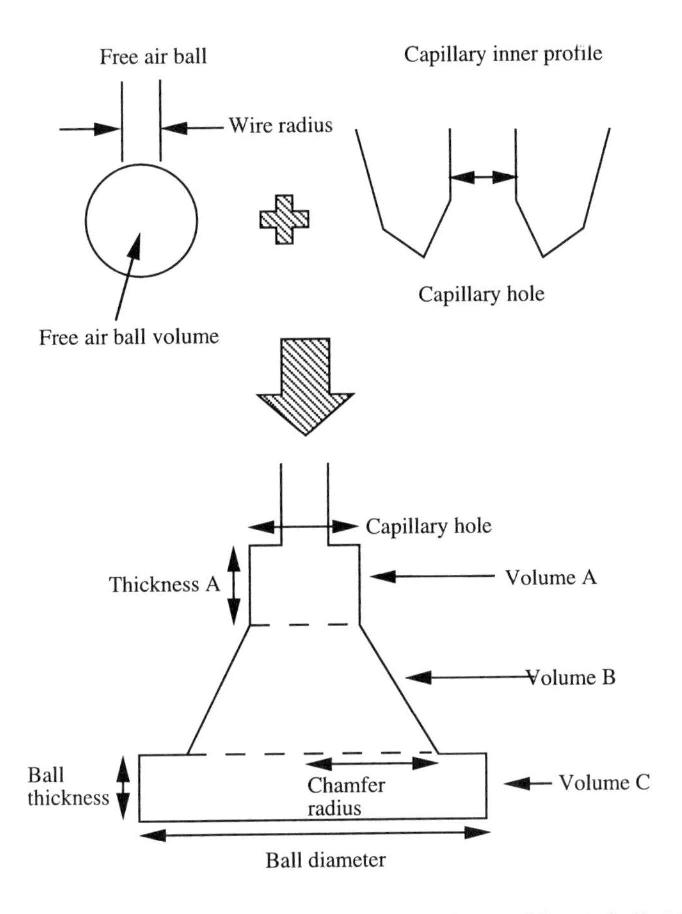


**Figure 3.7** *Capillary chamfer cone redesign for nail-head ball [Abdullah 1993]*

$$V_{cham} = \frac{1}{3}\pi(R_{cham}^2 H - r_{ch}^2 h) \tag{3.3}$$

where

$$h = \frac{r_{ch}}{tan\ \alpha}; \quad H = \frac{R_{cham}}{tan\ \alpha} \tag{3.4}$$

and where $\alpha$ is the chamfer angle [Abdullah 1993].

The area of ball contact with the substrate in a conventional ball bond is typically 75% of the ball diameter. A nail-head bond, because of its flatter profile, utilizes most of the ball for bonding. Thus, for the same ball diameter, a nail-head bond will have a larger bonded area but take up less bonding room than a conventional ball bond, and its shear strength will be 20 to 30% greater. The higher values of bonding pressure and ultrasonic energy used in nail-head bonding, reduce the probability of void formation in gold-aluminum bonding [Uno et al. 1993].

***Low-profile wirebonding.*** Low-profile bonding is achieved by special wire compositions and bonding head motions. The wire compositions were discussed in Chapter 2; the special wire tool motions are discussed in this section. During low-profile bonding, the trajectory of the capillary during its ascent from the first bond is designed to address two objectives including: wire should be bent at appropriate locations and in desired directions; and precise length of the wire is allowed to feed through the capillary. The bond head envelope for descent of the bonding head from the top of the loop to the second bond is described in Figure 3.8. Deviation of the bonding head from the envelope will result in broken wire or buckling of the bonding wire. Specialized bond head motions are incorporated at the second bond to eliminate the conventional catenary loop of the wire.

***Ultrafine-pitch wirebonding.*** The bond strength for ultrafine (≈10-μm-diameter) wirebonding decreases with an increase in the product of ultrasonic power and force and the product of ultrasonic power and time times force and time. Yamashita et al. [1991] found that as the product of ultrasonic power and time increases, the bond-pull strength increases and the deformed wire length decreases. Wedge bonding was found to be a superior technique for fine pitch (approximately 40-μm) bonding compared to ball bonding.

The deformed wire width deviation and deformed wire width were characterized by Yamashita et al. [1991] to evaluate the process limits for ultrafine-fine-pitch bonding. The limit condition for formation of short circuit between two wedge bonds was represented as

$$P > W + \sqrt{2\,(T_{wirebond}^2 + M^2 + F^2)} \tag{3.5}$$

and for short circuit between the wedge bond and the adjacent pad is

$$P > (W/2 + B/2) + \sqrt{(T_{wirebond}^2 + M^2 + R_{acc}^2 + F^2)} \tag{3.6}$$

where $P$ is the bonding pitch, $W$ is the deformed wire length, $F$ is the deviation of deformed wire length, $T_{wirebond}$ is the teaching accuracy of the wirebonder, $M$ is the mechanical accuracy of the wirebonder, $R_{acc}$ is the recognition accuracy,

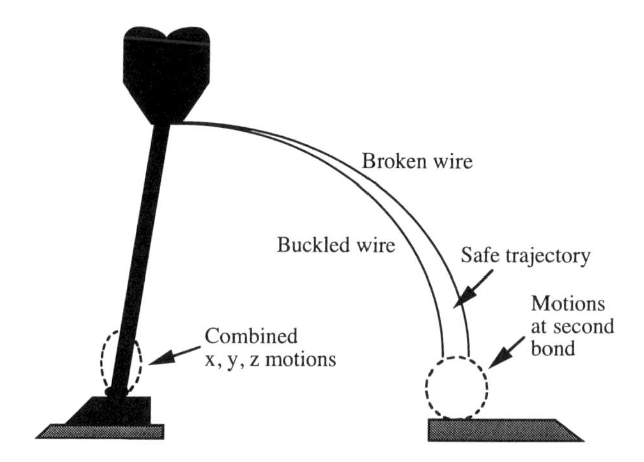


**Figure 3.8** *Bond-head motion envelope, patented by Kulicke & Soffa Industries Inc. [Levine and Sheaffer 1993]*

and $B$ is the pad width. Based on Equations (3.5) and (3.6), Figure 3.9 shows the allowable area for fine pitch wirebonding. The wire selected for 40-μm-pitch wirebonding is 10 μm in diameter (producing a deformed wire width of 19 to 21 μm).

The quality of the wedge bonds is judged by the deformed wire width of the wedge bond. The bonding load has to be sufficient to transfer the bonding load to the joining surfaces. Too large a bonding load will form a large non welded area at the central part of the wire and result in a weak bond. Yamashita et al. [1991] while developing the fine-pitch bonding process for 10-μm, 13-μm, and 18-μm gold wire, found that the wire pull strength attained its maximum value for deformed wire widths of 1.5 to 2 times the wire diameter and beyond that decreased for larger deformed wire widths. The optimum bonding conditions in terms of deformed wire widths were found to lie in a range of 19 to 21 μm for 10-μm-diameter wire; 22 to 24 μm for 13-μm-diameter; and 31 to 33 μm for 18-μm-diameter (Figure 3.10).

Yamashita et al. [1991] evaluated the effects of bonding tools on the positional accuracy of the bonding head. The tools evaluated include standard tool and the guide-type tool. With the standard tool the wire is inserted at the heel of the tool, whereas for the guide-type tool the wire is inserted the rear of the tool and guided to the surface. The guide-type tool with its smaller bore diameter was found to better positional accuracy than the standard type tool (Figure 3.11).

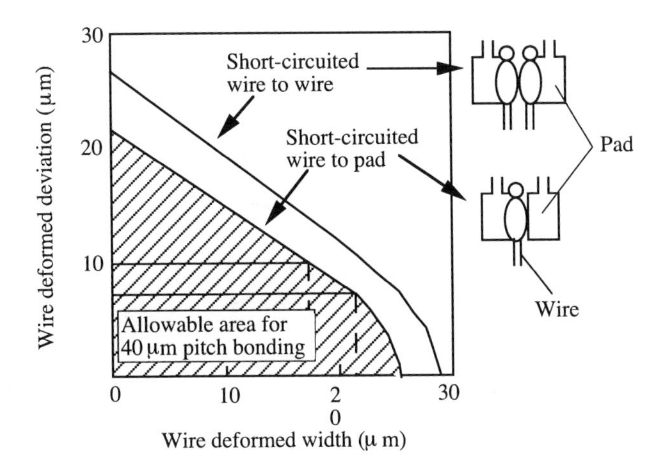


**Figure 3.9** *Relationship between wire deformed width and wire deformed deviation to achieve 40µm-pitch wirebonding [Yamashita et al. 1991]*

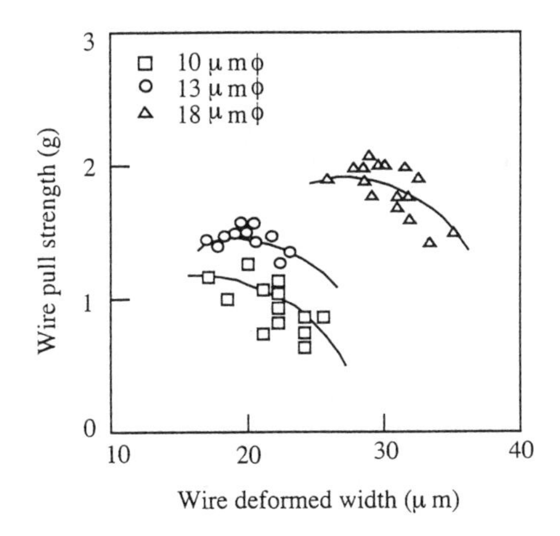


**Figure 3.10** *Wire deformed width and wire pull strength [Yamashita et al. 1991]*

In an effort to develop ultra fine-pitch wirebonding (≈ 40 µm), Yamashita et al. [1991] addressed the need for high-strength wire materials, the establishment of a bonding technology using these materials, and the prevention techniques for wire-wire contact. Wedge bonding was found to be a superior technique for fine-pitch (approximately 40 µm) bonding compared to ball bonding. The wire selected for 40 µm pitch wirebonding was 10-µm-in diameter (producing a deformed wire width of 19 to 21 µm).

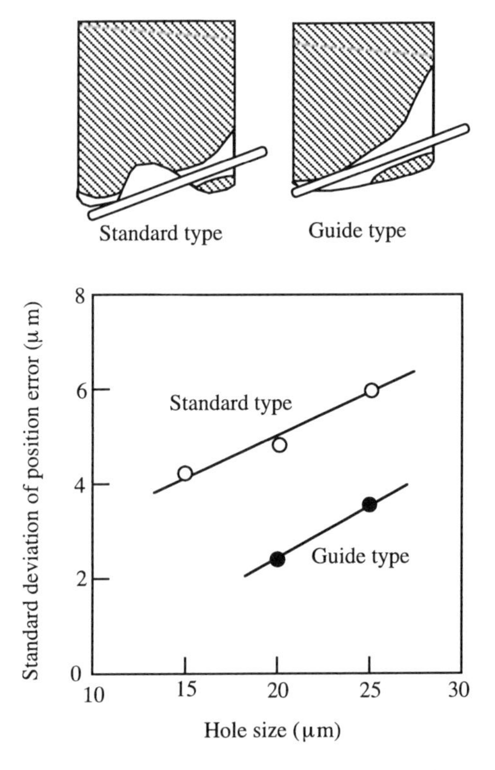


**Figure 3.11** *Bonding position error and wedge tool shape [Yamashita et al. 1991]*

## 3.4 TAPE AUTOMATED BONDING

Tape automated bonding (TAB) is an automated surface-mount method that can provide interconnection for chips with large numbers of input-output terminals (≤ 500). Figure 3.12 shows the flow of a TAB process.

A continuous polymer tape is fabricated with fine-pitched metal leadframes spaced along its length. A window is made in the center of each leadframe where the chip is to be placed. The leads of the leadframe are then bonded to the chips, on which bonding platforms, or "bumps," have been deposited by electrolytic or electroless gold plating and liftoff photolithography. The bonds to the leadframe fingers are typically made by thermocompression. The chip face may then be coated with liquid plastic encapsulant and cured. The reel of radial-spread, coated chips is then sent to the substrate (or circuit card) assembly line, where individual chips are excised from the tape, the leads are formed, and the package is bonded to the substrate.

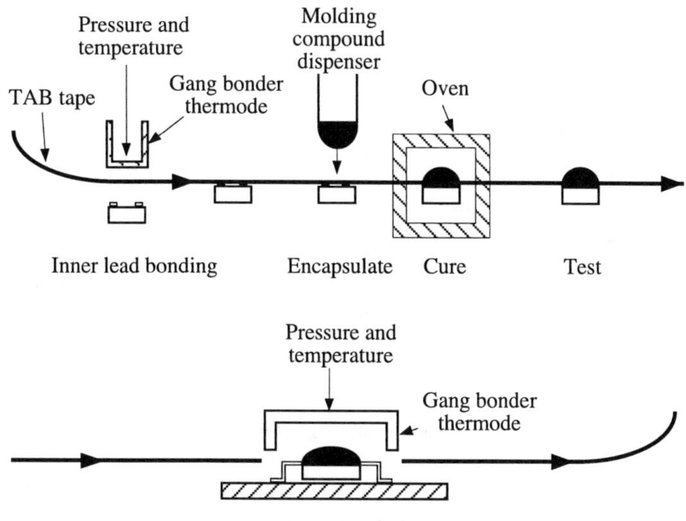


**Figure 3.12** *Outline of a complete TAB process using liquid plastic encapsulants*

The selection of the carrier tape (usually polyimide) for a particular package depends on cost, I/O count, and in-process testability. One-layer tape is the least expensive, but two- and three-layer tapes can accommodate high I/O counts and in-process testability. In principle, two-layer tape is capable of a higher I/O count than three-layer tape because of a semiadditive plating metallization process used in the former, versus the subtractive process used in the latter. Also, due to the chemical etching that defines the windows in the polyimide of two-layer tape versus the punching used for three-layer tape, the two-layer tape can provide greater design flexibility in the shape and size of the polyimide windows. However, under temperature-humidity stress conditions, three-layer tape with polymeric interlayer adhesive may provide stronger metal-polymer adhesion [Holsinger and Sharenow 1987] (See Chapter 2).

In the TAB process, inner-lead bonding is accomplished by a tool that feeds the tape continuously past a bonding stage, where chips are placed, aligned with the frame of tape, and bonded with a hot thermode. This thermode may be a solid piece of metal heated to a set temperature, or it may consist of four blades that can be heated by pulsing. The pulsed thermode offers more control of key bonding parameters, but is usually limited to perimeter-pad bonding. Although gold to tin thermocompression bonding and Au-Sn eutectic bonding can both achieve very high average pull strengths of 60 to 70 g, eutectic bonding, in which the tape has a thin layer of tin, allows the use of lower bonding forces and temperatures [Kawanobe et al. 1983].

Outer-lead bonding follows plastic encapsulation and testing. Each package is first excised from the tape with a specially designed punch and die, and leads (typically gull-wing) are formed. The package is aligned with the pads and bonded eutectically to a substrate on which solder is predeposited. If required, a heat sink may be attached to the package at this stage. Additional coatings are sometimes added after mounting to enhance reliability. A detailed presentation of tape automated bonding technology can be found in the book by Lau et al. [1992].

## 3.5 ENCAPSULATION PROCESS TECHNOLOGY

Several molding techniques are available for manufacturing plastic packages. The most widely used is the transfer molding process; others include injection molding and reaction-injection molding. A comparison of these processes is shown in Table 3.4.

### 3.5.1 Transfer Molding

After die-attach and wirebonding, the die-leadframe assembly is encapsulated. The transfer molding process is the most popular encapsulation method for essentially all plastic packages in integrated circuit technology. Transfer molding is a process of forming components in a closed mold from a thermosetting material that is conveyed under pressure, in a hot, plastic state, from an auxiliary chamber, called the transfer pot, through runners and gates into the closed cavity or cavities. Thermosets are polymers that are plastic or fluid at low temperatures and react irreversibly when heated to form a cross-linked network no longer capable of being melted. Figure 3.13 depicts the encapsulation process.

Transfer molding offers various advantages over other methods of molding. The process involves molding intricate parts with inserts. Loading times are shorter because fewer and larger preforms are used. Tool and maintenance costs are lower because deep loading wells are not necessary with transfer molds; and mold sections can be thinner because the stresses involved during closing of the mold are lower. The wear of the molds is lower in transfer molding and also the tendency toward breakage of pins is less. As the components are produced in closed molds, which are subjected to less mechanical wear and erosion by the molding material, closer tolerances on all molded dimensions are possible in transfer molding. The flash in a transfer molded component is small or absent in properly designed mold with adequate clamping pressure.

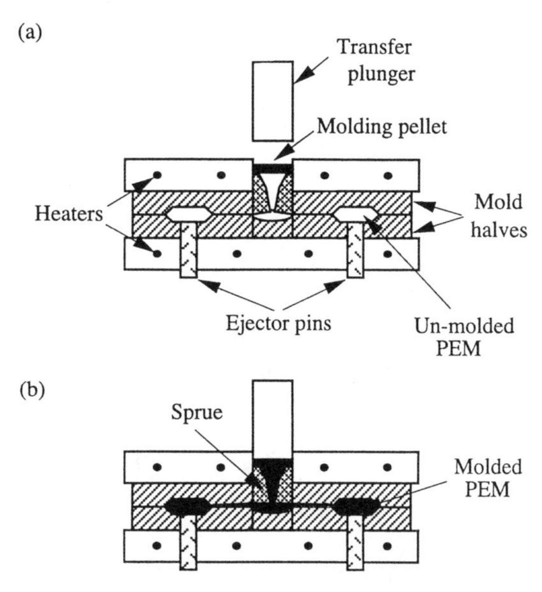


**Figure 3.13** *Encapsulation by transfer molding [Kovac 1989]*

One of the main limitations of the transfer molding process is the loss of material. The material left in the pot or well and also in the sprue and runner is completely polymerized and must be discarded. This is unavoidable and for small articles it represents a sizable percentage of the weight of the pieces molded. However in the fully automatic transfer molding presses employed for electronic parts, the savings elsewhere, including mold costs and finishing costs, usually offsets the loss of material. Another limitation is that the process is limited to use of standard metal lead frames. Extension of the process to include package designs such as TAB is still not clear.

***Molding equipment.*** Transfer molding requires four key pieces of capital equipment: a preheater, a press, the die mold, and a cure oven. The transfer molding press is normally hydraulically operated. Auxiliary ram-type transfer molds are commonly used in transfer molding. This mold has a built in transfer pot separate from the cavities. The molding compound is placed in the pot. Both the volume and the size of the molding compound preforms have to be appropriately selected for the press capacity. The mold is then clamped using the clamping pressure. The transfer plunger is activated to apply the transfer pressure to the molding compound. The molding compound is driven through the runners and gates into the cavities.

**Table 3.4 Comparison of molding processes**

| Molding type | Advantages | Disadvantages |
|---|---|---|
| Transfer molding | • multiple cavities, high yield<br>• relative material savings<br>• short cycle time<br>• low tool maintenance costs | • high molding pressure<br>• high viscosity<br>• restricted to the packaging of leadframes |
| Injection molding | • good surface finish<br>• good dimensional control | • poor material availability<br>• extremely high pressure<br>• rapid tool wear (screw and barrel) wear<br>• high capital investment |
| Reaction-injection molding | • energy efficiency<br>• low mold pressure<br>• good wetting of chip surface<br>• adaptability to TAB | • few resin systems available for electronic packaging<br>• requires good mixing<br>• high capital investment |

Since the 1980s, aperture-plate and multiplunger molds have been the dominant approaches to PEM molding. Table 3.5 compares the features of these molding methods. Aperture-plate molds are a patented transfer molding technology (U.S. Patent No. 4,332,537, 1 June 1982) exclusively developed for PEMs. An aperture-plate mold is constructed by assembling a series of stacked plates. The leadframes form the aperture plates. The top and bottom of the body are formed by separate plates. The bottom body-forming plate contains the runner system, while the top plate is finished for either laser or ink marking. The gates are positioned between the runners, parallel to the bottom of the body; the aperture plate cavities can be formed anywhere along this intersection, and their width can be any fraction of this length of intersection. This flexibility of gate positioning, along with the much lower pressure drop across the gate in an aperture plate, results in negligible wire sweep and paddle shift during molding. These molds are also highly adaptable for different package types and pinouts.

Multiplunger molds, also called gang-pot molds, have a number of transfer plungers, feeding one to four cavities from each transfer pot. They are highly automated and can be easily set and optimized for a new molding compound. However, their productivity is much less than conventional molds, and the use of low preform-preheat temperatures in manual tools can lead to some molding-

**Table 3.5 Comparison of molding tools**

| Feature | Aperture-plate molds | Multiplunger molds |
|---|---|---|
| Number of cavities | • several thousands possible | • from 2 to 100 |
| Flexibility of package types | • extremely flexible | • relatively inflexible |
| Process setup and control | • moderate | • straight forward for automated tools |
| Flow-induced stress problems | • minimum | • intermediate |
| Flash and bleed | • minimum | • average |
| Molding compound waste (%) | • 20 to 40 | • 10 to 25 |
| Package yield per cycle | • high | • very high |
| Ejection | • external ejection | • ejector pins in each cavity |
| Flow time and cavity material uniformity | • good | • excellent |
| Packing pressure (MPa) | • 1.4 to 2.8 | • variable on number of cavities fed from a pot (1.4 to 4.1) |
| Temperature profile and stability | • requires care | • excellent |
| Automation susceptibility | • intermediate | • very high |
| Capital cost | • low | • high (lower for small tools) |
| Labor cost | • high | • low |
| Maintenance cost | • medium | • high, due to automation |

compound temperature-related problems. Figure 3.14 shows a multiplunger mold used for simultaneous encapsulation of DIPs and quad flat packs. Figure 3.15 shows a multiplunger mold with each pot feeding just one cavity.

In general, the mold consists of two halves, the top and bottom. The mating surface of these two halves is called the parting line. Figure 3.16 shows the different parts of a transfer mold mounted on the molding machine. Platens are massive blocks of steel used to bolt the two halves to the molding press. Guide pins ensure proper movement of the two halves. The ejector pins aides the

**Figure 3.14** *Multiplunger mold used for simultaneous encapsulation of DIPs and quad flatpacks [Courtesy of National Semiconductor 1994]*

ejection of the component after the mold has opened. Gates are located where they can be easily removed and buffed if necessary. Properly designed gates should allow proper flow of material as it enters the mold cavity. Gates should be located at points away from the functioning parts of the molded component. Vents are provided in all transfer molds to facilitate the escape of trapped air. The location of these vents depend on the part design, and locations of pins and inserts. The vent is sufficiently small so that it allows the air to pass through but negligible amount of molding compound can pass through it. Vents are often placed at the far corners of the cavity, near inserts where a knit line will be formed, or at the point where the cavity fills last. Figure 3.17 shows a mold used for encapsulation of quad flatpacks. Here one transfer pot feeds more than ten cavities. The top of the figure shows the leadframe with the encapsulated packages. The bottom of the figure shows the cavities, as seen the material fills the runner and gates. Figure 3.18 shows a close-up of the cavities. As seen the material fills the runners and gates. The embossing done on the cavity to imprint the logo is clearly visible. The air vents on the cavities are visible.

**Figure 3.15** *Multiplunger mold with each pot feeding just one cavity [Courtesy of National Semiconductor 1994]*

***Transfer molding process.*** Figure 3.19 shows the various stages of a typical transfer molding process. In this process, leadframes are loaded (six to twelve in a row) in the bottom half of the mold. For both plate and cavity-chase molds, this is done at a workstation separate from the molding press. A cavity-chase mold uses a loading fixture; most molding operations have automated leadframe loaders. Both the moving platen and the transfer plunger initially close rapidly, but the speed reduces as they close. After the mold is closed and clamping pressure is applied, the preform of molding material (which has usually been preheated to 90 to 95°C, which is below the transfer temperature) is placed in the pot, and the transfer plunger or ram is activated. Preheating of the molding compound is done by a high-frequency electronic method that works on a principle similar to microwave heating. The transfer plunger then applies the transfer pressure, forcing the molding compound through the runners and gates into the cavities. This pressure is maintained for a certain optimum time, ensuring proper filling of the cavities. The mold then opens, first slowly. This step is known as the slow breakaway. Sometimes it is desirable to have the transfer plunger move forward so that it pushes out the cull, or the material

**Figure 3.16** *Different parts of a transfer mold [Courtesy of National Semiconductor 1994]*

**Figure 3.17** *A mold used for encapsulation of quad flatpacks [Courtesy of National Semiconductor 1994]*

remaining in the pot. Finally the component is ejected using the ejector system in the mold. In an aperture-plate mold, the plates themselves are loaded with the leadframe strips, as they shuttle in and out of the molding press.

The parameters controlled by the press include the temperatures of the pot and the mold, the transfer pressure, the mold-clamping pressure, and the transfer time to fill the mold cavities completely. The mold temperature should be high enough to ensure rapid curing of the part. However, precautions should be taken to control the mold temperature, because too high a mold temperature may result in "precure" or solidification of the molding compound before it reaches the cavities. Electric heating is the most commonly used method for heating the molds. Multiple electric heating cartridges are inserted both in the top and bottom halves of the molds; positioned to supply heat to all the cavities. The applied pressure ensures the flow of material into all parts of the cavity. Clamping pressure applied on the mold ensures the closure of the mold during polymerization or cure and also against the force of the material entering the cavities. Packing pressure, usually higher than the transfer pressure, is applied once the mold has been filled. While it is being transferred into the mold, the molding compound is reacting. Consequently, the viscosity of the material

**Figure 3.18** *Close-up of the cavities [Courtesy of National Semiconductor 1994]*

increases, at first gradually. As the reacting molecules become larger, the viscosity increases more rapidly until gelation occurs, at which point the material is a highly cross-linked network.

The pressure applied by the transfer plunger is critical. It should be sufficient to force the material through the runners and gates into the cavity and hold the material until polymerization. For high throughput, it is desirable to transfer and react the material as rapidly as possible, but decreasing the transfer time requires an increase in the transfer pressure to fill the mold. The high transfer pressures, typically as high as 170 MPa (25 ksi), can cause damage within the package, such as wire-sweep short or, in extreme cases, wirebond lift-off, shear, or fracture. These problems can be avoided by careful design of the package size and the transfer-gate size to minimize shear rate and flow stresses during cavity filling. In general, factors that tend to exacerbate these problems include high wire-loop heights, long wirebonds, bond orientation perpendicular to the advancing polymer flow front, rapid transfer times with corresponding high transfer pressures, high-viscosity molding components, and low elastic modulus wire.

An important way to decrease flow-induced stresses and improve molding yield is by velocity reduction. Velocity reduction in epoxy molding compounds reaches its limit before the need for flow-induced stress reduction is satisfied. In constant-pressure programmed transfer-rate control, high flow-induced stresses are created in the cavities nearest to and farthest from the transfer pot, while the middle cavities experience the lowest velocities and stresses. Mold mapping of flow-induced defects shows lowest yield in extreme-positioned cavities. A transfer-rate profile that is slow at the start, then higher over the middle cavities, and slow again when the last cavities are filling helps in reducing the flow-induced stresses.

Different mold designs and molding compounds require distinctive transfer profiles. Velocity-reducing tool changes are permanent for a particular molding compound and package design. Mold designs aim at balanced mold filling by maintaining nearly uniform pressure fronts through all segments of the mold; in these, channel cross-sectional areas are controlled for volume-flow and pressure drop uniformities in all cavities. However, in all cases a velocity surge occurs when the molding machine switches from the transfer pressure to the packing pressure. These transient high velocities, along with the rapid compression of any remaining voids, can cause wire sweep. Using a programmable pressure controller to profile this pressure transition is the best approach to minimizing this problem.

After 1 to 3 min at the typical molding temperature of 175°C, the polymer is cured in the mold. Following curing, the mold is opened, and ejector pins remove the parts. The molded packages are ready for ejection when the material is resilient and hard enough to withstand the ejection forces without significant permanent deformation. After ejection, the molded leadframe strips are loaded into magazines, which are postcured in a batch (4 to 16 hr. at 175°C) to complete the cure of the encapsulant. Postcure normally involves holding the part in an oven at a temperature somewhat lower than the mold temperature but well above the room temperature for several hours [Rubin 1990]. In some instances, postcuring is performed after code marking to eliminate an additional heat-cure cycle. The most important consideration in postmold cure analysis is the development of thermomechanical properties. As shown in Figure 3.20 the glass transition temperature of an epoxy resin increases significantly as the degree of conversion approaches 100%, indicating the value of proper postcure treatment in realizing full material properties.

***Simulation.*** The transfer-molding process is more complicated and difficult to treat analytically than either thermoplastic extrusion or injection molding, because

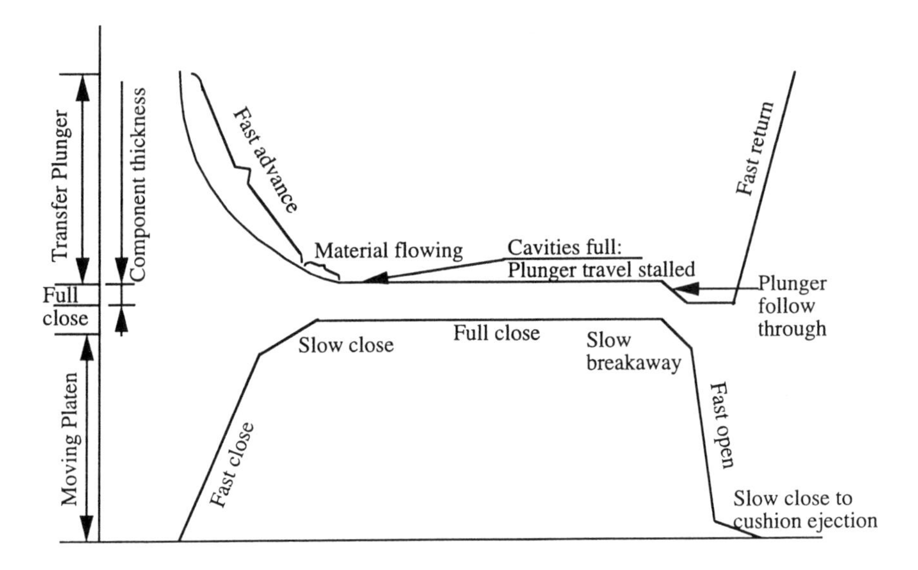


**Figure 3.19** *Various stages of a typical transfer molding process [Rubin 1990]*

of the time-dependent behavior of the molding compound, the irregular cross sections of the runners, and the common presence of inserts in the cavities. However, once a good quantitative model of the transfer-molding process has been proven, substantial time and cost can be saved. Long and expensive experimental runs no longer need to be carried out by trial and error to debug a mold or qualify a new compound.

Modeling the dynamics of transfer molding have met with moderate success [Manzione et al. 1989, Nguyen 1993]. The approach typically involves formulating a chemorheological description of the epoxy molding compound and coupling it with a network flow model. The gating factor for a successful simulation is a good characterization of the compound to determine its kinetic and rheological behavior. Most often, the curing reaction of the epoxy can be described by an autocalytic expression. The dynamic viscosity, as measured generally through a plate and cone viscometer, is reasonably well covered by the Castro-Macosko model [Nguyen 1993]. By incorporating this material information into a flow model that includes the geometrical intricacies of a multi-cavity mold, an understanding of the filling characteristics of a particular compound can be obtained.

With a rheological model in hand a variety of scenarios can be simulated to optimize processing conditions and mold design [Nguyen et al. 1992, Han and Wang 1993]. For instance, different temperatures, pressures, and packing settings can be estimated to provide optimized filling profiles. To reduce wire sweep, the

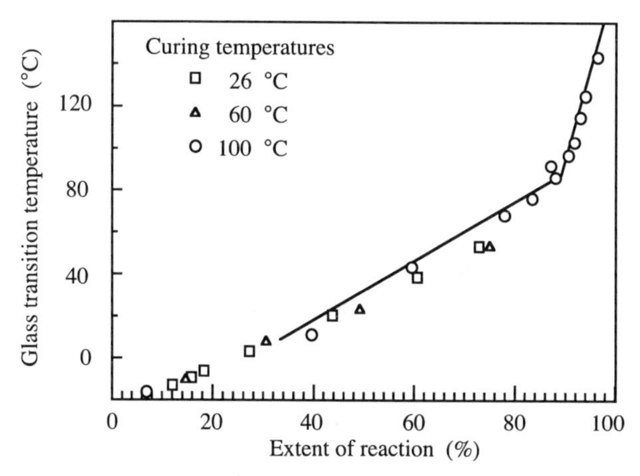


**Figure 3.20** *The glass transition temperature of an epoxy resin plotted against the extent of reaction for three different cure temperatures; the solid line indicates predictions of glass transition temperature from a thermodynamic property model [Matsuoka et al. 1989]*

dominant-yield loss concern for fine-pitch packages, an analytical model of wire deformation can be coupled with rheological and kinetic information to improve process conditions and package layout design rules [Nguyen et al. 1992]. Once a mold has been translated into its geometrical equivalent, various combinations of runner sizes, gate locations, gate dimensions, and cavity layout can be evaluated.

### 3.5.2 Injection Molding

Injection molding was initially developed for thermoplastic materials, but with slight modifications in the process it can also be used for thermosetting plastics and rubbers. It is a technique used for making large volumes of molds at low costs. Injection-molded parts are inexpensive because of the automated process and low cycle time; however, the molds, which are extremely complex and withstand high pressures, are costly.

The mold cavity is made up of two halves that are clamped together. Plastic pellets are fed into a screw from a feed hopper and are subjected to several heating zones so that when the melt comes out of the screw it is completely melted, de-aired, and dried. High pressure is also applied to the melt so as to fill the melt in the mould at the required speed. The plastic melt that comes out of the nozzle located at the other end of the screw is in a viscous state compressed to 3000 kg/cm$^2$. High pressure is then applied to the melt so as to fill the mold at the required speed. Extra melt may be required so as to compensate for shrinkage. After the component has cooled (with chilled water) and its shape set, the mold is opened and ejected.

Moldings made by injection molding usually do not require finishing, if good practice is followed in the manufacturing phase. Small components can be manufactured in large quantities at reasonable costs. Injection molding was applied to semiconductor packaging in the 80s with thermoplastic materials such as polyphenyle sulfide (PPS). The process was a novelty at the time, since it provided a faster alternative to transfer molding, but poor reliability quickly tempered the enthusiasm. The high viscosity of the PPS melt generated a lot of wire sweep due to the high injection pressure, which is typically an order of magnitude larger than the packing pressure of transfer molding. Furthermore, an almost total lack of adhesion between the thermoplastic and the copper leadframes was discouraging. Injection molding dropped out of the limelight once the cost of expensive capital investment was considered.

Despite these discouraging results with PPS, internal efforts were continued at some European companies to evaluate injection molding as an option. Since thermosets such as polyesters were already commercially processed using injection molding for some applications, the extension to epoxies was obvious. Results indicated that the reliability of packages produced by the two methods was similar, since the same material was used in both cases. Although wire sweep in injection-molded parts is generally greater than in transfer-molded parts, proper process optimization can minimize such defects. The differences in the methods occur mainly in process conditions and the required preventive maintenance; cleaning, parts wear, and residue removal are much more troublesome with injection-molding equipment.

### 3.5.3 Reaction-Injection Molding

Reaction-injection molding is different than other molding processes in that the starting materials are liquids at room temperature. Polyurethanes are usually associated with the process, but if polyesters and epoxies are used as resins then the process is known as resin transfer molding. The ingredients of the melt are polyol, polyisocyanate, and a promoter for the chemical reaction that takes place during the course of molding. The reactive liquid components are prepared separately and pumped into a mixing head in fine stream liquid form where they are thoroughly mixed. From here the mixed liquid resins are pumped into the heated mold for the part to cure.

One major advantage of reaction-injection molding is that low clamping pressures are required, which means large parts can be made at low costs.

Reaction-injection molding suffers from the same fate plaguing injection molding, namely, the lack of an appropriate material and expensive capital outlay. The absence of good candidate materials was not for lack of trying. Efforts by resin suppliers such as Shell and Dow Chemical did not produce a two-part epoxy system of sufficiently high purity and reactivity to compete with epoxies tailored for transfer molding. Foreign competition gained the upper hand, and both companies are now no longer active in this area. Ciba Geigy was partly successful in creating such a system in the early 1990s, although the reliability of the molded parts was still inferior to that of the control samples made by transfer molding. Once more, when the need to retrofit molding presses with reaction-injection molding equipment was considered, there was little incentive to put more development resources into the process.

### 3.5.4 Glob-topping

As in all plastic packages, TAB chips require encapsulation to provide both mechanical and environmental protection. Because of the delicacy of the TAB leadframe, transfer molding is not generally considered an option for TAB encapsulation. The most common approach is thus glob-topping, or radial-spread coating, a chip-surface coating process in which the coating is of nonuniform thickness across the chip surface and is usually much thicker than the coatings produced by any of the molding processes.

In glob-topping, the surface to be coated is positioned under a dispensing nozzle connected to a liquid-encapsulant reservoir. The bonded chip surface is generally heated to facilitate spreading of the encapsulating compound to the tip and exposed areas. Material flowing out of the nozzle impinges against the surface and fans outward (thus the term radial-spread coating). Process control parameters include reservoir pressure, resin viscosity, clearance between the nozzle and surface, dispensing pressure pulse, and physical design of the nozzle.

After the deposition, the coated package is cured by heating in a resistance, microwave, or radiant-heated enclosure or oven. Postcuring of the coating may be performed off-line in a batch process, if necessary.

To ensure chip reliability, the choice of coating is critical. Low-viscosity epoxies, silicone-modified epoxies, and silicone materials are the most widely used polymers for this application. To avoid bubble formation, coatings must be solvent-free and low-stress after cure.

Compared with transfer molding, radial-spread coating has two advantages: the lack of molding pressure and the absence of molding compound flow-related problems. Therefore, it is adaptable to TAB. However, manufacturing disadvantages include a long cycle time and a narrow range of low-viscosity encapsulating compounds. Because of the one-sided coating produced by the process, the packages are also more susceptible to moisture and high shrinkage stress leading to die fracture. (See Robock and Nguyen [1989] and Owczarek [1993]).

A version of radial-spread coating is often used in encapsulating plastic pin-grid arrays (PGAs). The plastic pin-grid array is a variation of the ceramic PGA that offers similarly high I/O capability and equivalent reliability at a lower cost, because of the simplicity of its manufacture. The substrate is a printed wiring board (PWB) in which fiberglass-reinforced epoxy or polyimide polymer sheets are laminated to copper foil. The copper foil is etched to provide bonding pads and a fanout pattern (as in a leadframe) that ends in copper-plated through-holes. Pins are soldered into the plated through-holes and the circuit is gold-plated for wirebonding. The chip is attached with polymer adhesive on this flat or recessed substrate, cured at a low temperature (about 100°C), and connected to the gold-plated circuit leads on the substrate by wirebonding. A plastic frame is placed around the chip and wirebonds, and the area enclosed by the frame is filled with an epoxy or silicone encapsulant by radial-spread coating. A plastic or metal cap is then attached with a polymer adhesive to cover the entire face of the substrate. Good reliability can be obtained in plastic PGAs as long as the formation of water films is prevented [Otsuka et al. 1987]. Figure 3.21 shows a cross-section of the fully fabricated plastic PGA.

## 3.6 DEFLASHING

During the molding process, molding compound can flow through the mold parting line and onto the leads of the device. In its thinnest form, this material is known as resin bleed. A thicker bleed of material is known as flash. If this material is left on the leads, it will cause problems in the downstream operations of lead trimming, forming, and solder dipping and/or plating.

The deflashing process usually consists of mechanical abrasion to remove both light and heavy flash material from the leadframe. If only a thin resin bleed is present, it may be chemically softened followed by mechanical removal of the residue [Zecher 1985]. Media deflashers use a mixture of pressurized air and an abrasive to mechanically remove the material from the surface of the leadframe. In many cases, a plastic granular medium is used to remove the flash and slightly abrade the leadframe. The use of natural media, such as walnut shells and apricot pits is not recommended, since these materials tend to leave an oily residue behind that hinders solder dipping. After removing the resin, the plastic

**Table 3.6 Granulated media specifications**

| Material | Mohs hardness | Specific gravity |
|---|---|---|
| Melamine | 4.0 | 1.5 |
| Urea | 3.5 | 1.5 |
| Acrylic (cross-linked) | 3.5 | 1.2 |
| Acrylic (thermoplastic) | 3.5 | 1.19 |
| Polyester | 3.0 | 1.15 |
| Walnut/apricot | 3.0 | 1.0 |

medium will also matte-finish the leadframe. This roughening enhances the adhesion of the solder material to the leadframe, increasing the durability of the solder adhesion during the steam aging test. The degree of leadframe roughening is dependent on the hardness of the blast medium and the pressure (See Table 3.6). Another method of deflashing uses a slurry mixture of water and abrasive material and high-pressure water to remove the resin from the leadframe; however, this does not abrade the leadframe in any way to enhance soldering.

In some cases, media deflashing can be used to remove the plastic junk from between the dam bar and package body. If this is not accomplished during deflashing, the device must be fed through a dejunking operation to prevent damage in the trim-and-form operation.

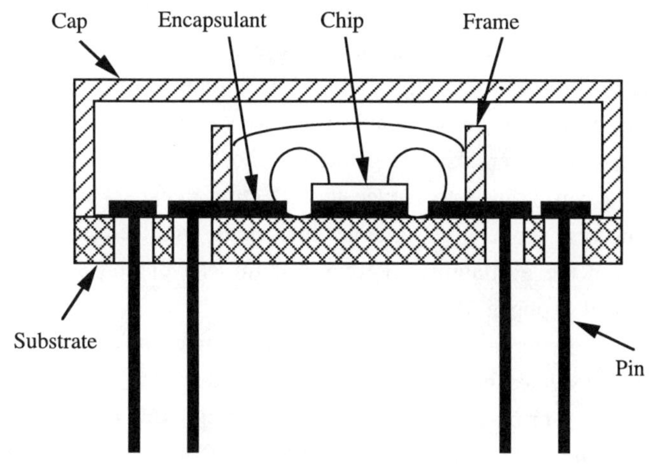


**Figure 3.21** *Cross section of a typical plastic pin-grid-array package*

## 3.7 LEAD FINISH

Lead finish — the protective coating applied to the leads to increase corrosion resistance and to enhance lead solderability — is typically applied after leadframe encapsulation. The finishing process can be either a plating or a coating process. Plating is typically an electrolytic process in which the leadframes are immersed in an electrolytic solution of a salt of the plating metal followed by thorough rinsing. Coating, also called hot dipping, involves dipping cleaned leads first in a flux, then in molten solder, and finishing the process with a hot-water rinse. These processes are performed after device encapsulation because the previous lead finishes cannot withstand encapsulation temperatures of up to 185°C.

Electroplating, has several advantages. It results in a uniform thickness lead finish, while hot-dipped coatings are nonuniform and tend to be thicker in the center. Nonuniform coating thickness is caused by the surface tension of molten solder, which tends to pull away from the ends. The thinner coating at the lead edges makes the leads susceptible to corrosion and loss of solderability. Electroplating can sometimes cause dog-bone plating — that is, plating that is thicker at the lead corners than at the center — since the higher current density results in charge concentration at the corners. Dog-bone plating can be avoided by agitating the electrolytic solution to nullify the effects of charge concentration [Harris 1992]. The biggest drawback of the plating process is that the package is exposed to an electrolytic solution that can introduce ionic contaminants into the package and increase susceptibility to corrosion failures.

Electroplating can be used with fine-pitch devices without the danger, inherent in hot dipping, of solder bridging across adjacent leads. Moreover, it does not subject the leadframe to the high molten-solder temperatures that accelerate the formation of brittle and nonsolderable copper-tin intermetallics at the leadframe-finish interface, and can also cause thermal shock loads on the package. Thus, electroplating can be used to reduce the effective thickness of the lead plating. Earlier concerns about high organic inclusions in electroplating have been alleviated with the availability of improved tin-lead plating solutions based on kysulfonic acid [Foppen 1993].

Some of the material handling systems are similar for electroplating and hot dipping. Encapsulated cut strips are typically loaded onto an endless carrier belt and automatically gripped by holders. The holders provide temporary mechanical attachment while the strips undergo pretreatment, rinsing, finishing, post-treatment, and drying. If the process is electroplating, the holders also provide electrical contact with the carrier belt. The desirable features of the loading system for leadframe plating include the ability to accommodate a variety of leadframe geometries, quick changeover from one product type to another (less than 5 min), capacity for handling large volumes of leadframes (up to 2400 strips per hour), minimal operator attention, computer controls using standard software, and incorporation of a process monitoring system [Foppen 1993].

In the electroplating process, the distance between consecutive leadframe strips on the carrier belt can affect plating uniformity. The leadframe strips can be placed very closed to each other (the minimum distance is 1.5 mm with a tolerance of 0.75 mm), which makes the system similar to a reel-to-reel plating system, which avoids excessive plating thickness at the strip ends [Foppen 1993]. If successive leadframe strips are further apart, the problem of higher current density at the leadframe edges has to be addressed by lowering the overall current, which slows the plating deposition rate and hence the throughput.

Figure 3.22 (a) illustrates the process flow for tin-lead plating; Figure 3.22 (b) illustrates the process flow for the hot-dip process. In both cases, the initial steps of fixturing the encapsulated leadframes, pretreating, and rinsing are common. Pretreatment is a chemical cleaning process to remove stamping oil and light soil common in leadframe plating lines during silver spot plating. Pretreatment also includes a surface activation step, which reduces surface oxides and provides good adhesion for the plating. Hot dipping requires a drying step after rinsing; the electroplating process does not require this step, because the leadframe is immersed in an aqueous solution (electrolyte). Before hot dipping, flux is applied to the leads, which improves solder flow and lead wettability, ensuring complete coverage of the lead by solder. The package is typically preheated, to prevent thermal shock, before dipping the leadframe in hot solder. After the coating or plating is applied, the process steps of rinsing followed by drying, are the same for both plated and hot-dipped leads.

As in any manufacturing process, controls are essential for the lead-finishing operation. The following parameters need to be monitored for plating [Intel 1993]:

- condition of plating — electrodes and fixtures;
- physical arrangement of electrodes and fixtures, including distance to the leadframes;
- chemical composition of plating and pretreatment solutions for all constituents and contaminants, including by-products;
- physical parameters during plating, including temperatures, rinse flows, voltage and current density, rate of plating deposition, and process times.

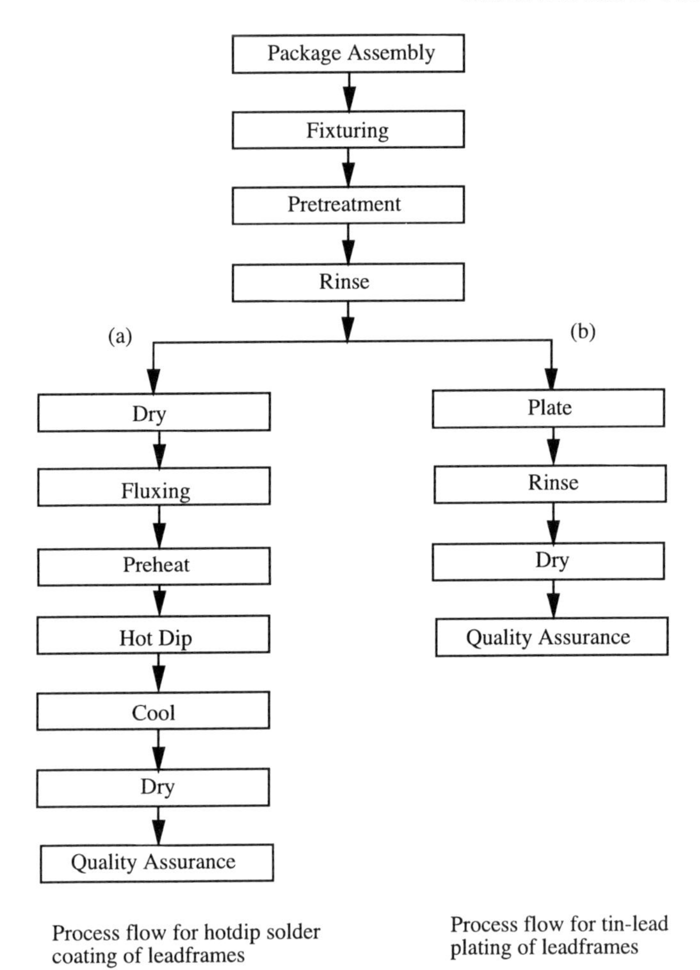


**Figure 3.22** *The process flow for (a) tin-lead plating and (b) the hot-dip process [Intel 1993]*

The parameters to be monitored for hot-dip process control are similar, include [Intel 1993]:

- temperature of the solder in the solder bath;
- height of solder in the bath and condition of fixtures;
- chemical analysis of the pretreatment solution and solder in the solder bath, especially to control the amount of copper that has gone into solution, forming a copper-tin intermetallic;
- operational parameters such as conveyor speed, flux density, preheat temperature, solder temperature, thermal profile of the package during solder dipping, and rinse flow;
- plating uniformity and solder bridging.

Palladium plating is similar to other plating processes, except it requires application of a relatively thick, low-porosity, nickel-rich barrier. After cleaning and rinsing bare leadframes, the surface is activated by chemically reducing surface oxides, producing a surface compatible for deposition of a nickel barrier layer. The nickel layer must be ductile so that it does not crack during lead forming. A second activation step produces a surface ionically similar to the palladium plating bath, ensuring good adhesion of palladium to the nickel barrier layer. The palladium layer is about 0.076 μm (3 microinches). Overall coverage of the lead depends on the surface of the underlying nickel barrier layer. The palladium plating can withstand encapsulation temperatures and can be applied even before the die is attached to the leadframe.

Palladium is suitable for die attachment and wirebonding, and eliminates the need for selectively plating the die paddle and silver spot-plating of inner leads for wirebonding. This results in some operational advantages. Unlike silver plating, palladium plating does not involve cyanides, eliminating the safety issues and waste treatment costs. Further, palladium plating is done over the entire leadframe, eliminating the cost of special spot tooling costs and also reducing product changeover time.

The parameters to be monitored for palladium-plating process control are similar to those for tin-lead and tin plating processes:

- plating thickness for both nickel and palladium;
- plating temperatures, rinse flows, voltage and current density, rate of plating deposition, and process times;

- condition of the plating electrodes and fixtures;
- physical arrangement of electrodes and fixtures, including distance to the leadframes;
- chemical composition of plating and pretreatment solutions for all constituents and contaminants, including by-products;

The advantages of palladium plating reflect the disadvantages of solder dipping. It is difficult to control the lead finish thickness with hot dipping surface-mount components with a lead pitch of less than 1.27 mm (50 mils). With finer lead pitches, solder bridging invariably occurs. If lead-tin or tin plating is used, the plating process has to be carried out after leadframe encapsulation, thus exposing the package to highly corrosive chemicals in the plating bath. Finally, plating the leadframe prior to encapsulation at competitive prices, offers advantages because of inherent controls. The reader is referred to Abbot et al. [1991] and Koford et al. [1981] for further details.

## 3.8 LEAD TRIMMING AND FORMING

The lead trimming and forming operations cut the peripheral leadframe metal that joins all the leads and gives the leads one or more bends to define specific lead shapes. Because the leads are plated, extreme care is required to expose the leads to minimum stress and abrasion to avoid damage to the lead finish.

Leadformers are typically pneumatic presses provided with different die sets for specific lead shapes. The package is held in a fixture and the die set cuts and forms the leads in what appears as a single step. The die sets must hold the package precisely to prevent excessive shock and strain to the leads due to misalignment, and also to fix the required height of the package within close tolerances. The die blocks are typically made of aluminum with replaceable brass bushings that last for over a year [Soldering Technology 1991]. The die set itself is made of tool hardened steel, heat treated, and reground to increase life. The bottom nest or supporting structure for the package usually slides out for ease of part removal. Because PEMs are almost always produced on a mass scale, the lead-forming operation tends to be highly automated.

An important issue in lead forming of surface-mount packages is maintaining lead coplanarity. Leads are said to be noncoplanar if the lead-joining surfaces do not lie in the established seating plane of all the component leads. Lack of coplanarity causes poor solder-joint coverage where package leads contact the circuit boards. Generally, the maximum allowed deviation from the seating plane is 100 μm (4 mils) [JEDEC 1989, Linker et al. 1993] although manufacturers are often able to achieve lead coplanarity within 50 μm (2 mils) of the seating plane [Nguyen 1993, Fancort 1988]. For surface-mount packages, manufacturers recommend leaving a minimum of 1250 μm (50 mils) of the lead beyond the encapsulation for clamping before bending [Fancort 1988]. This distance can be reduced to 750 μm (30 mils), but the more lead surface is available for clamping, the greater will be the precision in lead coplanarity from component to component.

Lead noncoplanarity is due to two main sources: handling damage and molding-induced shrinkage stress [Nguyen et al. 1994]. Handling damage occurs during leadframe loading and unloading from one workstation to the next — for example, from wirebonder to the subsequent die-coating stage or molding platform. In the typical production environment, such steps are sufficiently automated to minimize the unpredictable human factor. Thus, lead damage from mishandling is currently almost nonexistent, although some manufacturers occasionally report lead damage during electrical testing. This arises from the improper fixturing of the plastic loading tubes in the tester; slight misalignment of the tubes or deformation of the tube walls from high clamping forces can cause a variety of lead bending and twisting problems.

Molding-induced shrinkage stress, on the other hand, is an intrinsic processing problem. Right after molding, the shrinkage of the epoxy polymerization, combined with the thermal mismatch between the plastic and the metal leadframe, can cause warpage on the entire strip. Warpage, which affects lead integrity, was not a concern in the days of low pin-count packages made from thick leadframe stock (250 to 300 μm [10 to 12 mils]). However, with the current trend toward thinner and finer-pitch packages, the leadframes have been redesigned to minimize the longest uninterrupted lengths in order to prohibit buckling [Nguyen et al. 1994].

## 3.9 MARKING AND INSPECTION

After deflashing, trimming, and forming, the packed devices are code-marked on the top surface. The process involves the production of indelible, and legible alphanumeric characters and logos, including information on the manufacturer, country of origin, and device code, for the purposes of product identification and traceability. The methods used for this purpose include mechanical engraving, sand blasting, photolithography, etching, printing, and manual scribing. Printing is done either with laser or with a rubber stamp dipped in a polymer-based ink.

Screen and pad printing are also used. Most inks require heat or ultraviolet radiation cure after application. These applications are generally semiautomated, with speeds much greater than one per second.

Laser marking uses a TEA $CO_2$ laser, pulsed or Q-switched Nd:YAG laser, or an excimer laser to produce an indelible mark that is more reliable than ink [Yuoff et al. 1994, Matsunawa 1991, Pecht 1991, Ueda at al. 1990, Zeltner 1990, Parnas 1989, Basting 1988, Poulin et al. 1988, Perry and Huffines 1986, Seth and Scaroni 1986]. The marks are achieved by mask-marking or stroke-marking. Mask-marking involves marking onto the package by a high-energy pulse-emitting laser through a relay lens of a mask containing the necessary alphanumeric characters and logos. In stroke-marking, a computer controls the movement of either the worktable with respect to a stationary laser beam, or a pair of galvanometer-driven mirrors which deflect the laser beam onto the surface of the specimen. A mixture of both processes, for example, the line-scanning method, is sometimes used. For more information on laser marking, see Willis [1990], Basting [1988], Benhard [1987], and Bernard [1983].

The quality of laser marking is assessed by the contrast between the exposed area and the background, the mark width, the mark depth, the severity of spattering, and the evidence of microcracks. These characteristics are dependent on a number of factors, including the laser beam characteristics, the focal length of the lens, the focal position, the type and pressure of the gas used, and the thermal properties of the workpiece material. In general, the laser marking produces marks of low contrast. The contrast can be improved by incorporating a small volume fraction of a metal oxide to the molding compound. However, tinkering with the epoxy formulation is not desirable, because of both the inherent cost of material customization and the larger inventory stock numbers needed to keep track of different tools.

Laser marking is rapidly becoming the preferred marking method in the plastic packaging industry today, although it has a relatively slower speed (1 mark per second) compared to ink marking. According to a survey conducted in 1991 by the Nanyang Technological University, about 44% of all laser materials processing systems in Singapore, and 24% in Japan, were used for laser marking in the microelectronics industry. Most of these industries formerly used the ink-marking technique [Tam et al. 1991, Matsunawa 1991].

High-volume package production generally favors the pad-printing technique. Simple in concept, the method still requires careful developmental effort to achieve mark permanency. Most inks adhere rather well to standard mold compounds; however, with low-stress and the newer ultra low-stress molding compounds special tricks of the trade need to be employed. In this case, the problem concerns the migration of silicone or other stress-modifying additives to the surface under long exposure to high temperatures (e.g., postcure at 175 to 185°C). If marking is done prior to postcure, the moving layer of additives will lift off the ink and destroy legibility; if marking is done after postcure, proper cleaning procedures must be introduced. Surface conditioning generally involves burning the layer with a hydrogen flame-off, followed by vigorous brushing.

Pretest visual inspection is usually conducted after trimming, forming, and code marking. This is also a semiautomated operation, with packaged devices moving on a conveyor — possibly the same line that brought them through the code-marking station. During pretest inspection, lead position and formation and defects in the molded body and code mark are critically noted. Inspection equipment may vary depending on need. Custom or semicustom inspection lines, which involve gauges for sensing lead positions, are often used. Visual systems for detecting bent leads or for checking lead planarity are also employed.

For more information on marking techniques, see Miller [1990], Parnas [1989], Chua et al. [1988], and Choudhury [1986].

## REFERENCES

Abbott, D.C., Brook, R.M., McLelland, N., and Wiley, J.S. Palladium as a Lead Finish for Surface Mount Integrated Circuit Packages. *IEEE Transactions on Components, Hybrids, and Manufacturing Technology* 14 (September 1991) 567-572.

Abdullah, M. A. Nail Head Ball Bonding. *Third International Microelectronics and Systems '93 Conference* (August 1993) 25-43.

Basting, D. Excimer Lasers for Industrial Processing: Results and Applications *J. Laser Application*, 1:1 (1988) 9-16.

Benhard, H. K. State of the Art in Laser Marking and Engraving: Lasers in Motion for Industrial Applications, *Proc. SPIE* 744 (1987) 185-189.

Benhard, H. K. Review of Laser Marking and Engraving: *Laser Marking is the Most Cost-Effective Method and of Permanent Marking, Lasers & Optoelectronics* 7, 9 (1988) 61-67.

Bernard, B. Laser Marking Techniques. Lasers in Materials Processing. *American Society for Metals* LA (1983) 48-52.

Choudhury, M. A. Making Your Mark. *Electronic Packaging and Production* 26, 7 (1986) 47-48.

Chua, M.C.K., Rao, S., Tan, S. H., and Chua, S. J. Computer Controlled Optical System for Nd: YAG Laser Plastic Integrated Circuit Package Marking. *Proc. Asia Pacific Conf. on Optics Technology.* Institute of Physics, Singapore (1988) 1-18.

Doane, D.A., and Franzon, P. Electrical Design of Digital Multichip Modules. In *Multichip Modules: Technologies and Alternatives*, Van Nostrand Reinhold, New York (1993) 368-393; Wire Bonding 525-568.

Dumoulin, P., Seurin, J.P., and Marce, P. Metal Migration Outside the Package During Accelerated Life Tests. *IEEE Transactions on Components, Hybrids and Manufacturing Technology* (1982) 479.

Edwards, D.R. Shear Stress Evaluation of Plastic Packages. *Proceedings of the 37th Electronic Components Conference, IEEE* (1987) 48.

Fancort Lead Forming of Surface Mounted Components. Company literature, Fancort Industries, Inc., 31 Fairfield Place, West Caldwell, NJ 07006 (1988).

Foppen, R. Tin-Lead Plating on Encapsulated Semiconductor Leadframes. *Metal Finishing* (January 1993) 27-31.

Fujita, K., Onishi, T., Wakamoto, S., Maeda, T., and Hayakawa, M. Chip-Size Plastic Encapsulation on Tape Carrier Package. *International Journal of Hybrid Microelectronics* 8, No. 2 (1985) 9-15.

Han, S., and Wang, K.K. A Study of the Effects of Fillers on Wire Sweep Related to Semiconductor Chip Encapsulation. *ASME Winter Annual Meeting* (1993) 123-130.

Harper, C.A. *Electronic Packaging and Interconnection Handbook.* McGraw Hill, New York (1991).

Harris, D.B. Westinghouse Electric Corp., Baltimore, MD. Phone interview (April 1992).

Hirota, J., Machida, K., Okuda, T., Shimotomai, M., and Kawanaka, R., The Development of Copper Wire Bonding for Plastic Molded Semiconductor Packages. *35th Proceedings of IEEE Electronic Components Conference* (May 1985) 116-121.

Holsinger, S., and Sharenow, B. Advantages of a Floating Annular Ring in Three-Layer TAB Assembly. *IEEE Transactions CHMT-10*, No. 3 (1987) 332-334.

Intel. *Packaging Handbook.* Intel Literature Sales, P.O. Box 7641, Mt. Prospect, IL 60056-7641, Intel Order No. 240800 (1993).

Iscoff, R., "Ultrathin Packages: Are They Ahead of Their Time?", Semiconductor International, Volume 17, No. 5, pp. 48, May 1994.

JEDEC Coplanarity Test for Surface Mount Leaded Devices. *JC-14.1-88-33B, JEDEC Solid State Products Engineering Council*, Washington, DC (November 1989).

Kamijo, A., and Igarashi, H. Silver Wire Ball Bonding and Its Ball/Pad Interface Characteristics, *Proceedings of the 35th Electronic Components Conference* (1985) 91-97.

Kawanobe, T., Miyamoto, K., and Hirano, M. Tape Automated Bonding Process for High Lead Count LSI. *Proceedings of the 33rd Electronic Components Conference IEEE* (1983) 221-226.

Kessler, L.W., Semmens, J.E. and Walter, K. Nondestructive Inspection of Ag/Glass Die-Attach Bonds. *Proceedings of the 37th Electronic Components Conference* (1987) 55-60.

Kew, T. G., Tow, C. K., and Fuaida, H. Low Loop Wire Bonding for Thin Small Outline Package (TSOP) *Third International Microelectronics and Systems '93 Conference* (August 1993) 68-76.

Khan, M., Tarter, T., and Fatemi, H. Aluminum Bond Pad Contamination by Thermal Outgassing of Organic Material From Silver-Filled Epoxy Adhesives. *IEEE Transactions* CHMT CHMT-12, No. 4 (1987) 586-592.

Koford, S. Lamczyk, R., and Buszklewicz, B. Palladium Plating: A Replacement for Gold. *IEEE Electronic Components Conference* (1981) 405-414.

Krumbein, S.J. Metallic Electromigration Phenomenon. *33rd Meeting of the IEEE Holm Conference on Electrical Contacts*, Chicago, IL (September 1987).

Lau, J., Erasmus, S.J. and Rice, D.W. Overview of Tape Automated Bonding Technology in *Electronic Materials Handbook:* Packaging, ASM Intl (1989), 274-296; also *Handbook of Tape Automated Bonding*, ed. by John Lau, Van Nostrand Reinhold, New York (1992).

Levine, L., and Sheaffer, M. Wirebonding Strategies to Meet Thin Packaging Requirements, Part I. *Solid State Technology* (March 1993) 63-70.

Linker, F., Levit, B., and Tan, P. Ensuring Lead Integrity. *Advanced Packaging* (1993) 20-23.

Mahalingam, M., Nagarkar, M., Lofgran, L. Andrew, J., Olsen, D., and Berg, H. Thermal Effects of Die Bond Voids in Metal, Ceramic and Plastic Packages. *Proceedings of the 34th Electronic Components Conference IEEE* (1984) 469.

Manzione, L.T., Osinski, J.S., Poslzing, G.W., Crouthamel, D.L., and Thierfelder, W.G. A Semi-empirical Algorithm for Flow Balancing in Multi-cavity Transfer Molding. *Polymeric Engineering Science* 29, No. 11 (1989) 749.

Matsunawa, A. Present and Future Trends of Laser Materials Processing in Japan. *Industrial Applications of High Power YAG and $CO_2$ Lasers, Proc. SPIE* 1502 (1991).

Matsunawa, A., and Ohnawa, T. Interaction in Laser Material Processing. *Transactions of JWRI, Welding Research Institute,* Osaka University, Japan, 1 (1991) 9-15.

Matsuoka, S., Quan, X., Bair, H.E. and Boyle, D.J. A Model for the Curing Reaction of Epoxy Resins. *Macromolecules* 22 (1989) 4093-4098.

Miller, J. Laser Marking Enhances IC Identification. *Evaluation Engineering* 29, 1 (1990) 20-22.

Minges, M.L., et al. *Electronic Materials Handbook,* 1, ASM Intl., Materials Park, OH (1989).

Mori, S., Yishida, H., and Uchiyama, N. The Development of New Copper Ball Bonding-Wire. *Proceedings of the 38th Electronic Components Conference, IEEE* (1988) 539-545.

Nguyen, L.T. Reactive Flow Simulation in Transfer Molding of IC Packages. *Proceedings of the 43rd Electronic Components and Technology Conference* (1993) 375-390.

Nguyen, L.T. National Semiconductor Corp., P.O. Box 58090, Santa Clara, CA 95052, Phone interview (April 1993).

Nguyen, L.T., Danker, A., Santhiran, N., and Shervin, C.R. Flow Modeling of Wire Sweep During Molding of Integrated Circuits. *ASME Winter Annual Meeting* (1992) 27-38.

Nguyen, L.T., Wallberg, R.L., Chua, C.K., and Danker, A. Voids in Integrated Circuits Plastic Packages from Molding. *Joint ASME/ISME Conference on Electronic Packaging* (1992) 751-762.

Nguyen, L.T., Chen, K.L. and Lee, P. Leadframe Designs for Minimum Molding-Induced Warpage. *Proceedings of the 44th Electronic Components and Technology Conference* Washington, DC (1994).

Nguyen, M. Johnson Matthey Electronics. Personal communications (1994).

Olson, T., and Owens, N. Surface Pick-up Tools, Solder Die-attach, and Packaging Reliability as SIP Applications. *Proceedings of the 1991 International Symposium on Microelectronics* (October 1991) 302-306.

Onuki, J., Masateru, S., Iizuka, T., and Okikawa, S. Study of Aluminum Ball Bonding for Semiconductors. *Proceedings of the 34th Electronic Components Conference* (1984) 7-12.

Otsuka, K., Takeo, Y., Ishida, H., Yamada, T., Kuroda, S., and Tachi, H. The Mechanisms That Provide Corrosion Protection for Silicone Gel Encapsulated Chips. *IEEE Transactions Components Hybrids Manufacturing Technology* CHMT-12, 4 (1987) 666-671.

Owczarek, J.A. Coating of Electronic Components, I: Physical Model of the Deposition Process Determined by Drop Tests, *ASME Transactions, Journal Electronic Packaging,* 115 (1993) 233-239; II: Determination of the Residual Yield Stress After Deposition, and Method to Produce a Drop of Required Cured Skin Thickness and Diameter. *ASME Transactions, Journal of Electronic Packaging* 115 (1993) 240-248.

Parnas, S. J. Indelible coding With Lasers: Variable Information Marking On-the Fly. *Surface Mount Technology* 3, 8 (1989) 27-28.

Pecht, M. *Handbook of Electronic Package Design,* Dekker, New York (1991).

Pecht, M., *Integrated Circuit Hybrid and Multichip Module Design Guidelines: A Focus on Reliability*, Wiley, New York (1993).

Perry, D. V., and Huffines, W. H. Quality Evaluation of $CO_2$ and Nd: YAG Laser Marking of Ceramic Chip Capacitors. *IEEE Proc. 36th Electronic Components Conference* (1986) 655-658.

Poulin, G. D., Eisele, P.A., and Znotins, T. A. Advances in Excimer Laser Material Processing. *International Congress of Optical and Science Engineering,* Germany (1988).

Robock, P.V., and Nguyen, L.T. Plastic Packaging. *Microelectronics Packaging Handbook* ed. Tummala, R. R. and Rymaszewski, E. J. Van Nostrand Reinhold, New York (1989) 556-557.

Seth, A., and Scaroni, J. Ceramic Component Marking with YAG Lasers. *Semiconductor International* 9, 10 (1986) 80-83.

Soldering Technology. *Lead Formers*, Company literature, Soldering Technology International Inc., 555 W. Allen Ave., Suite 10, P.O. Box 3217, San Dimans, CA 91773 (1991).

Tam, S. C., Yeo, C. Y., Yusoff, M. Noor, Lim, L. E. N., Jana, S., Yang, L. J., and Lau, M. W. S. Applications of Laser Machining in Singapore. *International Conf. on Productivity '91* (1991) 194-218.

Ueda, M., Sitoh, Y., Hachisuka, H., Gokoh, Y., and Mantani, H. Studies on $CO_2$ Laser Marking. *Optics and Lasers in Engineering* 12, 4 (1990) 245-249.

Uno, T., et al. Void Formation and Reliability in Gold-Aluminum Bonding. *Nippon Micrometal Corporation*, Kawasaki, Japan (1992); quoted from Mohammed Azli Abdullah, Nail Head Ball Bonding, paper given at *Third International Microelectronics and Systems '93 Conference*, Malaysia (August 1993) 25-43.

van Kessel, C.G.M., Gee, S.A., and Murphy, J.J. The Quality of Die-Attachment and Its Relationship to Stresses and Vertical Die-Cracking. *Proceedings of the 33rd Electronic Components Conference, IEEE* (1983) 237.

Willis, J. B. Techniques and Applications of Laser Marking. *Laser Lines Limited*, Banbury, Oxon, England (1990).

Yamashita, T., Kanamori, T., Iguchi, Y., Arao, Y., Shibata, S., Ohno, Y., and Ohzeki, Y., Development of Ultra Fine Wire and Fine Pitch Bonding Technology, *IEICE Transactions* E 74, No. 8 (August 1991).

Yuoff, Md. Noor, Tam, S. C., Lim, L. E. N., and Jana, S. A Review of Nd: YAG Laser Marking of Plastic and Ceramic IC Packages. (Paper accepted for publication in Journal of Materials Processing Technology) (1994).

Zecher, R. F. Deflashing Encapsulated Electronic Components. *Plastic Engineering* (June 1985) 35-38.

Zeltner, H. Coding at the Speed of Light. *Journal of Packaging Technology* 4, No.3, (1990) 12-14.

# 4

# ASSEMBLY ONTO PRINTED WIRING BOARDS

Plastic-encapsulated microcircuits (PEMs) are assembled on printed wiring boards (PWBs), along with other parts (capacitors and resistors), to form functional circuits. Although there are many variations in the order of assembly steps, the common sequence includes parts placement, solder application, solder reflow, cleaning, and, in some cases, conformal coating. Many manufacturers also have inspection, testing, and rework procedures, although these are not value-added activities. Figure 4.1 shows a generic printed wiring board assembly flow.

Because of the small geometries, high parts count, and stringent quality required of today's electronic assemblies, significant attention must be paid to the development and control of assembly processes via a structured, proactive approach.

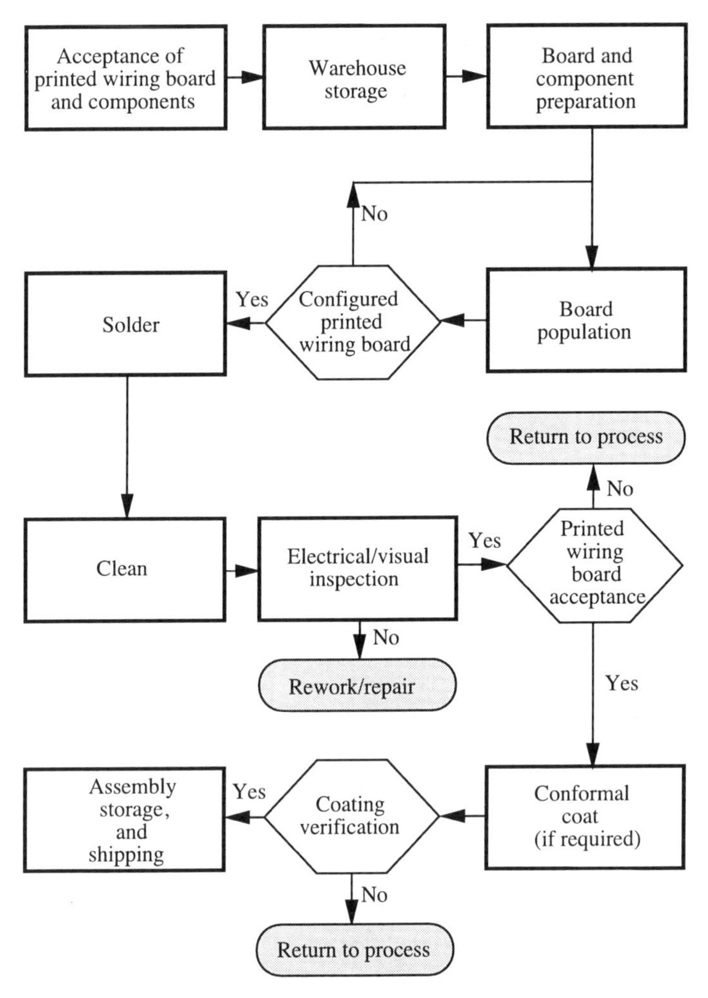


**Figure 4.1** *A generic printed wiring board assembly flow*

Understanding the constraints imposed by the parts used is critical to high-quality assembly.

## 4.1 ASSEMBLY TECHNOLOGIES

Based on the attachment configuration, three types of assembly processes are possible: through-hole assembly, surface-mount assembly, and mixed through-hole and surface-mount assembly. Through-hole assembly involves inserting component leads into plated through-holes in the board and soldering them. The PEMs (i.e., dual in-line and plastic pin-grid arrays) may have their leads trimmed to the proper length prior to assembly or, in a few cases, the leads may be trimmed after assembly and soldering. During the soldering process, molten solder rises through, and fills, the plated through-hole, forming a mechanical and electrical connection between the printed wiring board and the component. The most common through-hole soldering process is wave soldering (see Section 4.2.1). Other, less common through-hole assembly processes include fountain soldering and solder dipping. See Manko [1992], Rahn [1993], and Hwang [1992] for more information on these processes.

Until the 1980s, through-hole assembly was the dominant interconnection method for printed wiring board assemblies. However, by the end of the decade, the need for increased functionality, higher speed, reduced size, higher lead counts, and lower costs placed demands on printed wiring board assemblies that could no longer be met by through-hole designs. Surface-mount technology provided the capabilities to meet these requirements.

In surface-mount assembly, the component leads are soldered to the surface of the printed wiring board. The solder joint, the sole mechanical and electrical connection between the printed wiring board and the component, bears a significant amount of the stress from any vibrational or thermomechanical loads on the board. During assembly, surface-mount packages are placed on pads patterned on the printed wiring board and pretreated with solder paste. Components are kept in place by the inherent tackiness of the paste. The board is then heated, during which the molten solder paste reflows around the leads. Subsequent cooling of the assembly solidifies the solder joint. The most common surface-mount soldering process is reflow soldering (see Section 4.2.2). The interested reader is referred to Pecht [1993a], Hwang [1992], and Capillo [1990] for more information on surface-mount assembly.

Mixed assembly combines through-hole and surface-mount components on one PWB. For mixed assembly, manufacturing engineers must determine the optimal process sequence, considering, for example, which major soldering processes should occur first, and whether or not to include such common steps as cleaning. The manufacturing design team must decide whether to put all through-hole components on one side of the printed wiring board and surface-mount components on the other, or place the surface-mount components on both sides of the board. These decisions impact cost, quality, reliability, and production scheduling.

## 4.2 SOLDERING

The solder joint provides the mechanical attachment for the device, as well as electrical continuity and, in some cases, a path for heat transfer. As a metal-joining technique for electronic assemblies, soldering offers such important attributes as low-temperature manufacturing, good electrical and thermal conductive paths, strength to withstand thermomechanical and vibrational loads, and ease of automation for mass production of a wide variety of assemblies. The geometry of the solder joint depends on the type of package (through-hole, leaded, or leadless surface-mount) being soldered to the board (see Chapter 1).

Soldering process classifications are based on the method of applying the solder and the heating processes involved. Solder can be applied to all joints on one side of the board simultaneously by passing the board over a jet of molten solder; this is called wave soldering. Reflow soldering involves applying solder, typically in the form of a paste, to pad patterns on the printed wiring board, placing the components at those sites, and heating (reflowing) the solder. The solder paste melts, wets the surfaces of the component leads and pads, and upon cooling, solidifies to form a joint.

### 4.2.1 The Wave Soldering Process

Wave soldering includes fluxing, preheating, passing the board over a solder wave, and removing excess solder. Figure 4.2 depicts the passage of the populated, fluxed, and preheated PWB over the solder wave.

Fluxing consists of applying a flux (a chemical compound with activators, solvents, and detergents) to the solder joint locations to chemically clean the surfaces to be joined before passing them over the solder wave.

Preheating reduces thermal shock to the components and the PWB, activates the flux, and evaporates any flux solvents to prevent the formation of blowholes (voids) in the solder joint.

During soldering, the board is passed over a jet of molten solder. The hydrostatic pressure of the solder from the nozzle wets the underside of the PWB

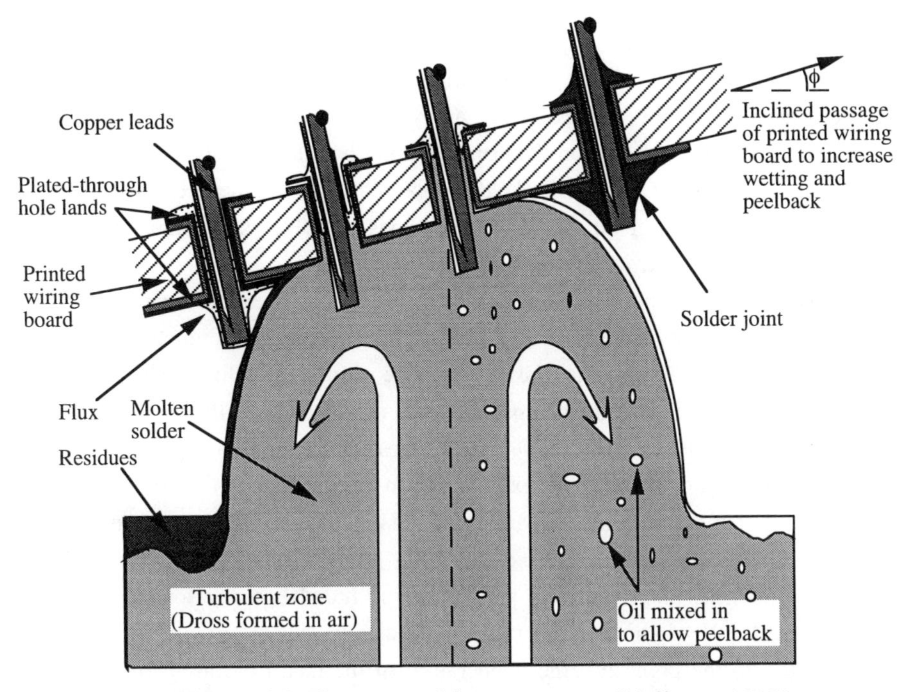


**Figure 4.2** *The wave soldering process [Malhotra 1993]*

and pushes the solder slightly up the plated through-hole. Subsequently, capillary action resulting from the surface tension between the solder and the walls of the plated through-hole forces the molten solder to rise up through the hole. Upon emerging on the other side of the board, the solder wets the upper lands of the through-hole. The top portion of the risen solder then cools down and solidifies into a cap, assuring that the solder stays in the plated-through hole during cooling.

The printed wiring board is usually inclined with respect to the solder wave to improve wetting. In the molten state, excess solder generally peels back from the board and falls into the solder pot; to improve the effectiveness of this process, an air knife may be used to force a jet of air on the underside of the PWB that blows away any excess solder. Soldering can be conducted in an inert atmosphere to avoid the formation of unwanted oxides.

For a detailed discussion on wave soldering processes and equipment, see Malhotra [1993] and Woodgate [1988].

### 4.2.2 The Reflow Soldering Process

The reflow soldering process consists of solder-paste dispensing, component placement, preheating, drying, reflowing, and cooling. Reflow involves remelting (reflowing) solder previously applied to a PWB joint site (pad) in the form of a preform or paste. No solder is added during reflow.

The first step in the process is to apply solder paste and any necessary adhesives to the land patterns on the printed wiring board. Adhesives are needed only if the components cannot be held in place by the tackiness of the solder paste. More information on the various methods of solder paste and adhesive dispensing can be found in Hwang [1992].

Surface-mount components are designed to save space and increase the potential functionality of the electronic assembly. However, the reduction in size and weight also increases the risk of damage during handling. Placement accuracy is critical because the component leads must line up with the pads (land area) on the boards to prevent solder bridges or incomplete solder joints. Computer inspection defines the exact location of the land and monitors the alignment. For more information on surface-mount component placement, see Kou [1991], DeCarlo [1991], Arnone and Tong [1991], and Pecht [1993b].

After component placement, the printed wiring board is ready for soldering. The populated PWB is transported to the preheater to evaporate solvents in the solder paste. In the preheater, the temperature of the board is raised to 100° to 150°C at a rate low enough to prevent solvent boiling and the formation of solder balls; this rate, established at 2°C/sec. for wave soldering of ceramic capacitors, has been accepted as an industry norm for reflowing all sensitive components [Linman 1991]. A short preheat time, followed by oxygen-free reflow, improves solder wetting, allows the use of low-flux solder paste, and effectively reduces the amount of subsequent cleaning required.

The second stage in the reflow process is slow heating, which increases the temperature to the solder melting point and activates the flux in the solder paste. The activated flux removes oxides and contaminants from the surfaces of the metals to be joined. The heating is kept short to allow the moisture in the paste to evaporate without splattering the solder, which could lead to shorts between metallization traces if the solder bridges two metal conductors.

The third stage in the process, reflow, consists of melting and fillet formation. During the melting stage, the temperature of the solder paste is raised to just above its melting point. When the solder melts, it replaces the liquid flux formed in the previous drying process. The temperature is held at the melting point while the solder coats the surfaces to be joined to form a good bond. Higher temperatures increase the fluidity of the solder, causing it to move away from the desired joint location along either the component lead or the PWB, depending on whether convection or radiation heating is used. The total time that the solder is above the melting point, or wetting time, is critical to reliable solder joint formation. Excessively long periods at the wetting temperature may lead to intermetallic formation and embrittlement. Fillet formation is the most critical part of the reflow process. The temperature must be regulated to allow the melted solder particles to coalesce and then cool. The cooling of the joint produces surface tension in the formation of a fillet around the component lead.

The last stage of the reflow soldering process involves cooling the solder joint, either by conducting the heat through the board layers or by natural or forced convection to the ambient air. Forced convection is not recommended because it excessively stresses the boundary layers in the solder joint, reducing joint reliability [Woodgate 1988]. However, if the board assembly is complex enough, heat trapped within the board layers may slow down the cooling rate and cause poor joints.

Reflow methods differ according to the method of heat transfer to the reflow site. Conductive, convective, and radiative heat transfer are variously used for this purpose. Conductive heat-transfer methods include the soldering iron, hot-bar reflow, and conductive-belt reflow. Radiative heat-transfer methodologies include infrared, laser, and optical-fiber systems. Convective heat-transfer is used in vapor-phase (or condensation) reflow systems. Each method has its own merits and limitations. The most common technique for mass reflow soldering is infrared heating, because it allows temperature control throughout the reflow soldering cycle. However, vapor-phase soldering is considered the best way to limit the maximum temperature reached during reflow, since the temperature in the oven never rises above the temperature of the saturated vapor. Discussions on the types of reflow soldering processes are presented in Malhotra [1993] and Hwang [1992].

The effect of the time-temperature relationships for soldering on the reliability of plastic packages and their integrated circuits must be known before setting up a soldering facility. Figure 4.3 shows the acceptable operating region for solder reflow. Important concerns include the rate of increase in temperature to the solder melting point, the total time that the devices and components are subjected to the soldering temperature, and the impact of the maximum temperature on the potential failure mechanisms of a plastic-encapsulated microcircuit.

***The reflow temperature profile.*** Temperature variation must be optimized and controlled throughout the reflow soldering process to ensure a high-quality, high-yield solder joint. Figure 4.4 is a typical infrared reflow temperature profile for quality solder-joint production, showing the temperature of the workpieces. The temperature of the infrared panels and emitters in the soldering machine must be much higher than the desired workpiece temperature. Usually, infrared reflow

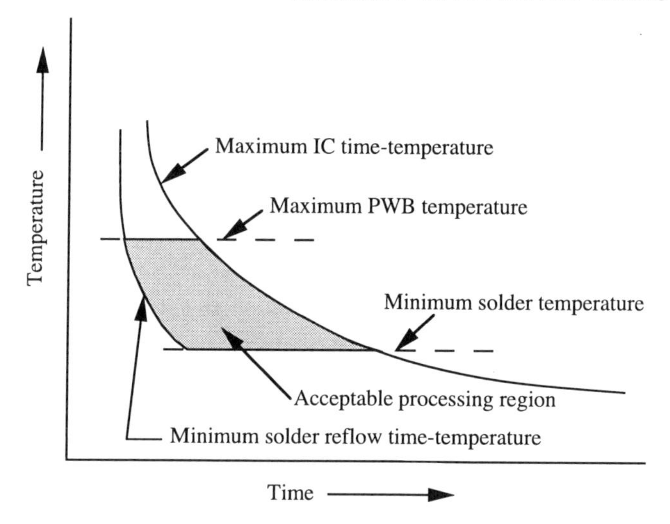


**Figure 4.3** *Time-temperature relationships in reflow soldering [Hutchins 1990]*

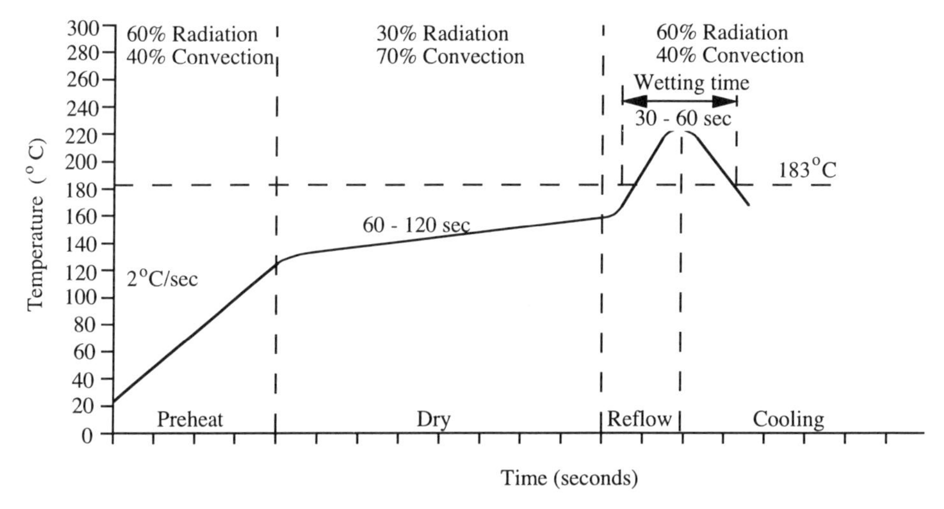


**Figure 4.4** *A typical infrared-reflow soldering time-temperature profile [Malhotra 1993]*

ovens are provided with both top and bottom heater controls so that different board geometries can be accommodated.

Reflow profile bands depict the temperature operating range within which process quality is acceptable. Figure 4.5 is an example of the soldering profile band for the reflow profile in Figure 4.4. The figure shows a typical band size of 25°C; the actual bandwidth depends on the solder paste, the types of components, the board material, and the board loading throughput.

Three common reflow profiles are depicted in Figure 4.6. Type A, depicting a fast heating rate, is the most common, with a moderate peak temperature typical of surface-mount assemblies with eutectic solder pastes that do not require predrying. Type B shows a slow heating rate for a moderate peak temperature, typical of assemblies containing small ceramic chip capacitors or components sensitive to thermal shock. Type C depicts a fast heating rate with a high peak temperature, common for high-lead solders with high melting points (in excess of 280°C).

Figure 4.7 is a vapor-phase reflow soldering temperature profile. Because the vapor temperature is held constant, the stabilization period and thermal uniformity after preheating differ significantly from the other profiles. Typical vapor-phase temperatures vary from 205 to 215°C; the upper temperature limit of 215°C ensures that the flux is easy to clean later. This is much lower than the 230 to 240°C for infrared reflow and 260°C for solder dipping and wave soldering, making vapor-phase reflow a preferred process for any temperature-sensitive package.

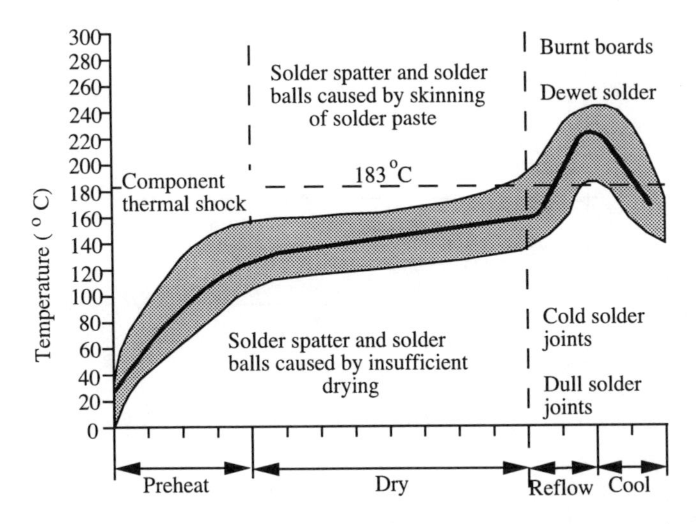


**Figure 4.5** *A typical reflow soldering profile band and potential problems that can arise at various stages in the manufacturing process [Malhotra 1993]*

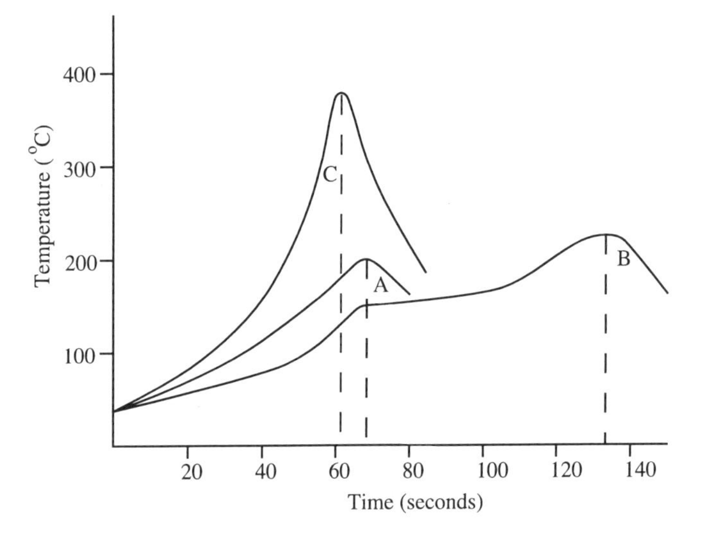


**Figure 4.6** *Types of reflow profile [Hwang 1992]*

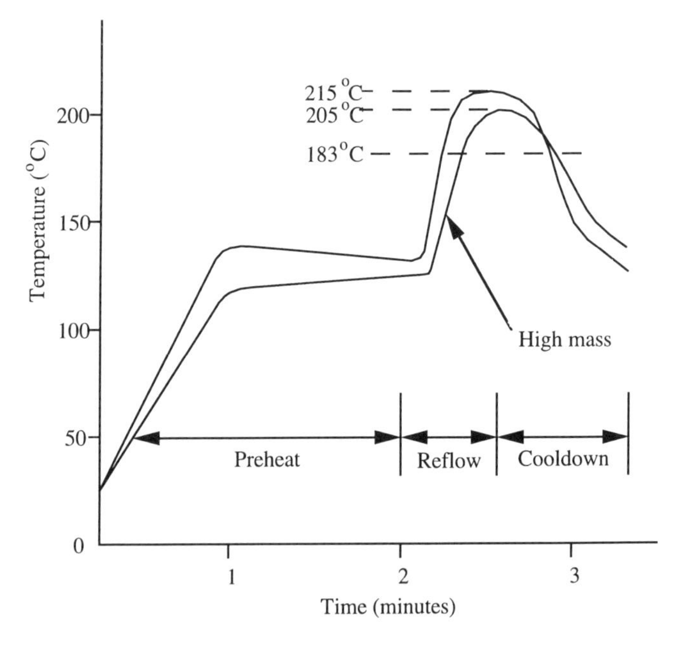


**Figure 4.7** *A typical vapor-phase-reflow soldering profile [Linman 1991]*

***Preheating rate.*** With plastic packages, the maximum package temperature during solder reflow and the rate of temperature change must be controlled to avoid some unique assembly-related failure mechanisms. Solder temperatures above 220°C will increase the probability of failure [IPC-SM-786].

The recommended rate of temperature change is between 2 and 10°C/sec [IPC-SM-786]; an accepted norm for preheating the infrared reflow process is 2°C/sec [Hutchins 1990], derived from studies conducted during the wave soldering of ceramic capacitors. These rates are used in reflow soldering because integrated circuits are less susceptible to thermal stress damage than ceramic capacitors. The ramp-up time from room temperature to solder melting temperature is highest for the vapor-phase reflow process — from 6 to 8°C/sec still less than the maximum allowable rate of 10°C/sec.

## 4.3 POPCORNING

Soldering processes expose the PEM to temperatures of 210 to 260°C for periods of 30 sec to 5 min; these temperatures are above the glass-transition temperature (140 to 160°C) of most molding compounds. Moreover, at such temperatures, ingressed moisture in the molding compound can vaporize and expand the package outline, leading to package cracking, a phenomenon known as popcorning. Cracks may be further propagated by rework procedures to remove defective components, and by temperature cycling during operation.

Studies on popcorning failures have shown that cracks initiate due to delamination of the molding compound from the die, the leadframe, or the die paddle interface [Ilyas and Poborets 1993, Tay et al. 1993]. Delamination is initiated by the coefficient of thermal expansion mismatch between die and encapsulant materials, which can cause a buildup of unwanted stresses at the boundary between the attached materials. In extreme cases, these stresses can cause deadhesion or delamination at the material interface.

The greatest CTE mismatch occurs in the in-plane direction between the molding compound and the die paddle in the leadframe. The in-plane coefficient of thermal expansion of plastic packages varies with temperature [Panchwagh et al. 1993]. The plastic body of the package has a CTE of 12 to 14 ppm/°C below the glass transition temperature (approximately 140 to 160°C) and about 50 ppm/°C above it (see Figure 4.8). The temperature changes that occur during soldering cause both the encapsulant and the die paddle to expand at different rates, resulting in unwanted stresses at the interface.

The mechanism of delamination has been visually observed by Tay et al. [1993] in a small-outline J-lead package with a transparent encapsulant.

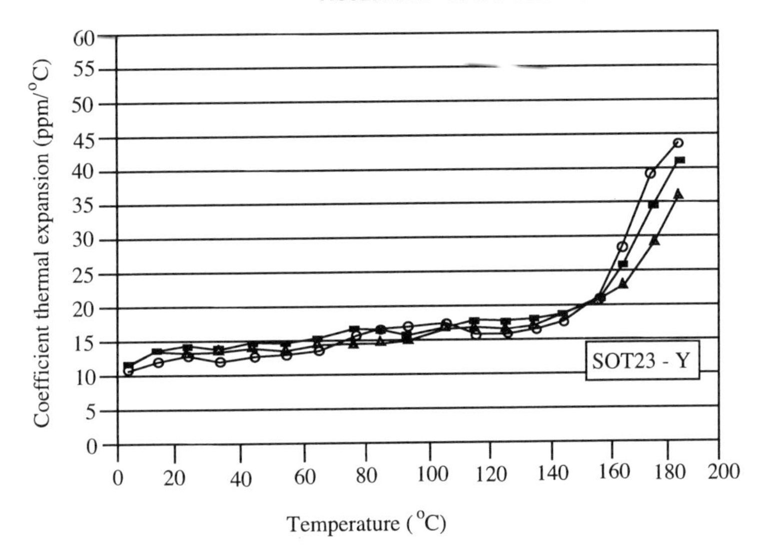


**Figure 4.8** *In-plane coefficient of thermal expansion of a SOT23 as a function temperature [Panchwagh et al. 1993]*

Delamination initiates at the middle of the long side of the die paddle and propagates until the die paddle is fully delaminated from the molding compound. Next, a side crack occurs on the long side of the die and propagates, causing delamination between the die and the die paddle. The resulting deformation is usually less than that caused by the delamination between the die paddle and the molding compound [Kawamura et al. 1993]. In certain instances, delamination occurs between the die surface and the molding compound without any further crack propagation. In these cases, moisture and contaminants extracted from the molding compound are trapped in the interfacial void, creating the possibility of long-term corrosion.

In less extreme cases, the delamination may not be severe, and internal package cracks may not reach the surface. The electrical characteristics of the PEM remain stable even after cracking occurs [Fukuzawa et al. 1985, Ilyas and Poborets 1993]. Nevertheless, reliability risks are introduced, since the plastic is free to move relative to the die surface during temperature excursions. Moreover, delamination may concentrate stress on the bonds, leading to sheared or cratered ball bonds that can cause intermittent electrical failures.

Figure 4.9 illustrates the adhesion strength and percent delamination for groups of 68-lead plastic-leaded chip carriers encapsulating dummy chips, after preconditioning at 85°C and 85% relative humidity followed by three one-minute

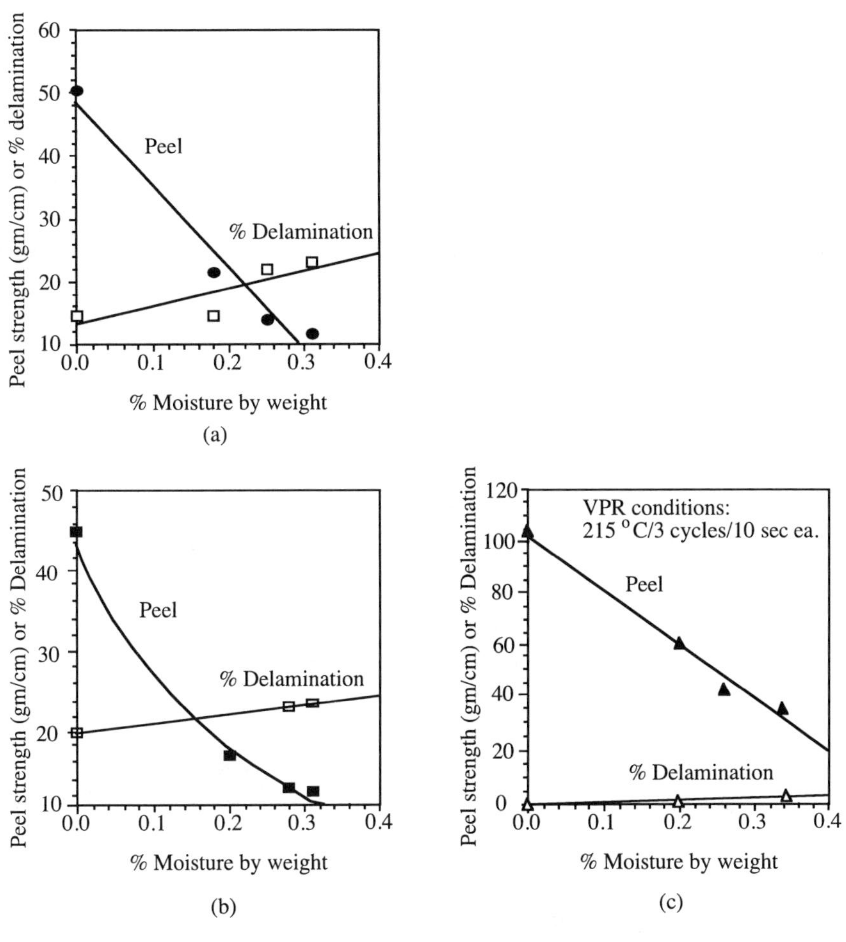


**Figure 4.9** *Adhesion strength and percent delamination after three 1-min cycles of vapor-phase reflow at 85°C/85% RH for groups of 68- lead plastic-leaded chip carriers encapsulating dummy chips [Moore et al. 1991]*

cycles of vapor-phase reflow [Moore et al. 1991]. The packages were examined nondestructively using a C-mode scanning acoustic microscope. Figures 4.9a and b show packages with low adhesion strengths, while Figure 4.9c depicts the percentage delamination for a package fixed with adhesive. The graphs show a definite increase in delamination and a reduction in peeling strength with an increase in moisture content. However, for the package in Figure 4.9c, the increase in percentage delamination is small, suggesting that increasing the adhesion improves the resistance of the package to delamination caused by moisture ingress during reflow soldering.

Tay et al. [1993] studied the occurrence of delamination to determine the critical stress intensity factor, $K_{IIC}$, corresponding to the maximum strength of the bond between the molding compound and the die; $K_{IIC}$ was used to predict the initiation of delamination in a plastic-encapsulated package. In the study, plastic packages were preconditioned in four different environments: 24 hrs. at 121°C and 100% relative humidity, 168 hr at 85°C and 85% RH, 168 hr at 85°C and 65% RH, and 168 hr at 85°C and 40% RH. After preconditioning, the packages were subjected to an adhesion test to determine the force required to pull a strip of the leadframe material from the molding compound (see Figure 4.10). The nodal displacements obtained from a finite element analysis of the test specimen were used to determine the stress intensity factor. Since, delamination is purely a shear phenomenon, $K_{IIC}$, is the critical stress intensity value for delamination and can be used as the design criterion for preventing it.

In thin packages soldered at low temperatures, a bulge appears at the bottom after soldering [Fukuzawa et al. 1985]. At high soldering temperatures, the same package shows cracks in the encapsulation beneath the die pad. The cracks are too small to be detected by the naked eye, so the packages are inspected using such nondestructive evaluation techniques as optical, X-ray, and acoustic microscopy (see Chapter 6). Figure 4.11 depicts acoustic microscopy on a popcorn-cracked 84-lead plastic quad flatpack after solder-reflow [Gallo 1993]. The top part of the picture shows delamination on the bottom of the die paddle, and the bottom part gives a side view. There are two cracks in the package: one extends to the bottom surface, and the other remains internal (the package is shown upside down).

The wave-soldering process, normally used with through-hole packages, exposes only the leads of the package to the solder temperature. Surface-mount soldering processes, on the other hand, subject the entire molded body to elevated temperatures and can, therefore, promote popcorning. The greatest temperature stresses, however, are imparted by solder-dipping, a liquid-immersion process that exposes the entire component to molten solder. Among reflow heating processes, infrared reflow and vapor-phase soldering impose a significant amount of thermomechanical stress, while laser heating and thermode heating, which locally heat the component leads, cause the lowest thermally induced stress. These reflow heating techniques, however, have far lower production rates than solder dipping, infrared heating, and vapor-phase reflow, and are not used commercially.

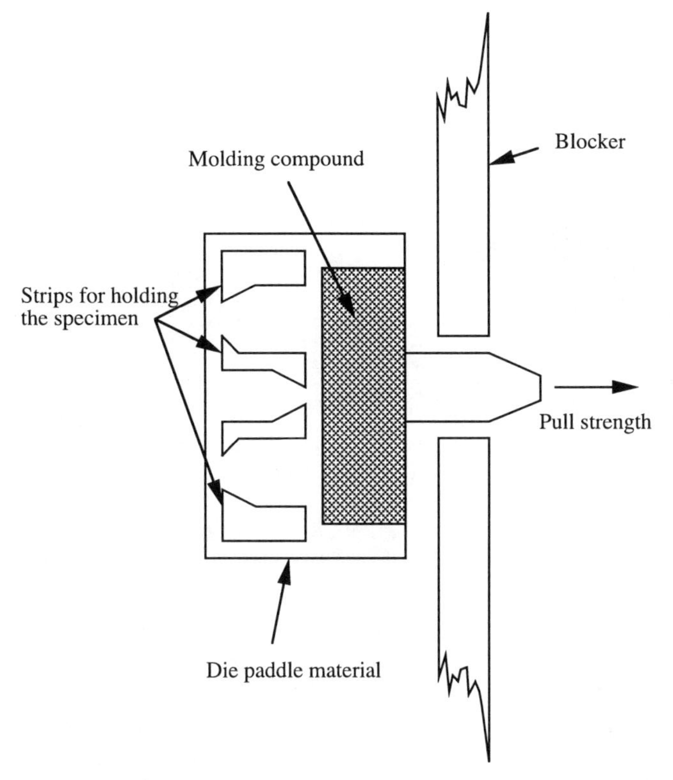


**Figure 4.10** *Experimental setup to determine the adhesion strength between the molding compound and the leadframe in a plastic encapsulated package [Tay et al. 1993]*

### 4.3.1 Types of Package Cracks

Package cracks resulting from soldering stress can be divided into three types, depending on the origin and direction of crack spread [Omi et al. 1991]. In type I, the crack originates at one edge of the die pad and spreads to the bottom of the package. In type II, the crack originates on one edge of the die pad and spreads to the top surface. In type III, the crack spreads from an edge of the chip to the top surface of the package. The predominance of one type over another depends on the following factors:

- package structure: chip size, relative resin thickness on the chip and under the die pad, leadframe material, and elastic modulus of the die-bonding material;
- soldering conditions: soldering temperature and time;

**Figure 4.11** *Acoustic microscope scan of a plastic quad flatpack cracked during solder reflow [Courtesy of Dexter Corp 1993]*

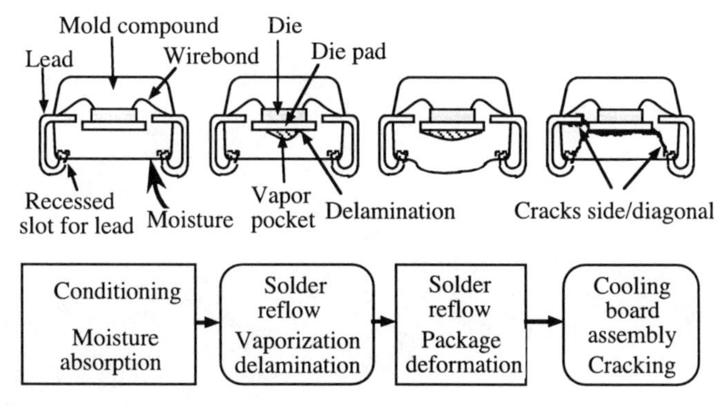


**Figure 4.12** *Conditions leading to package popcorning during assembly [Nguyen 1993]*

- moisture absorption: ratio of moisture absorbed in the package to the maximum possible absorption.

Figure 4.12 depicts the mechanism of popcorn cracking; the lower pad surface delaminates completely from the die, effectively concentrating the thermal shrinkage forces on the paddle edges [Nguyen 1991]. The cracked package is susceptible to corrosion and contamination of the bond pad and die metallization by the migration of external ions along the surface of the cracks [Cergel 1989]. Figure 4.13 indicates the end results for molded packages, with cracks emanating from the die pad and propagating either upward or downward.

### 4.3.2 Analysis of Popcorning

Variables that influence popcorning include

- the material properties of the encapsulant, the leadframe, the die and the die paddle;
- the leadframe design;
- the ratio of the paddle area to the minimum encapsulant thickness surrounding the paddle;
- the adhesion of the encapsulant to the die paddle and the leadframe;
- the moisture absorbed in the encapsulant;
- the contamination level; the voids in the encapsulant; and
- the reflow process parameters [Nguyen 1991].

(*a*)

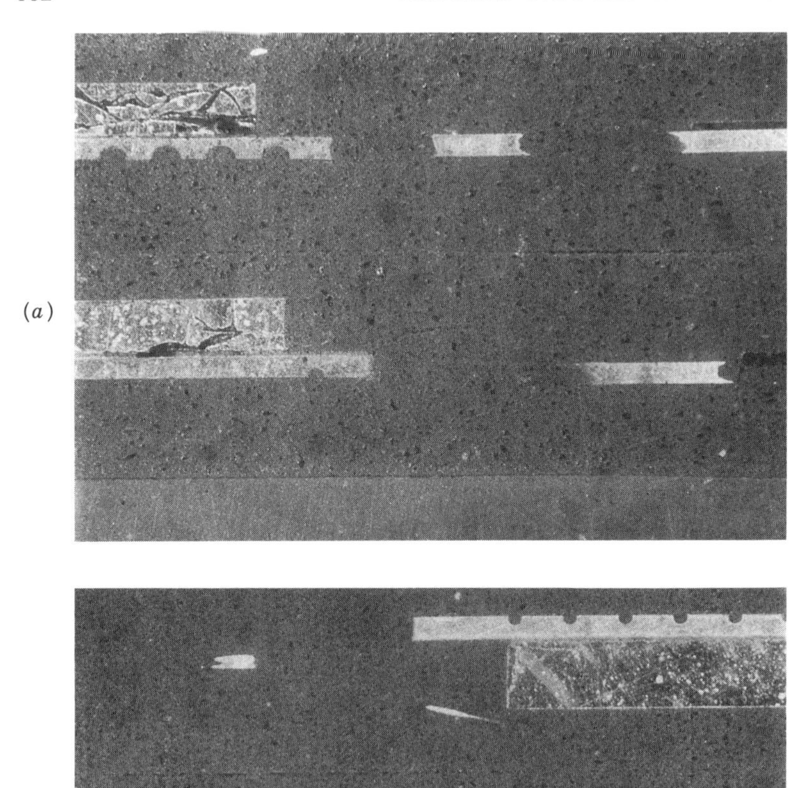

(*b*)

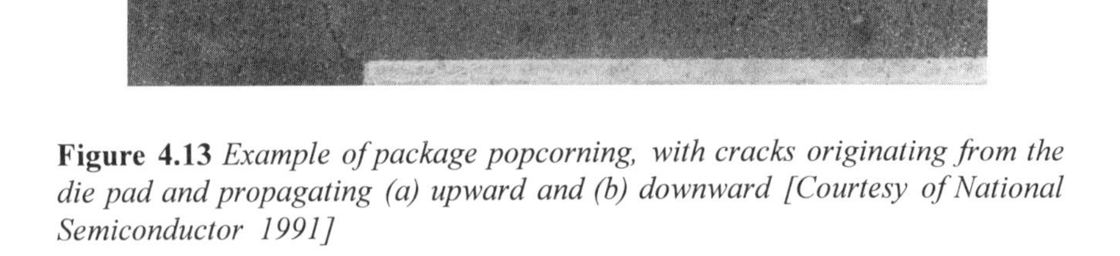

**Figure 4.13** *Example of package popcorning, with cracks originating from the die pad and propagating (a) upward and (b) downward [Courtesy of National Semiconductor 1991]*

Encapsulant material properties that affect package cracking are the coefficient of thermal expansion, the stiffness (e.g., flexural modulus), the mechanical strength, the moisture diffusion and absorption, the adhesion strength, and the fracture toughness. Defects, such as burrs on the die paddle and sharp edges on the die, can also accelerate popcorning. Both finite element simulation and experimental results show package cracks initiating at the sharp edges and corners of the leadframe.

Kitano et al. [1988] analyzed the package cracking phenomenon during 215°C vapor-phase soldering by analytically expressing moisture distribution in the encapsulant according to Fick's second law. The initial conditions specified were that

$$t = 0 \quad \text{and} \quad O(t) = O_0(x) \tag{4.1}$$

where $O_0(x)$ is the initial distribution of moisture in the encapsulant and $x$ is the perpendicular distance from the die paddle towards the bottom of the package. The boundary conditions specified were that

$$\begin{aligned} x &= 0,\ O(t) = p_{pe} O_{sol} \\ x &= h_{package},\ O(t) = RH\, p_s O_{sol} \end{aligned} \tag{4.2}$$

where $p_{pe}$ is the vapor pressure in the pad-encapsulant space, $O_{sol}$ is the solubility coefficient of moisture in plastic, $h_{package}$ is the thickness of plastic below the chip pad, and $RH$, $p_s$ are the relative humidity and the saturated vapor pressure in the ambient, respectively. The saturated moisture content of the plastic is assumed to follow Henry's law, that is, it is proportional to $p_s$. Fick's first law of diffusion describes the mass of water diffusing into the delaminated space per unit area per unit time. Integrating this function with time, the accumulation of moisture in the interface is obtained as

$$m = m_o + A \int_o^t D \left( \frac{\delta c}{\delta x} \right)_{x=o} d\tau \tag{4.3}$$

where $m_o$ is the initial mass of water in the space, and $A$ is the area of the chip pad. Under delamination and bubble formation conditions during soldering, the amount of moisture at the interface can be used to obtain the vapor pressure in the bubble, the volume of the bubble, and the maximum clearance, $\delta$, between the chip pad and the plastic. The diffusion and solubility coefficients of moisture are experimentally determined for a moisture-saturated epoxy molding compound slab by mass change during desorption, and are expressed as

$$D = D_0 \exp\left(-\frac{E_d}{kT}\right) \tag{4.4}$$

$$S = S_0 \exp\left(\frac{E_s}{kT}\right) \tag{4.5}$$

where $D_0$ is the diffusion coefficient at 0°C, $O_0$ is the solubility coefficient at 0°C, $E_d$ and $E_s$ are the activation energies for the diffusion and solution phenomena, respectively, $k$ is Boltzmann's constant, and $T$ is the absolute temperature.

The modulus of elasticity and bending strength of the molding compound are experimentally determined as a function of temperature. Using these experimental data, popcorning of small-outline J-leaded (SOJ) packages under infrared heating was correlated with the maximum measured clearance $\delta$, between the chip pad and the encapsulant and calculated as a function of the percentage moisture saturation of the package during both absorption and desorption cycles. The calculated interface moisture saturation of 60 to 63% in an 85°C/85% RH ambient agreed with the experimental data, demonstrating that popcorning indeed depended on the moisture content at the interface of the die paddle and the molding compound.

A complementary study on the relationship between popcorning and the physical properties of four specially formulated epoxy molding compounds was conducted by Ito et al. [1991]. Along with conventional molding compounds, these authors studied compounds with low glass transition temperatures, ultra-low stress, and super ultralow stress. Using mathematical and experimental methods to determine moisture diffusivity, moisture solubility, flexural modulus, and flexural strength (as did Kitano et al. [1988]), they obtained a relationship between the calculated maximum stress in the package due to moisture evaporation and the mechanical flexural strength of the encapsulants at soldering temperature for popcorning in an 80-lead (80L) PQFP. For the two packages studied, Table 4.1 shows the physical properties of the four molding compounds and their experimental flexural strength versus the calculated maximum popcorn-producing package stress at the indicated soldering temperatures. The maximum stresses generated during soldering are directly related to the moisture absorption and diffusion characteristics in the molding compound and the thickness of the encapsulant. In general, only the low glass transition temperature compound can

**Table 4.1 Relationship of molding compound properties to interface popcorn-producing stress [Ito et al. 1991]**

| Molding compound | Diffusion parameters $D=D_0 \exp(-E_d/kT)$ ($cm^2/sec$) | Solubility parameters $S=S_0 \exp(E_s/kT)$ ($mg/cm^3/kg/mm^2$) | $T_g$ (°C) | Flexural strength ($kg/mm^2$) | | Maximum stress in the package (1 cm x 1 cm pad) ($kg/mm^2$) | |
|---|---|---|---|---|---|---|---|
| | | | | 240°C | 260°C | Plastic quad flatpack, soaked, 30°C/85%RH /20 hr, 0.3 mm plastic, 260°C, 20 sec dip soldering | Plastic-leaded chip carrier, soaked, 85°C/85%RH /72 hr, 1.4 mm plastic, 240°C, 30 sec IR soldering |
| Low glass transition temperature anti-popcorn | $D_0= 2.00 \times 10^{-3}$<br>$E_d= 0.344$ eV | $S_0=4.00 \times 10^{-3}$<br>$E_s= 0.379$ eV | 115 | 0.90 | 0.90 | 0.92 | 0.65 |
| Ultralow stress | $D_0= 8.57 \times 10^{-2}$<br>$E_d= 0.425$ eV | $S_0= 0.98 \times 10^{-3}$<br>$E_s= 0.439$ eV | 165 | 0.90 | 0.74 | 3.5 | 2.75 |
| Super ultralow stress | $D_0= 1.48 \times 10^{-2}$<br>$E_d= 0.400$ eV | $S_0= 1.75 \times 10^{-3}$<br>$E_s= 0.427$ eV | 153 | 1.00 | 0.90 | 2.6 | 1.5 |
| Conventional | $D_0= 2.43 \times 10^{-2}$<br>$E_d= 0.424$ eV | $S_0= 1.97 \times 10^{-3}$<br>$E_s= 0.426$ eV | 162 | 1.05 | 0.90 | 2.0 | 1.35 |

prevent popcorning during soldering, even for a 0.3-mm (12 mil) thick plastic encapsulant saturated with moisture at 30°C. The addition of stress-relieving agents to generate low-stress molding compounds increases both moisture solubility and diffusivity in such encapsulants, hampering their ability to withstand popcorning [Ito et al. 1991]. Similar studies on phenomenological and mathematical analyses of popcorning were also done by Oizumi et al. [1987].

The molding compound properties required for popcorn resistance are different from those for thermal shock or cycling resistance in surface-mounted devices, as shown by Ito et al. [1991]. The key failure mechanism associated with thermal shock is passivation cracking. The relationship used to calculate the number of passivation cracks is

$$N_p = c_6 n (I_s)^{7.0} x^{2.9} \tag{4.6}$$

where $N_p$ is the number of passivation cracks, $c_6$ is the constant for a particular device and package type, $n$ is the number of thermal shock cycles, and $I_s$ is the stress index, given by

$$I_S = \int_{T_1}^{T_2} (E_B \alpha)\, dt \tag{4.7}$$

where $t_{package}$ is package thickness, $E_B$ is the flexural modulus of the encapsulant, $T_1$ is the lower limit of temperature cycling, $T_2$ is the upper limit of temperature cycling, and the exponents and constants are specific to an 80-lead plastic quad flatpack. For all encapsulant thicknesses studied, passivation cracks determined experimentally for all four epoxy molding compounds (Table 4.1) were correlated with the ultra low-stress molding compound demonstrating the most thermal shock cycling resistance or the fewest top passivation cracks. In summary, an encapsulant with the antipopcorn properties of the low glass transition temperature molding compound and the anti-thermal shock properties of the ultra-low stress molding compound (coefficients of thermal expansion of 14 ppm/°C) will have the highest reliability under use conditions.

Matsushita Electric Industrial Corporation [1993] found no correlation between popcorning and moisture absorption- and desorption-induced weight change in a plastic quad flatpack at any fixed water concentration level. The researchers then considered the level of water concentration at the leadframe-plastic interface for plastic quad flatpacks subjected to dip soldering at 260°C, finding that all packages cracked at a water concentration level between 0.17 moles per liter (up to 9 hr desorption at 125°C storage for samples exposed for 100 hr to 85°C/85%

RH) and 0.2 moles per liter (up to 17 hrs. absorption in 85°C/85% RH for samples baked for 15 hr at 125°C). They also found the critical condition for package cracking as $\sigma_P \geq \sigma_R$, where $\sigma_P$ is the maximum stress loading at the edge of the die pad and $\sigma_R$ is the flexural strength of the encapsulant at the soldering temperature. The relationship for calculating $\sigma_P$ is

$$3\sigma_P = c\left(\frac{a_{max\,dp}}{h_{package}}\right)^2 p_{water} \tag{4.8}$$

where $c$ is a constant, $p_{water}$ is the vapor pressure of water at the soldering temperature, $a_{maxdp}$ is the longer dimension of the die pad, and $h_{package}$ is the package thickness under the die pad. Figure 4.14 shows the correlation of 260°C dip-soldering-induced plastic quad flat pack cracking with varying die pad sizes, obtained by Matsushita using Equation 4.5 with their interface water concentration data. Figure 4.15 shows the correlation with the distance from die pad edge to the package end and the thickness of the encapsulant under the die pad (for underside crack or side crack), for the same soldering process applied to SOJ packages. The interfacial water concentration level for side-crack generation in SOJ packages was the same as in plastic quad flat packs during desorption, but only 0.13 moles per liter after 11 hr of absorption for samples baked in 85°C/85% RH storage for 15 hr at 125°C. Figure 4.16 depicts the moisture degradation curve for plastic-encapsulant surface-mount packages and shows the "safe region" where popcorning can be avoided.

According to standard specifications, the probability of package cracking is strongly dependent on the ratio of the die-paddle size (short side) to the minimum plastic thickness of the package, as illustrated in Figure 4.17. The minimum plastic thickness of a package is defined as the distance from the die-attach paddle, die surface, heat spreader (if used), or die coating (if used) to the nearest exterior of the package, and from the tip of the lead fingers to the die paddle. As the ratio of the die-paddle area to the minimum plastic thickness increases, the probability of package cracking rises. This ratio is only an indicator of the susceptibility to package cracking, not a rule. In Figure 4.17, the cross-hatched area represents the uncertainty inherent in classifying a package safe or crack-sensitive, according to this indicator.

A conflicting view is presented by Ilyas and Poborets [1993] in their study of susceptibility of various encapsulants to ambient moisture and consequent popcorning and package cracking in eight types of plastic packages. A flow chart of their experiment is shown in Figure 4.18. The results of the study

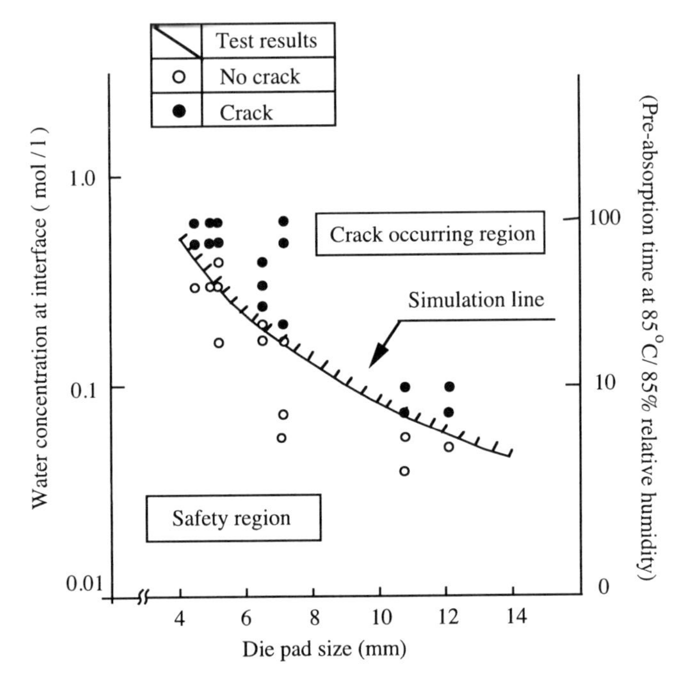


**Figure 4.14** *Correlation of 260 °C dip soldering induced plastic quad flatpack cracking with varying die-pad sizes [Matsushita Electric Industrial Corporation 1993]*

indicated that neither the ratio of die-paddle thickness to package thickness nor the ratio of die-thickness to package thickness were significant. Rather, it was shown that the thickness of the molding compound under the die and the amount of moisture in the package distinguished a cracked from an uncracked package. The study concluded that package cracking during solder reflow does not occur in packages with a moisture content less than 0.1% by weight and a molding compound thickness greater than 1 mm (40 mils) under the die; packages with a moisture content greater than 0.2% and a molding compound thickness less than 1.75 mm (70 mils) were highly susceptible to cracking.

Electrical testing on the packages often showed that delamination did not necessarily cause electrical failure of the component. Of the packages showing delamination, some passed the electrical test.

### 4.3.3 Preventing Package Cracking

The simplest way to reduce package failures during surface mounting is to keep the molded devices in a sealed, moisture-barrier bag with desiccant until they are based on the 0.11% moisture content limit have been questioned for their usefulness and cost-effectiveness. Instead, parts are classified based on their

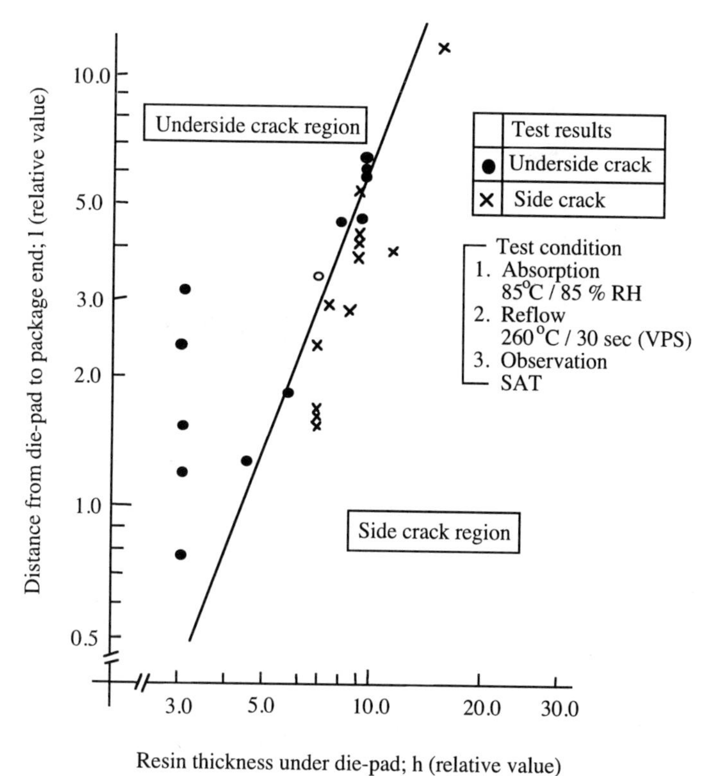


**Figure 4.15** *Correlation of 260 °C dip soldering-induced small-outline J-lead package cracking with the distance from die-paddle edge to package and the thickness of encapsulant under the die paddle for underside cracks or side cracks [Matsushita Electric Industrial Corporation 1993]*

attached to the circuit board. However, this imposes restrictions on component handling in the factory and can reduce efficiency for very-low-volume assembly plants. Ensuring that the PEMs have a low moisture content is done by baking them immediately before the mounting process.

A temperature bake is often used to reduce the moisture content of the plastic packages before mounting and eliminate moisture-related cracking. According to Lin et al. [1987, 1988], the safe allowable moisture content is around 1100 ppm, or approximately 0.11% by weight. Baking the components at 125°C for 24 hr drives off a sufficient amount of the absorbed moisture. A resulting residual moisture of 0.08% by weight can thus be obtained, allowing the manufacturer and user sufficient shelf life. However, integrated circuit packages have been shown to withstand damage due to package cracking at moisture levels as high as 0.3% by weight [Hickey 1994]. Therefore, dry-bagging precautions,

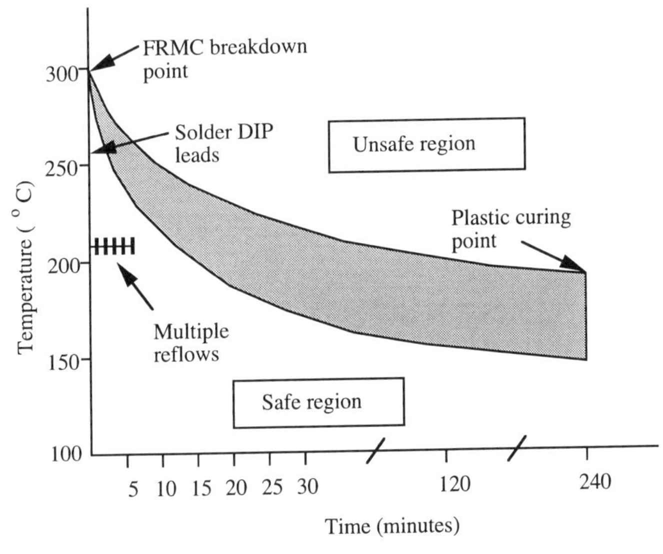


**Figure 4.16** *Degradation curve for plastic surface-mount components [Hutchins 1990]*

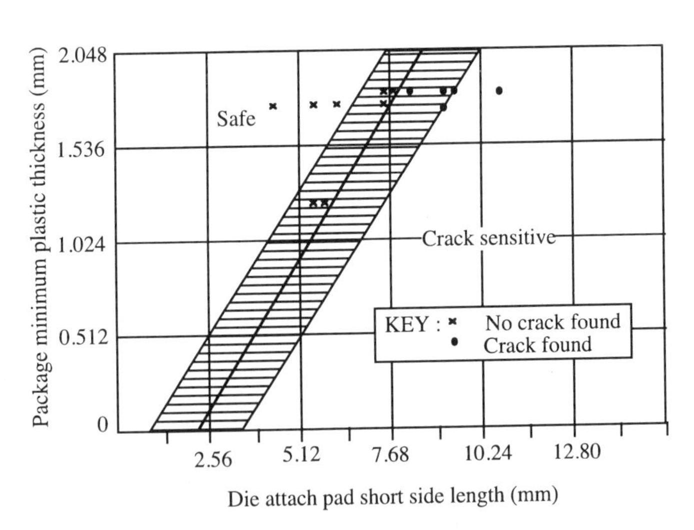


**Figure 4.17** *Probability of package cracking as a function of die-paddle ratio; 1 mil (0.001 inches) is approximately 40 mm [IPC-SM-786]*

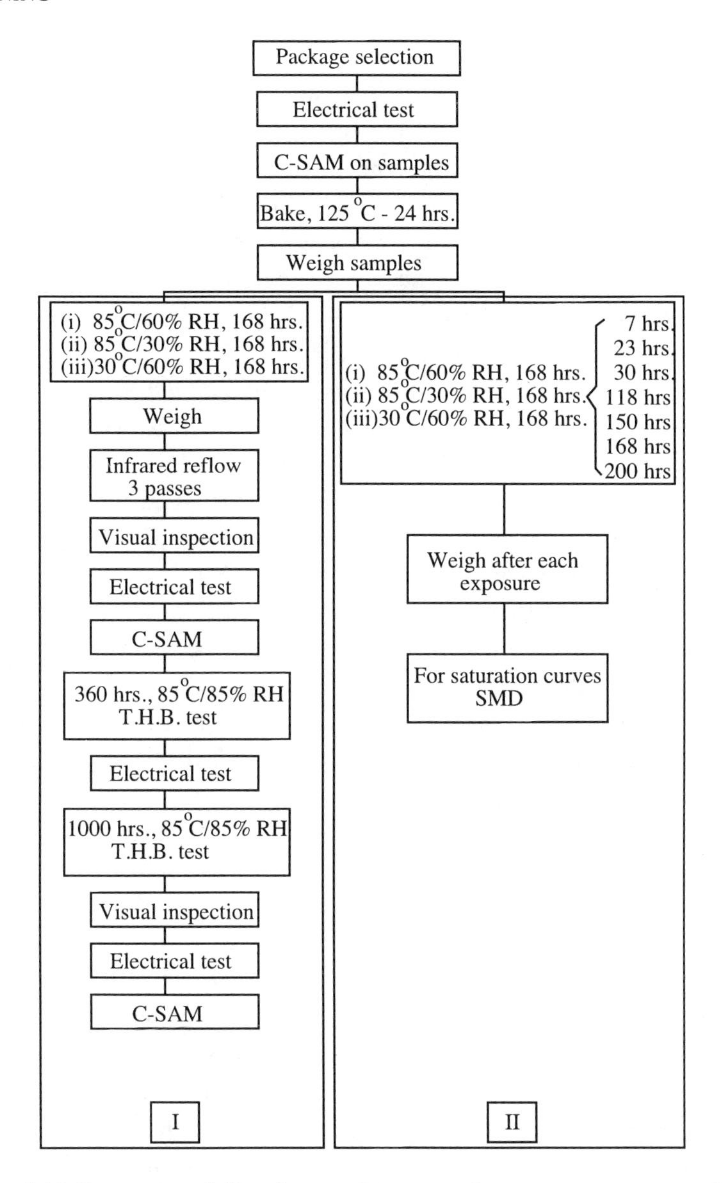


**Figure 4.18** *Experimental flowchart to determine the moisture sensitivity levels of various plastic package [Ilyas and Poborets 1993]*

storage floor life (out of bag) at the board assembly site. The parts are classified into 6 levels, and appropriate precautions are taken during handling. The classification scheme is discussed in Section 5.3.

Additional techniques for preventing popcorning include using molding compounds with improved resistance to soldering heat, putting a polyimide coating on the back side of the die paddle, adopting an improved leadframe design, preventing voids in the molding compound, and formatting a vent hole in the underside of the package.

***Molding compound with improved resistance to soldering heat.*** This technique uses a molding compound with improved adhesion strength, mechanical strength, and moisture absorption ratio [Omi et al. 1991]. The glass transition temperature of the molding compound is increased until it nears the solder reflow temperature. Although epoxy molding compounds are capable of reaching this level of glass transition temperature, the problem is maintaining post-encapsulation low die stresses. An opposite approach keeps the glass transition temperature below the solder reflow temperature to dissipate strain, while increasing the modulus of the molding compound in the rubbery region. This approach ensures that the material does not sustain enough water vapor-induced deformation to damage the package [Ito et al. 1991]. Increasing the strength of the molding compound above the glass transition temperature will prevent tearing of the encapsulant. Other methods to prevent package cracking include designing packages to decrease stress on the plastic due to water vapor and using plastic molding compounds with high bending strengths at soldering temperatures.

***Polyimide coating on the back side of the die pad.*** A polyimide film with excellent adhesion to both the die pad and the molding compound can be formed to suppress possible molding compound exfoliation from the die pad and to reduce the stress from moisture vaporization under solder heat [Omi et al. 1991]. Adhesion promoters, such as coupling agents or special plating, have been used to increase the adhesion of the encapsulant to the leadframe and the die paddle. Special leadframe designs intended to increase the "tooth," or roughness, of the leadframe can reduce or eliminate delamination of the encapsulant from the surface, particularly in the area of the die paddle [Nguyen and Michael 1992].

***Improved leadframe design.*** Cracks in plastic packages are usually initiated at the leadframe burrs and propagate from the burr through the thinnest plastic region. Burrs created in the leadframe stamping process can be stress

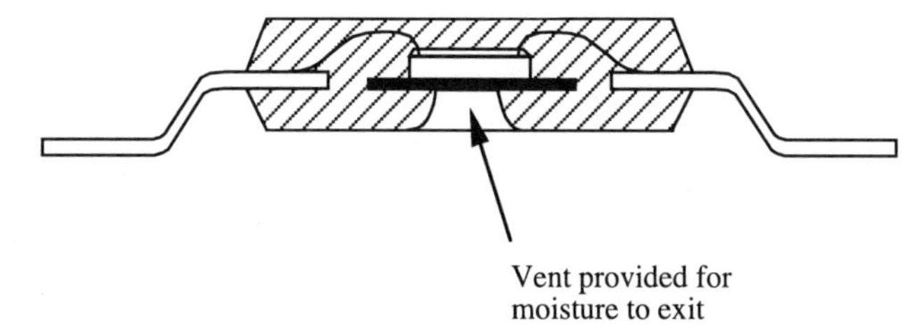


**Figure 4.19** *The provision of a vent hole in the plastic package to allow the vaporized moisture to exit the encapsulant [Fukuzawa et al. 1985]*

concentration points. Reversing the stamping direction to cause burrs at the top of the leadframe or etching the leadframe (coining) reduces the amount of cracking [Omi et al. 1991].

***Prevention of voids.*** Package cracks sometimes extend from die-bond voids at die edges, so a void-free die-bonding process is crucial. This technique requires forming the central portions of the longer sides of the die pad, where the maximum stress is concentrated, inward to the chip outline (anchoring).

***Formation of a vent hole.*** Fukuzawa et al. [1985] suggest providing a vent hole in the plastic encapsulant underneath the die pad to allow vaporized moisture to exit (see Figure 4.19). To be effective the hole must be drilled through the encapsulant into the die-attach material. Moisture in the plastic package tends to diffuse toward the largest void in the molding compound; as the largest void, the vent hole facilitates moisture diffusion out of the package. This is an economical method of preventing popcorning. Objections to this method arise because drilling a vent hole induces a defect in the package and may cause unforeseen reliability problems later.

A summary of common design and assembly remedies to avoid package cracking during solder reflow operations is given in Figure 4.20 and is discussed in detail in Chapter 9. When the plastic package is exposed to humidity, cracking depends on three issues: package design (e.g., thickness), leadframe design (e.g., burr-free etched leadframe versus stamped leadframe), and molding compound properties (e.g., flexural modulus and fracture toughness — factors that can be addressed only by the compound suppliers). Semiconductor suppliers, have some latitude in package outline selection (e.g., whether to include a pedestal) and leadframe configuration (e.g., whether to use a dimpled/anchored pad or a coined leadframe).

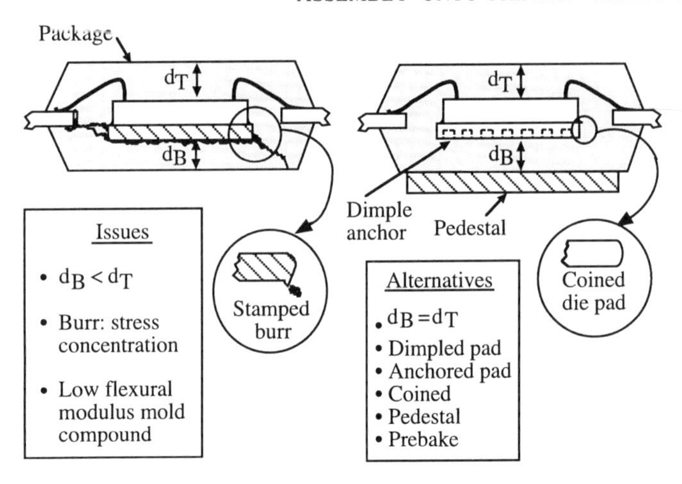


**Figure 4.20** *Common design and assembly remedies to avoid package cracking during reflow [Nguyen 1991]*

## 4.4 SOLDER JOINT FATIGUE

When using plastic-encapsulated microcircuits (PEMs), the coefficients of thermal expansion of the PEM and the board must be sufficiently matched so that expansion differences during reflow do not cause solder joint fatigue failures. Closely matched CTE of the component and circuit card help reduce thermally induced stresses.

With PEMs, the CTE of the components is from 20 to 23 ppm/°C, which corresponds to the coefficients of thermal expansion of FR-4 board (16 to 24 ppm/°C), BT board (8 to 16 ppm/°C), and CE board (6 to 13 ppm/°C) [Nelco 1994]. Care must be taken when using large PEMs on ceramic boards with coefficients from 5.4 to 8.3 ppm/°C. For more information on solder joint fatigue, see Lau [1991].

## 4.5 CLEANING

The soldering process tends to create by-products that can contaminate the plastic-encapsulated microcircuits and the printed circuit board. In general, a contaminant is any material—not just those that cause corrosion—that can degrade performance to unacceptable levels.

The principal source of soldering contaminants is the flux used during soldering. However, other sources include body oils, talcs, perfumes, dried skin, and dandruff from workers handling the boards; oils used by component placement machines; adhesives; solder balls; and the atmosphere. To reduce the chance of corrosion and increase the reliability of the finished product, the printed circuit board must be cleaned after soldering; if it is soldered several times, repeated cleaning may be necessary.

Many common solvents, such as chlorofluorocarbons (CFCs) and Freon 113 (trichlorotrifluoroethane), are used in cleaning electronics. The property that makes them attractive as cleaning agents is their stability. However, cleaning agents such as perchloroethylene and methyl chloride are now known to harm the environment.

Alternative cleaning strategies include semiaqueous cleaning and aqueous cleaning. One alternative being considered is a no-clean process in which eliminating the contaminating agents precludes the necessity for cleaning.

In semiaqueous cleaning, two solvents are used in succession to clean the printed circuit board. The first can be either a single compound designed to remove the nonpolar contaminants or an azeotrope that removes both polar and nonpolar contaminants. The second solvent is water, used to remove polar contaminants and the residue left by the first solvent. The final step in the process is drying the printed circuit board to prevent corrosion caused by any remaining water.

In aqueous cleaning, deionized water, along with a saponifier solution, is the only cleaning agent used. The boards are subjected to a prewash, a microdroplet wash, a rinse, and a drying stage.

No-clean processes tackle the problem of contamination where it occurs, i.e., at the time of soldering. They involve the use of no-clean fluxes that leave no residue after soldering. For further information on electronic assembly cleaning, see Hwang [1992], Curtis [1993], and Rahn [1993].

## 4.6 CONFORMAL COATING

To maintain the cleanliness of the board and prevent the entry of contamination and moisture from external sources, a conformal coating may be applied to an assembled board. The desirable characteristics of an uncured conformal coat include a long pot-life, easy applicability, secure adhesion to the printed wiring board, non-reactive nature with the board material, resistance to fungal growth, hydrolytic stability, environmental compatibility, safety, and ease of inspection [Waryold and Lawrence 1991]. In the cured state, the coating should effectively bar moisture and particle contaminant ingress, provide good electrical insulation over its temperature range, be chemically resistant, be flexible enough to absorb thermal shocks without cracking, and be repairable. During curing, shrinkage of the conformal coating can produce excessive stresses on delicate components

**Table 4.2 Typical physical and operating characteristics of cured conformal coatings (ratings are in decreasing order, with A being the optimum) [Waryold and Lawrence 1991]**

| | Acrylic | Polyurethane | Epoxy | Silicone |
|---|---|---|---|---|
| Physical characteristics | | | | |
| Humidity resistance | A | A | B | A |
| Humidity resistance (extended periods) | B | A | C | B |
| Abrasion resistance | C | B | A | B |
| Mechanical strength | C | A | A | B |
| Temperature resistance | B | C | C | A |
| Acid resistance | B | A | A | B |
| Alkali resistance | B | A | A | B |
| Organic solvent resistance | D | A | A | B |
| Operating characteristics | | | | |
| Application | A | B | C | A |
| Remove (chemically) | A | B | very difficult | C |
| Remove (burn-through) | A | B | C | stable |
| Pot-life | A | B | D | C |
| Optimum cure time: | | | | |
| Room temperature | A | B | B | B |
| Elevated temperature | A | B | B | C |

[Waryold and Lawrence 1991].

The common types of conformal coatings are acrylic, polyurethane, epoxy, and silicone. Tables 4.2 and 4.3 give typical physical, electrical, operating, and thermal characteristics of these coatings. Acrylics apply easily; cure in minutes; provide electrical, moisture, and fungus resistance; have a long pot-life; and readily adapt to existing manufacturing processes. An added advantage is that acrylics do not undergo an exothermic reaction during curing, which prevents damage to heat-sensitive components. Their susceptibility to chlorinated solvents

**Table 4.3 Typical thermal and electrical characteristics of cured conformal coatings [Waryold and Lawrence 1991]**

| Property | Acrylic | Polyurethane | Epoxy | Silicone |
|---|---|---|---|---|
| Resistance to heat, continuous (°C) | 135 | 125 | 125 | 200 |
| Linear coefficient of thermal expansion (microinches/in/°C) | 50 to 90 | 100 to 200 | 40 to 80 | 220 to 290 |
| Thermal conductivity ($10^4$ cal/sec/cm $^2$/cm) | 4 to 5 | 4 to 5 | 4 to 5 | 3.5 to 8 |
| Dielectric strength short time at 23°C, $10^6$ V/cm at 0.003 cm | 1.38 | 1.38 | 0.87 | 0.79 |
| Surface resistivity at 23°C, 50% RH, Ω-cm | $10^{14}$ | $10^{14}$ | $10^{13}$ | $10^{13}$ |
| Dielectric constant at 23°C, 1 MHz | 2.2 to 3.2 | 4.2 to 5.2 | 3.3 to 4.0 | 2.6 to 2.7 |

is used as an advantage during rework to remove the conformal coating.

Polyurethane coatings provide excellent protection from humidity and moisture. Their chemical resistance is also high, which often causes problems during rework; special strippers are required to remove the coating completely. The coating is very sensitive to the presence of moisture on the board and tends to blister. This can lead to moisture ingress and circuit failure.

Epoxy conformal coatings have short pot-lives, but good resistance to humidity, abrasion, and chemicals. However, the coating itself is difficult to remove. If rework must be done, the component must be removed by first burning through the epoxy with a knife or soldering iron, often destroying the component in the process. Moreover, epoxies take 1 to 3 hr to cure at elevated temperature, or four to seven days at room temperature.

Silicone conformal coatings are used in applications that require continuous exposure to elevated temperatures. Their advantages include excellent resistance to humidity, corrosion, and temperature changes. The fact that they withstand high temperatures makes them useful for power applications. Repair and rework, however, are not possible with silicone conformal coatings, since they do not dissolve easily.

## 4.7 SUMMARY

The assembly of plastic-encapsulated microcircuits onto printed wiring boards includes component placement, wave soldering or solder-paste deposition and reflow, cleaning, and conformal coating of the printed wiring board. The most important part of PEM assembly is the soldering operation, which subjects the assembly to temperatures ranging from 215 to 260°C. It is important to account for the effect on the components of exposure to high soldering temperatures and to observe the specifications for the proper use of PEMs.

Assembly process development must be efficient and cost-effective. Among the many factors involved are process parameters, material properties, and handling methods; process development basically consists of identifying and optimizing these.

Statistical process control (SPC) offers a means of identifying and controlling the key process parameters to ensure quality in the process and the resultant product. SPC assists in reducing defective products by preventing problems before they occur [Brust 1991].

Every aspect of the assembly operation is a potential source of defects. In addition to the soldering process, this includes board design, component and material selection, solder mask production, handling, cleaning, and other steps.

A typical SPC program includes defining a good product, identifying process parameters that can cause defects if not controlled, defining the appropriate corrective action required for each defect, and implementing and evaluating each solution [Brust 1991]. Once a list of defects is formulated, these are ranked, as in a Pareto chart, to identify the most critical ones. The next stage is identifying all the parameters controlling the defects and depicting them graphically, as in an Ishikawa diagram. These data must be monitored over a sufficient number of samples to determine the critical parameters for a particular defect. The sample size must be large enough to accurately monitor product quality and correct inconsistencies, while being small enough to use quickly and easily.

Typical parameters that require monitoring in a PEM assembly process include

- equipment parameters, including squeegee pressure, printed wiring board registration, reflow oven temperature, and conveyor belt speed;
- materials related parameters, including solder paste type, viscosity, powder size, flux type, printed wiring board composition, and molding compound properties;
- procedural parameters, including proper setup, implementation procedures,

and experimental procedures;

- operator related parameters, including setup time, training, and soldering technique;
- environmental parameters, including humidity and temperature profiles; and
- measurement parameters, including frequency of operation and accuracy of placement.

Future trends in soldering technology are addressing solder's vulnerability to temperature and stress and the health hazards of lead. A move in this direction is the use of lead-free solders and conductive adhesives. Some solder alternatives under consideration include eutectic-binary systems of tin-zinc, tin-magnesium, tin-aluminum, tin-barium, tin-calcium, tin-silver, and tin-copper. For more information on these and other solder alternatives, see Hwang [1993a, b, c], Brookes [1993], and Gilleo [1994a, b].

## REFERENCES

Arnone, D.J., and Tong, J.J. SMT Placement Tool Capability Analysis. *Proceedings of the Technical Program, NEPCON West* 3 (February 1991) 2154-2169.

Brookes, T.W. A Lead-free Alloy for Electronics Soldering. *Circuits Assembly* 4, 10 (October 1993) 58-60.

Brust, N.T. Statistical Process Control for Surface Mount Assembly. *Proceedings of the Technical Program, NEPCON West* 1, (February 1991) 85-93.

Capillo, C. *Surface Mount Technology: Materials, Processes, and Equipment,* McGraw Hill, New York (1990).

Cergel, L. Considerations About the Use of Surface Mount Technology in Automotive Electronic Applications. *7th International Conference on Automotive Electronics*, London (1989) 51-60.

Curtis, D.A. Cleaning and Contamination. In *Soldering Processes and Equipment*, edited by Pecht, M.G. Wiley, New York (1993).

DeCarlo, J.M. A Comparison of Alignment and Placement Methods for Fine Pitch Components. *Proceedings of the Technical Program, NEPCON West* 3, (1991) 2135-2141.

Dexter Corp. Private communication, Olean, NY (1993).

Fukuzawa, I., Ishiguro, S., and Nanbu, S. Moisture Resistance Degradation of Plastic LSIs by Reflow Soldering. *Proceedings of the 23rd International Reliability Physics Symposium* (1985) 192-197.

Gallo, A.A. Research Associate, Electronic Materials Division, Dexter Corporation. Personal communication (December 1993).

Gilleo, K. Evaluating Polymer Solders for Lead-free Assembly, Part 1. *Circuits Assembly*, 5, 1 (January 1994*a*) 52-56.

Gilleo, K. Evaluating Polymer Solders for Lead-free Assembly, Part 2. *Circuits Assembly*,5, 2 (February 1994*b*) 50-52.

Hickey, D.J. Project Engineer, Delco Electronics Corporation. Private communication (March 1994).

Hutchins, C.L. Time and Temperature Requirements for Surface Mount Soldering. *Proceedings of the Technical Program: National Electronic Packaging Conference West*, Anaheim, CA (February 1990) 288-297.

Hwang, J.S. *Solder Paste in Electronics Packaging*, Van Nostrand Reinhold, New York (1992).

Hwang, J.S. Can We Have Lead-free Solders, Part 1. *Circuits Assembly* 4, 10 (October 1993a) 32-39.

Hwang, J.S. Can We Have Lead-free Solders, Part 2. *Circuits Assembly* 4, 11 (November 1993*b*) 30-31.

Hwang, J.S. Can We Have Lead-free Solders, Part 3. *Circuits Assembly* 4, 12 (December 1993*c*) 30-32.

Ilyas, Q.S.M., and Poborets, B. Evaluation of Moisture Sensitivity of Surface Mount Plastic Packages. *Proceedings of the ASME Conference*, New Orleans, LA (1993) 145-156.

IPC-SM-786. *Recommended Procedures for Handling of Moisture Sensitive Plastic IC Packages.*

Ito, S., Nishioka, T., Oizumi, S., Ikemura, K., and Igarashi, K. Molding Compounds for Thin

Surface Mount Packages and Large Chip Semiconductor Devices. *Proceedings of the 39th International Reliability Physics Symposium* (1991) 190-197.

Kawamura, N., Kawakami, T., Matsumoto, K., Sawada, K., and Taguchi, H. Structural Integrity Evaluation for a Plastic Package During the Soldering Process. *Proceedings of the ASME Conference*, New Orleans, LA (1993).

Kitano, M., Nishimura, A., Kawai, S., and Nishi, K. Analysis of Package Cracking During Reflow Soldering Process. *Proceedings of the 26th International Reliability Physics Symposium* (1988) 90-95.

Kou, M.Y.F. Assessing Fine Pitch Placement Machine from a System Standpoint. *Proceedings of the Technical Program, NEPCON West* 3 (1991) 2125-2134.

Lau, J.H. *Solder Joint Reliability: Theory and Applications*. Van Nostrand Reinhold New York (1991).

Lin, R., et al. Control of Package Cracking in Plastic Surface Mount Devices During Solder Reflow Process. *Proceedings of the 7th Annual Conference of the International Electronics Packaging Society (IEPS)* (1987) 995-1010.

Lin, R., Blackshear, E., and Sevisky, P. Moisture Induced Package Cracking in Plastic-Encapsulated Surface Mount Components During Solder Reflow Process. *Proceedings of the 26th Annual International Reliability Physics Symposium* (1988) 83-89.

Linman, D.L. Vapor Phase for High Reliability Soldering. *Proceedings of the Technical Program: National Electronic Packaging and Production Conference West*, Anaheim, CA (February 1991) 278-285.

Malhotra, A. Wave Soldering and Reflow Soldering, in *Soldering Processes and Equipment*, edited by Pecht, M.G. Wiley, New York (1993).

Manko, H.H. *Solders and Soldering*, McGraw Hill, New York (1992).

Matsushita Electric Industrial Corporation. Personal communication (1993).

Moore, T.M., Kelsall, S.J., and McKenna, R.G. The Importance of Delamination in Plastic Package Moisture Sensitivity Evaluation. *Journal of Surface Mount Technology* (1991).

Nelco International Corporation. Personal communication with William Glover (March 1994).

Nguyen, L.T. Reliability of Postmolded Integrated Circuits Packages. *SPE RETEC* (1991) 182-204.

Nguyen, L.T. Reliability of Postmolded IC Packages. *Transactions of the ASME* 115 (December 1993) 346-355.

Nguyen, L.T., and Michael, M.M. Effects of Die Pad Anchoring on Package Interfacial Integrity. *IEEE Electronic Components and Technology Conference* (1992) 930-938.

Oizumi, S., Ito, S., and Suzuki, H. Analysis of Reflow Soldering by FEM. *Nitto Technical Report* (1987) 40-50.

Omi, S. Fujita, K. Tsuda, T., and Maeda T. Causes of Cracks in SMD and Type-Specific Remedies. *Electronic Components and Devices Conference*, GA (May 1991) 776-771.

Panchwagh, T.S., Wu, X., and Pecht, M. Determination of Temperature Dependent Coefficient of Thermal Expansion of Plastic Packages. *Report submitted to Hamilton Standards, Inc.* (1993).

Pecht, M.G. *Soldering Processes and Equipment*, Wiley, New York (1993*a*).

Pecht, M.G. *Placement and Routing of Electronic Modules*, Dekker, New York (1993*b*).

Rahn, A. *The Basics of Soldering*. Wiley, New York (1993).

Tay, A.A.O., Tan, G.L., and Lim, T.B. A Criterion for Predicting Delamination in Plastic IC Packages. *Proceedings of the 31st. International Reliability Physics Symposium* (1993) 236-243.

Waryold, J., and Lawrence, J. Selection Criteria for the Use of Conformal Coatings. *Proceedings of the Technical Program: National Electronic Packaging and Production Conference West*, Anaheim, CA (February 1991) 2174-2182.

Woodgate, R.L. *Handbook of Machine Soldering*. Wiley New York (1988).

# 5

# PACKING AND HANDLING

Packing and handling involves materials, containers, and precautions that ensure that the quality and reliability of plastic-encapsulated microcircuits are not degraded during shipping, handling, and storage. This necessary step between the manufacture of components and their mounting on printed wiring boards does not add value to the components. However, poor packing and handling can have a dramatically negative impact on board assembly yield, manufacturer's cost, and shipping schedules.

Plastic-encapsulated microcircuits (PEMs) must be adequately protected during shipping, handling, and storage. Precautions taken in packing can provide the requisite protection against high humidity conditions (which can lead to moisture ingress); against physical damage (which can cause bent or broken leads); and against damage due to electrostatic discharge. Automatic handling of packages is facilitated by providing special features in the packing.

While all the reasons for special handling procedures are discussed in this chapter, the focus is on issues unique to plastic packages, such as dry packing, providing a dry nitrogen environment, and baking before mounting packages on circuit cards.

## 5.1 CONSIDERATIONS IN PACKING AND HANDLING

Considerations in the packing of plastic-encapsulated microcircuits include prevention of moisture absorption, protection of leads from handling damage, prevention of lead solderability degradation, precautions against electrostatic damage, and packing for automatic handling.

### 5.1.1 Prevention of Moisture Absorption

Since plastic-encapsulated microcircuits are nonhermetic, they are susceptible to moisture-induced failures and latent defects. Molding compounds can absorb enough moisture to cause cracking in moisture-sensitive packages when exposed to temperatures typical of furnace-reflow soldering and wave soldering. The phenomenon is known as popcorning because the pop of the packages is audible. High temperature and temperature ramp rates, prime loads that cause popcorning [IPC-SM-786A 1994], are experienced by the packages during surface mounting on circuit cards. Of particular concern are solder reflow processes that heat the package body, such as infrared heating, hot-air furnace, and vapor-phase reflow. Solder reflow processes that selectively heat the leads and bond pads, but not the component body, such as hot-bar reflow, laser soldering, and manual soldering, are not problematic. Wave soldering may be a concern if components dip through the solder wave.

Through-hole mounted plastic-encapsulated microcircuits do not exhibit popcorning because the component body is protected from the solder heat by the board. However, popcorning may occur in through-hole components mounted on double-sided boards if the components come in direct contact with the solder wave.

Possible damage due to popcorning includes delamination of the encapsulant from adjacent metal structures such as the leadframe and die pad, damaged wirebonds, cratering of bond pads, and cracking of the encapsulant. Even if the plastic-encapsulated microcircuit continues to perform its electrical function after popcorning, its reliability is compromised because the cracked and delaminated encapsulant no longer provides protection. Cracks may also let contaminants reach the active circuit via capillary action.

Popcorning in moisture-sensitive plastic packages can be avoided by controlling moisture content with proper packing and handling procedures. The moisture content of a plastic-encapsulated microcircuit is defined as

$$moisture\ content\ (\%wt) = \left(\frac{wet\ weight\ -\ dry\ weight}{dry\ weight}\right) \times 100\% \tag{5.1}$$

Typical moisture absorption curves after a high-temperature bake are shown in Figure 5.1 as a function of temperature and relative humidity. The figure illustrates that temperature is the parameter determining the moisture absorption rate. The saturation limit is a function of both relative humidity and temperature, although relative humidity is the major factor determining the steady-state moisture content of the component. Weight gain by small outline packages as a function of exposure time in 85°C/85% RH and 85°C/30% RH environments is plotted in Figures 5.2(a) and (b), respectively [Altimari et al. 1992]. The plots show that the mold compound chosen has a significant effect on the amount of moisture absorbed by plastic-encapsulated microcircuits.

For a given package type, there may be safe moisture content which the package may not popcorn. Table 5.1 lists moisture content and package cracking for 68-leaded PLCCs; the safe moisture content is considered as 0.264% by weight for the PLCCs tested. The critical value at which the component exhibits cracking depends on the adhesion and interlocking between the encapsulant and metal surfaces, and on package size, architecture, and lead count. Generally, large and thin plastic packages are more susceptible to popcorning; larger packages generate higher stresses at the corner of the die pad, and thinner packages offer a shorter path for moisture ingress through the molding compound, leading to higher moisture concentration at the interface of encapsulant and metal leadframe. Table 5.2 [Switky 1988] documents cracking in small-outline packages with various pin-counts and die sizes after 24 hr in an autoclave and by three cycles of vapor-phase reflow.

At one time, the safe moisture content for plastic-encapsulated microcircuits was 0.11% by weight [ANSI/IPC-SM-786 1990]. A lower safe moisture content was recommended for thinner PEMs with finer lead pitches [Intel 1993]. Table 5.3 lists results of moisture absorption tests on PLCCs [Switky 1988]; these indicate that the safe moisture content for plastic-encapsulated microcircuits decreases as the pin count increases. The safe moisture content can be exceeded in a relatively short time under common manufacturing floor and warehouse environments, as shown in Figure 5.3.

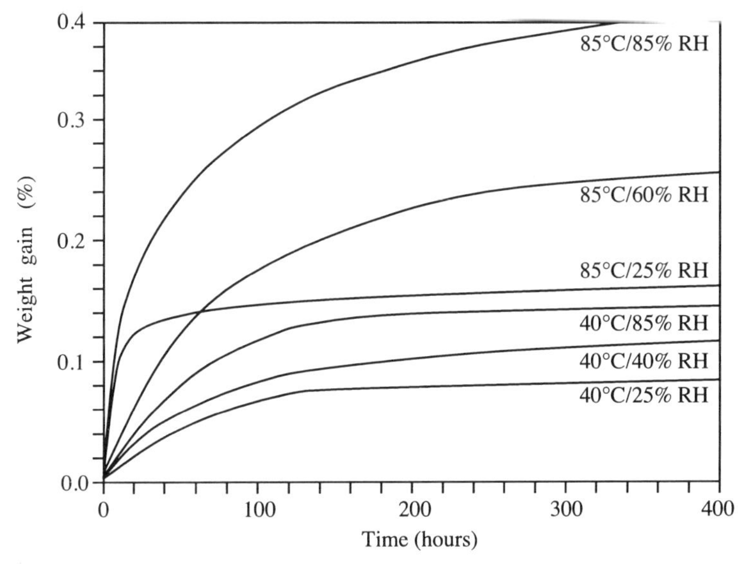


**Figure 5.1** *Typical moisture absorption curves, after high-temperature bake, shown as function of temperature and relative humidity [Intel 1993]*

**Table 5.1 Cracking of 68-leaded PLCCs as a function of moisture content after three cycles of vapor phase reflow (die size and pad size are 255 x 305 mils and 350 x 350 mils, respectively) [Switky 1988]**

| Moisture level (% increase in weight/dry weight) | Cracked samples/ sample size | Percent cracked (%) |
|---|---|---|
| 0.000 | 0/50 | 0 |
| 0.054 | 0/25 | 0 |
| 0.176 | 0/25 | 0 |
| 0.227 | 0/25 | 0 |
| 0.264 | 0/25 | 0 |
| 0.310 | 6/24 | 24 |
| 0.326 | 12/25 | 48 |
| 0.333 | 11/25 | 44 |
| 0.345 | 22/25 | 88 |
| 0.347 | 23/25 | 92 |
| 0.364 | 23/25 | 2 |

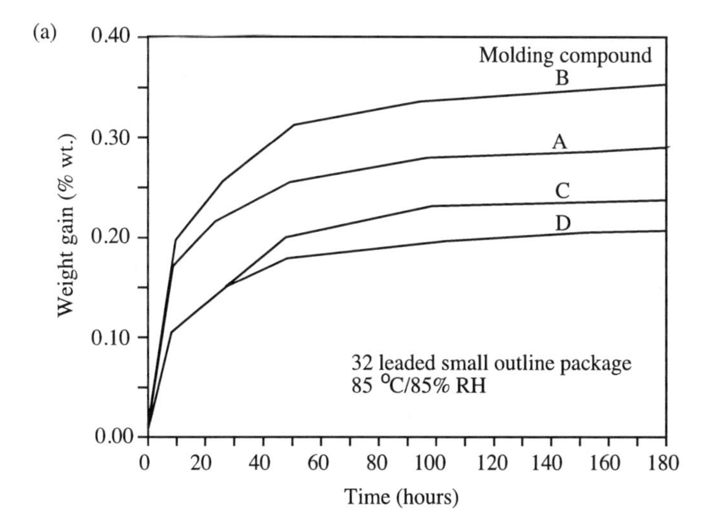


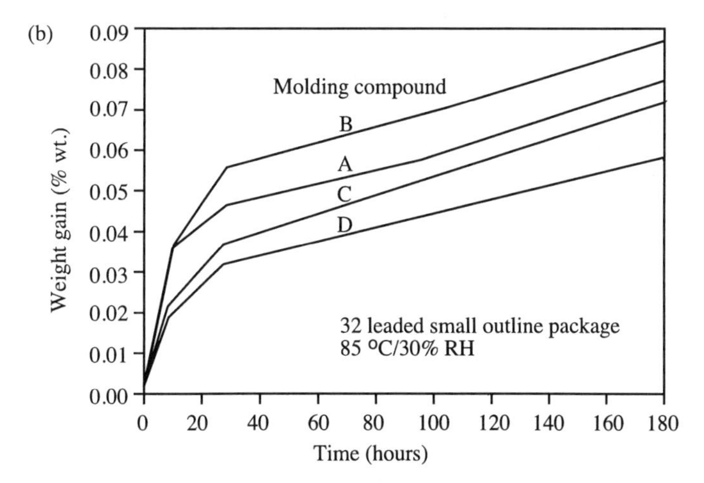


**Figure 5.2** *Weight gain by small-outline packages as a function of exposure time in 85 °C/85% RH and 85 °C/30% RH environments [Altimari et al. 1992]*

**Table 5.2 Package cracking in small-outline packages after moisture absorption in an autoclave for 24 hr and three cycles of vapor-phase reflow [Switky 1988]**

| Package type | Die size (mils) | Pad size (mils) | Moisture level (% wt) | Cracked samples/ sample size |
|---|---|---|---|---|
| 8-lead narrow | 43 x 74 | 80x80 | 0.43 | 0/50 |
| 14-lead narrow | 73 x 95 | 80 x 100 | 0.45 | 0/50 |
| 14-lead narrow | 62 x 60 | 80 x 100 | 0.49 | 0/50 |
| 20-lead wide | 105 x 77 | 130 x 150 | 0.40 | 0/50 |
| 24-lead wide | 75 x 142 | 140 x 160 | 0.40 | 0/10 |

**Table 5.3 Cracking of PLCCs as a function of lead count after moisture absorption in an autoclave for 16 hr and three cycles of vapor-phase reflow [Switky 1988]**

| Package leads | Die size (mils) | Pad size (mils) | Moisture level (% wt.) | Cracked samples/ sample size |
|---|---|---|---|---|
| 28 | 207 x 137 | 250 x 250 | 0.353 | 0/50 |
| 44 | 240 x 240 | 310 x 310 | 0.353 | 0/46 |
| 68 | 249 x 221 | 275 x 325 | 0.284 | 5/39 |
| 84 | 240 x 240 | 340 x 340 | 0.333 | 55/55 |

Dry packing must be used for shipping moisture-sensitive PEMs so they do not absorb more than the safe moisture content. The dry-pack process reduces the saturation moisture content in packages by keeping the relative humidity inside the moisture barrier bag below 20%. Other methods, such as storing in dry nitrogen and baking, may also be employed.

Three methods can be employed to dry components that have absorbed moisture: high-temperature bake, low-temperature bake, and storage in dry nitrogen. Generally, a package with a moisture level at or below 0.05% by weight is considered dry enough for dry packing. Components are dried to below the safe moisture content to provide the desired time for shipping and handling in a humid environment before the components are mounted on printed wiring boards.

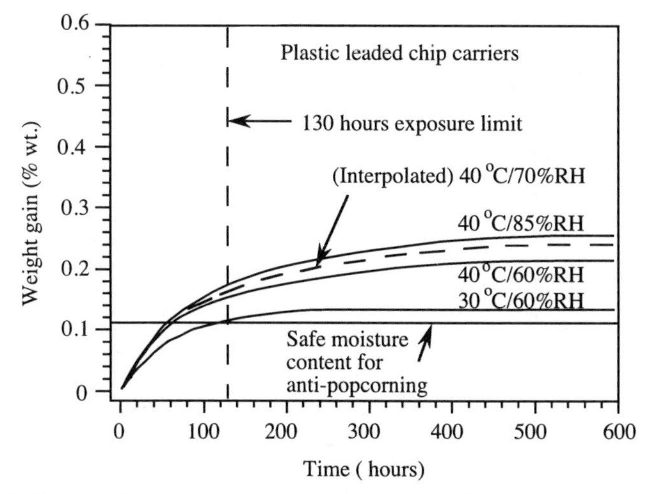


**Figure 5.3** *Safe moisture content exceeded in a short time under common manufacturing floor and warehouse environments for PLCCs [Intel 1993]*

***High-temperature bake.*** Moisture desorption is governed by diffusion. Higher temperatures help diffusion and reduce desorption time (a temperature above the glass transition temperature of the encapsulant is not recommended because package integrity may be jeopardized.) The temperature during high-temperature bake is typically kept at approximately 125°C. Figure 5.4 shows a moisture desorption plot for a 68-pin plastic-leaded chip carrier (PLCC) as a function of time at 125°C [Alger et al. 1987]; its characteristics are typical of plastic-encapsulated microcircuits. The chip carriers were saturated with moisture at 85°C/85%RH for 168 hr. An infinitely long time would be required to bake the packages 100% dry, because of the exponential nature of desorption; fortunately, an acceptable level of moisture for most components is reached in a more realistic time. The Institute for Interconnecting and Packaging Electronic Circuits recommends that the bake temperature not exceed 125 °C. Actual time of baking will vary with package type [IPC-SM-786A 1994]. Hitachi [1991] recommends a bake-out at 125°C for 16 to 24 hr.

The residual moisture content in a component must be carefully assessed, because moisture can be present at critical interfaces — such as between the metallization and the encapsulant — even when the moisture content of the whole component is low.

The moisture desorption time of a package is affected by the component architecture and encapsulant. Thin packages desorb quicker than thick packages because the path length for moisture to diffuse out of the package is shorter. Encapsulants that absorb moisture slowly also desorb slowly, as indicated by the desorption plots in Figure 5.5 for Intel's 32-lead small outline packages (SOPs)

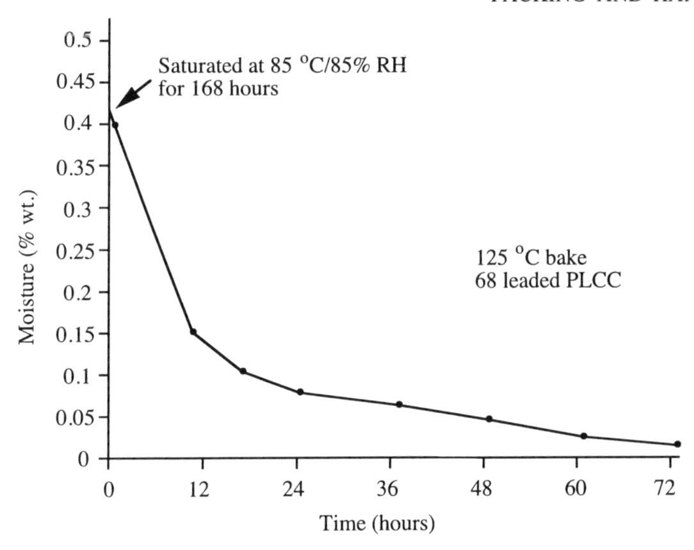


**Figure 5.4** *Moisture desorption plot for 68-pin plastic leaded chip carrier (PLCC) as a function of time at 125°C [Alger et al. 1987]*

[Altimari et al. 1992].

Although baking at a high temperature desorbs moisture rapidly, it can lead to temperature-accelerated intermetallic formation on the leads and to oxidation of the lead finish. Repeated high-temperature bakes can promote $Cu_6Sn_5$ intermetallic growth on copper leads, which inhibits solderability. No more than one 125°C bake for 24 hr. is recommended, unless postbake solderability has been verified [IPC-SM-786A 1994].

Most shipping containers are unstable at high temperature, and outgassing products condensing on component leads can degrade solderability; it is recommended therefore that packages be removed from the containers before they are baked. Unfortunately, the extra handling that may be required for baking components outside the shipping containers may adversely affect lead coplanarity and lead bend.

***Low-temperature bake.*** A low-temperature bake may be used instead of a high-temperature bake to desorb moisture from plastic-encapsulated microcircuits. The temperature and humidity for a low-temperature bake are limited to 40°C and < 20% RM, respectively; it is critical that low humidity be maintained at this low temperature. The Institute for Interconnecting and Packaging Electronic Circuits [IPC-SM-786A 1994] recommends baking at 40°C, with a dry nitrogen or dry air purge that maintains the RH below 20%. This may need to be continued for up to a month, depending on the initial moisture content of the packages.

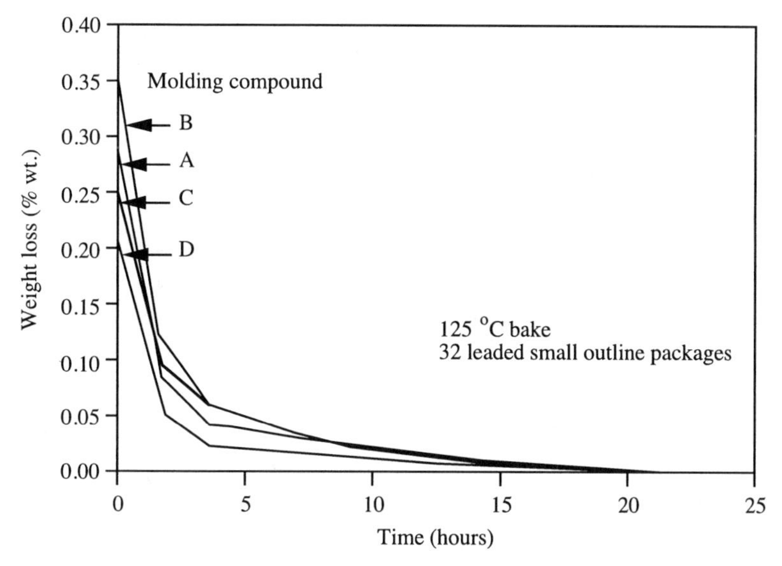


**Figure 5.5** *Intel's 32-lead small-outline packages (SOPs) desorption plot [Altimari et al. 1992]*

Intel recommends a bake at 40°C < 5% RH; this bake requires 192 hr. to reach a residual moisture level below 0.08% for 68-lead plastic-leaded chip carriers [Clifton and Pope 1988].

The advantage of a low temperature bake is that components can be baked without removing them from their shipping magazines, trays, or tape reels. This helps minimize handling that can damage leads. Components dried this way are equivalent to those dried high temperatures, but bake duration is much longer. Lead solderability is not affected because the formation of intermetallics is slowed. The postbake floor life of components is similar to that of high-temperature-baked components, assuming the distribution of residual moisture is the same in either case.

***Dry nitrogen environment.*** Packages may be kept in a dry nitrogen environment at room temperature to avoid moisture absorption or to eliminate the absorbed moisture. The packages lose moisture to the flowing nitrogen, which helps maintain a very low level of humidity inside the chamber; the isolated packages remain relatively dry. Lead solderability remains good after storage, because nitrogen is inert and prevents oxidation. Intermetallic formation is not accelerated at room temperature. The floor life of nitrogen-dried packages is comparable to temperature-baked packages, but it is finite and packages should be mounted when removed from the nitrogen flow.

### 5.1.2 Lead Damage and Solderability

Manual handling, together with drying methods, shock, and vibration, can cause damage to leads. Shipping containers are designed for automatic handling and for preventing lead bend, loss of coplanarity, and lead contamination.

Mishandling may occur at various process points, such as trim and form operations, loading of parts into the tester for electrical sorting, or removal of parts after burn-in. Lead deformation can also be induced by such process factors as high molding stresses that result in warpage, or the buckling of thin leadframe strips prior to singulation [Nguyen et al. 1993]. Regardless of the cause, damage to fragile leads can be extensive. Six typical defect modes have been characterized: coplanarity, sweep, tweeze, twist, body standoff, and center-to-center spacing [Linker et al. 1993]. Coplanarity is the vertical deviation of a lead from the seating plane of the device, and is the single most important parameter in assembly, packing, and handling.

Packing materials that include shipping trays, reels, tapes, and magazines are made of plastic materials, are often unstable at bake temperature, and outgas products that may condense on package leads, degrading their solderability. Outgassing of hydrocarbons and other organic contaminants from low-temperature stable packing materials is especially of concern during high-temperature bake. The container materials must be stable enough at bake temperature that any outgassing products do not degrade solderability. If solderability is a major concern, then the packages may be baked outside of the shipping containers.

Intermetallic growth is a time and temperature-dependent diffusion process. Long exposure to high temperatures during baking can form excessive intermetallics and degrade solderability. Often, leads are made of copper, which forms a $Cu_6Sn_5$ intermetallic with tin (Sn). The growth of $Cu_6Sn_5$ is one of the causes of lead-finish degradation leading to poor solderability [Alger et al. 1990]. Limiting the time of exposure, especially at high bake temperatures, is therefore crucial for package lead solderability. If the bake is repeated, the total time of exposure to high temperatures should be limited. Moreover, oxidation of lead surfaces can cause degradation of lead solderability; inert gas flow and a low relative humidity environment is preferred during bake to alleviate this problem.

### 5.1.3 Electrostatic Discharge Protection

Plastic-encapsulated microcircuits may be sensitive to electrostatic discharge (ESD). Electrostatic discharge can either cause overstress failure of the component (which can be exposed by an electrical function test), or induce damage leading to parametric shift and reduced reliability.

Electrostatic discharge occurs most readily in low-humidity environments. The low humidity levels inherent in baking increase the risk of ESD damage. Air ionizers may be used in baking ovens to keep devices from charging, but they cannot dissipate the charge in areas where the ionized air cannot reach because of the flow pattern.

The probability of electrostatic discharge occurring increases with increased handling, which may lead to tribologically produced charges. Charge-dissipative or electrically conductive shipping containers are employed to protect against electrostatic discharge. Charge-dissipative containers are made of electrically conductive materials, such as resin impregnated with conducting graphite fibers. Often, an antistatic coating of carbon is applied on the inner walls of shipping containers; the coating is water soluble and wears out due to friction as components slide around.

Evaluating ESD-preventive characteristics for both graphite-impregnated and antistatic-coated tubes indicates that either type of ESD protection is sufficient, even for low-humidity environments. There is no significant observed difference in component charging for graphite-impregnated tubes stored at ambient humidity and those stored at < 10%RH. Although there is a difference in charging of components stored in antistatic-coated tubes in < 10%RH, the jeopardy is considered low because the charging is 2.2 picocoulomb (pC)/lead — well below the critical level of 10 pC/lead [Pope 1988].

## 5.2 HIERARCHY OF PACKING

Dry packing is required to keep moisture-sensitive plastic surface-mount components dry during shipping and storage. There are generally three levels of packing for plastic-encapsulated microcircuits, as shown in Figure 5.6. The first level involves packing components in suitable shipping containers. The second level involves packing the containers along with desiccant pouches, a humidity indicator card, and dunnage in a moisture-barrier bag; the third level labels the shipping box with appropriate handling and shelf-life information.

### 5.2.1 First Level

The first level of packing requires placing plastic-encapsulated microcircuits in magazines, trays, or tape reels. Table 5.4 is a list of shipping containers available for various plastic-encapsulated microcircuits manufactured by Hitachi. ESD-preventive characteristics are built into the containers, either by making the

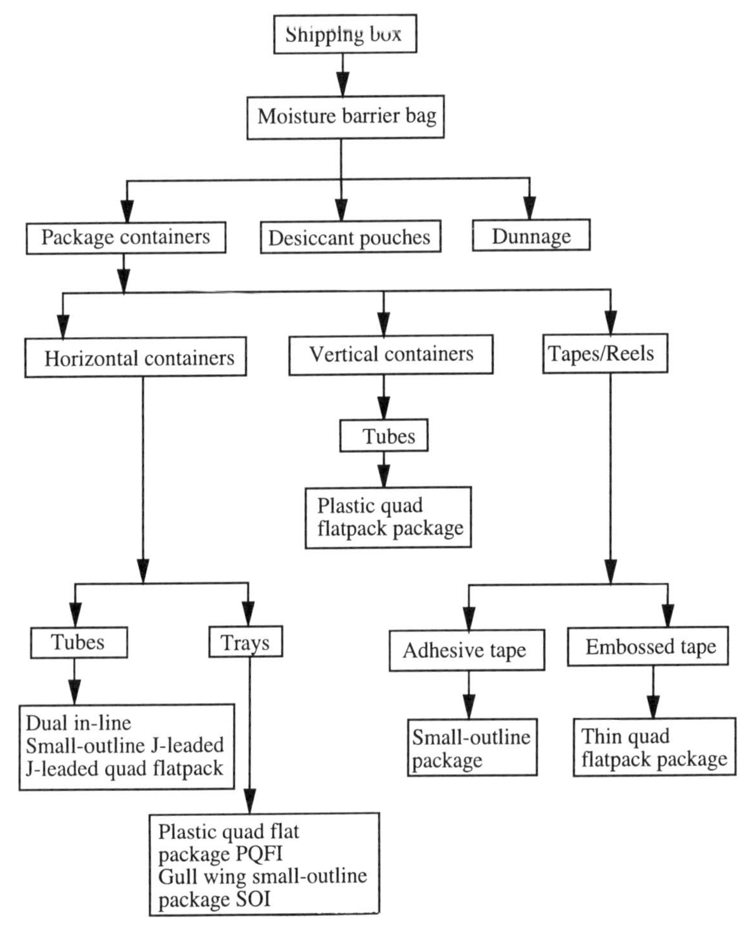


**Figure 5.6** *Examples of horizontal shipping magazines used by Hitachi [1991] and Nippon Electric Company [1988]*

chloride stoppers are inserted into the ends of the magazine to hold the packages in place.
container material electrically conductive or by coating with an antistatic agent.

Magazines are made of plastic or aluminum. Plastic magazines are treated with a water-soluble antistatic agent that wears out as a result of friction caused by package sliding or repetitive use. Magazines come in two forms, one for horizontal stacking of packages and the other for vertical stacking. Most types of plastic packages (for example, dual in-line packages, small-outline J-leaded packages, J-leaded quad flatpacks, and plastic-leaded chip carriers) can be stored in magazines. Dual in-line packages are the only plastic package type shipped exclusively in magazines. Examples of horizontal shipping magazines used by Hitachi and Nippon Electronic Company are shown in Figure 5.7. An example of a vertical magazine is shown in Figure 5.8 [Hitachi 1991]. Flexible polyvinyl

**Table 5.4 Shipping containers available for PEMs from Hitachi [1991]**

| Package type | Package types:(F=flat, M=square, C=chip carrier leads, D=dual leaded, N=skinny, S=shrink, T=fin) | Shipping container type | | | | |
|---|---|---|---|---|---|---|
| | | Magazine | | Tray | Tape/reel | |
| | | Horizontal | Vertical | | Adhesion | Embossed |
| Small-outline package | F-8D | ● | | | ● | ○ |
| | F-14D | ● | | | ● | ○ |
| | F-14DN | ● | | | | ● |
| | F-16D | ● | | | ● | ○ |
| | F-16DN | ● | | | | ● |
| | F-18D | ● | | | | ○ |
| | F-20D | ● | | | | ○ |
| | F-20DN | ● | | | | ○ |
| | F-24D | ● | | | | ● |
| | F-28D | ● | | | | ● |
| | F-32D | ● | | | | ● |
| Quad flatpack | F-44 | | | ● | | |
| | F-54 | | | ● | | |
| | F-60 | | | ● | | |
| | F-64 | | | ● | | |
| | F-80 | | | ● | | |

**Table 5.4 (cont.)**

| Package type | Package types:(F=flat, M=square, C=chip carrier leads, D=dual leaded, N=skinny, S=shrink, T=fin) | Shipping container type | | | | |
|---|---|---|---|---|---|---|
| | | Magazine | | Tray | Tape/reel | |
| | | Horizontal | Vertical | | Adhesion | Embossed |
| Small-outline package | F-8D | ● | | | ● | ○ |
| | F-14D | ● | | | ● | ○ |
| | F-100 | | | ● | | |
| | F-136 | | | ● | | |
| | F-168 | | | ● | | |
| Small-outline J-leaded package | C-20D | ● | | | | ● |
| | C-24D | ● | | | | |
| | C-28D | ● | | | | |
| | C-32D | ● | | | | |
| Quad flatpack with J-leads | C-18 | ● | | | | ● |
| | C-32 | ● | | | | |
| | C-44 | ● | | | | ○ |
| | C-52 | ● | | | | ○ |
| | C-68 | ● | | | | ○ |
| | C-84 | ● | | | | ○ |
| Small-outline butt-leaded package | M-26DT | | | ○ | | |

**Table 5.4 (cont.)**

| Package type | Package types:(F=flat, M=square, C=chip carrier leads, D=dual leaded, N=skinny, S=shrink, T=fin) | Shipping container type | | | | |
|---|---|---|---|---|---|---|
| | | Magazine | | Tray | Tape/reel | |
| | | Horizontal | Vertical | | Adhesion | Embossed |
| Small-outline package | F-8D | ● | | | ● | ○ |
| | F-14D | ● | | | ● | ○ |
| Quad flatpack with butt-leads | M-18, 18T | | ● | ● | | ● |
| | M-28, 28T | | ● | ● | | ● |
| | M-44, 44S, 44T | | ● | ● | | ● |
| | M-56S | | ● | ● | | ● |
| | M-68 | | ● | ○ | | |
| | M-84S | | ○ | ○ | | |

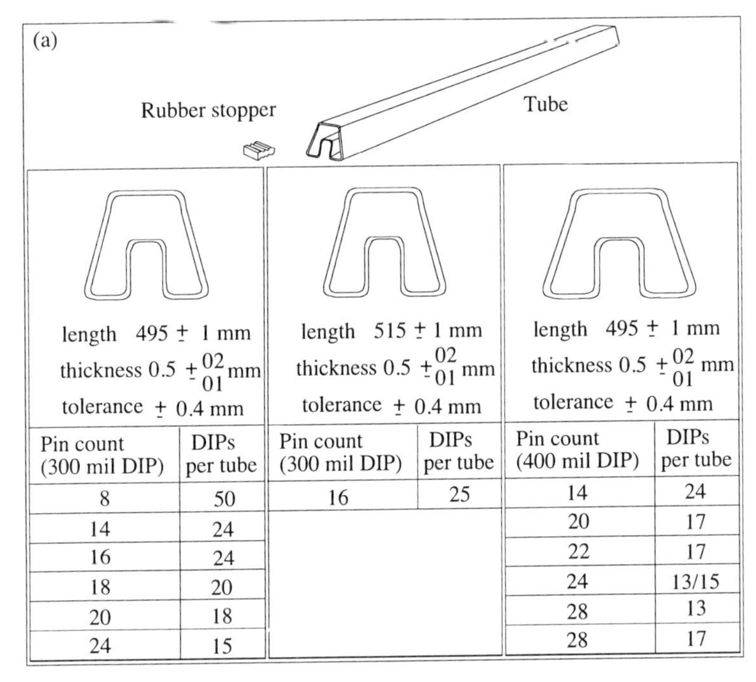


length 495 ± 1 mm
thickness 0.5 $^{+02}_{-01}$ mm
tolerance ± 0.4 mm

| Pin count (300 mil DIP) | DIPs per tube |
|---|---|
| 8 | 50 |
| 14 | 24 |
| 16 | 24 |
| 18 | 20 |
| 20 | 18 |
| 24 | 15 |

length 515 ± 1 mm
thickness 0.5 $^{+02}_{-01}$ mm
tolerance ± 0.4 mm

| Pin count (300 mil DIP) | DIPs per tube |
|---|---|
| 16 | 25 |

length 495 ± 1 mm
thickness 0.5 $^{+02}_{-01}$ mm
tolerance ± 0.4 mm

| Pin count (400 mil DIP) | DIPs per tube |
|---|---|
| 14 | 24 |
| 20 | 17 |
| 22 | 17 |
| 24 | 13/15 |
| 28 | 13 |
| 28 | 17 |

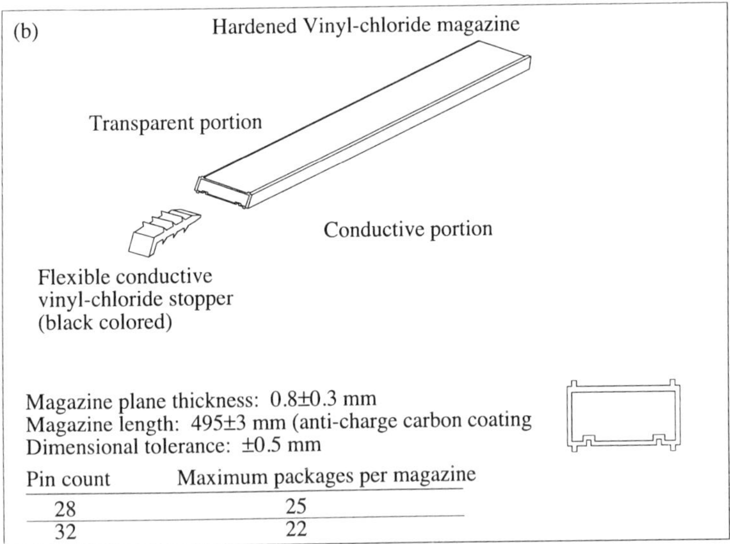


Magazine plane thickness: 0.8±0.3 mm
Magazine length: 495±3 mm (anti-charge carbon coating
Dimensional tolerance: ±0.5 mm

| Pin count | Maximum packages per magazine |
|---|---|
| 28 | 25 |
| 32 | 22 |

**Figure 5.7** *Examples of horizontal shipping magazines used by Hitachi [1991] and Nippon Electric Company [1988]*

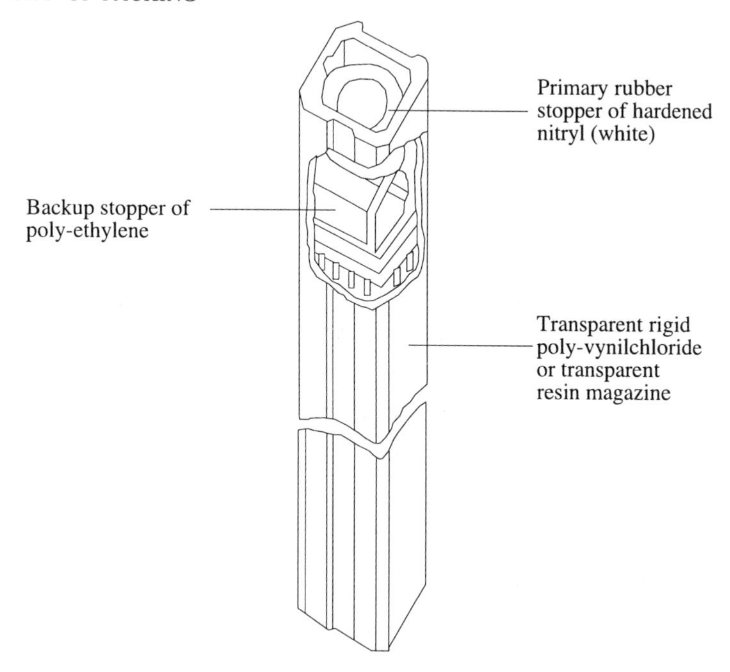


**Figure 5.8** *An example of vertical magazine [Hitachi 1991]*

Trays made from injection-molded plastic, such as polyvvynil chloride and polystyrene are used primarily for storage of quad flatpacks and bumped quad flatpacks. Antistatic characteristics are provided by impregnating conducting fibers with the injection-molded plastic, or by treating the trays with an antistatic agent, such as carbon. Trays made of polystyrene do not need an antistatic coating because sufficient fiber loading can be added to provide critical conductivity. Polyvinyl chloride trays should be stored between -25°C and +40°C because their shape and color may change above +55°C. As many as a hundred plastic packages can be packaged per tray. Figure 5.9 shows a diagram of a tray used for quad flatpacks.

Tape reels are used for smaller sized of plastic packages, such as small-outline plastic packages and small plastic-leaded chip carriers. Adhesive or embossed tapes are used. The adhesive tape holds the plastic-encapsulated microcircuits on carrier tape (see Figure 5.10); the embossed tape has depressed pockets in which the plastic-encapsulated microcircuits reside, and a cover tape ensures that they remain in place (see Figure 5.11). Many of these smaller plastic-encapsulated microcircuits are difficult to handle manually. Tape reels allow packing that is amenable to automatic handling equipment. The carrier tape can have sprocket holes for loading registration on one or both sides, depending on the size of the

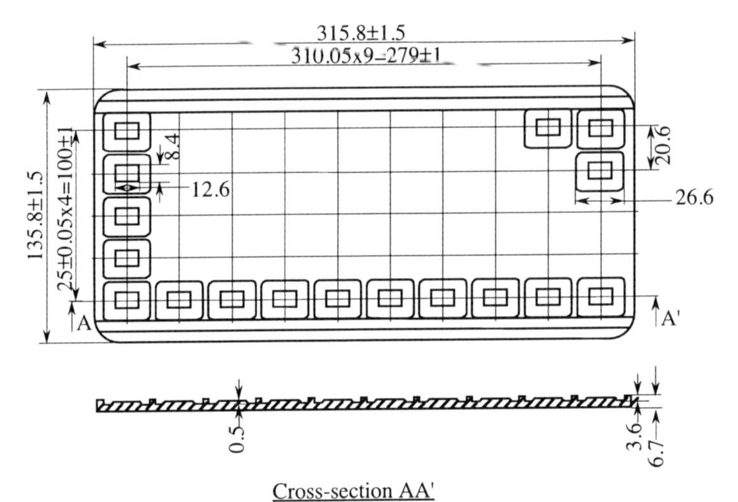


Magazine plate thickness: 0.7 +0.15, -0.25
Dimensional tolerance: ±0.5

**Figure 5.9** *A diagram of a tray used for quad flatpacks (the dimensions are in mm [NEC 1988]*

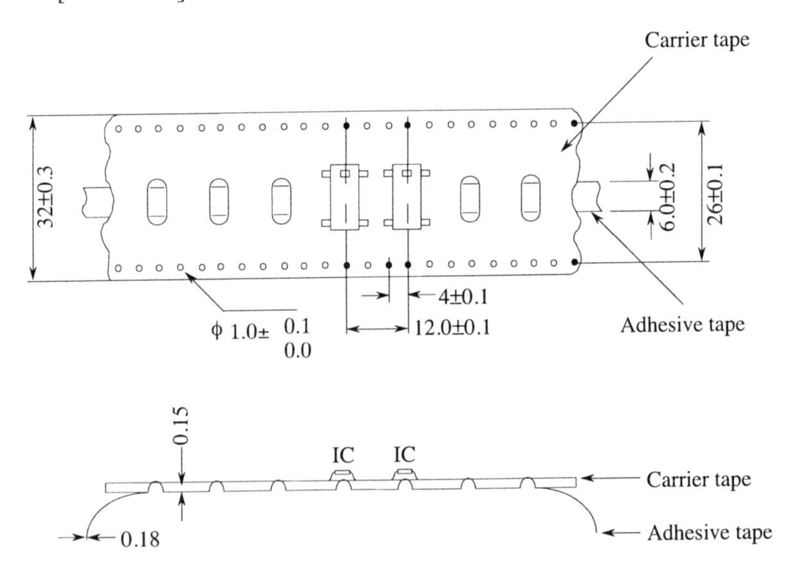


**Figure 5.10** *The carrier tape using an adhesive tape to hold plastic-encapsulated microcircuits (the dimensions are in mm) [Hitachi 1991]*

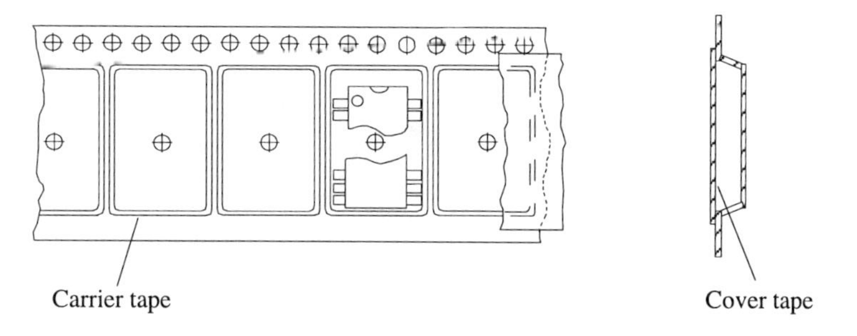


**Figure 5.11** *Embossed tape with depressed pockets in which the plastic-encapsulated microcircuits reside, and a cover tape to assure that it remains in place [Hitachi 1991]*

plastic-encapsulated microcircuits. Tape widths come in a variety of sizes, ranging from 12 to 44 mm. Many more plastic-encapsulated microcircuits can be stored in tape reels than in magazines and trays. Typical tape reels hold 500 to 3000 PEMs per reel. Nippon Electronic Company has a reel that holds 12,000 small transistors.

### 5.2.2 Second Level

The second level of PEM packing involves enclosing magazines, trays, or tape reels in bags. Moisture-barrier bags are used for dry-packing moisture-sensitive plastic-encapsulated microcircuits. At the time of dry-packing, devices should have a moisture content of less than 0.05% by weight. Not all plastic-encapsulated microcircuits are considered moisture-sensitive, so not all are shipped in drypacks. Table 5.5 lists some package types from Texas Instruments, the shipping container used, and whether they are shipped in dry packs.

Dry-packing uses a moisture barrier bag (MBB) in shipping containers with dried plastic-encapsulated microcircuits, a desiccant, and a humidity indicator card. The moisture-barrier bag is sealed and labeled with the seal date. Examples of box-and-bag second-level packaging are shown in Figure 5.12.

Moisture-barrier bags are generally made of Tyvek, a multilayered opaque material. MIL-STD-81705B [1989] classifies moisture-barrier bag materials into two types. Type I has a water vapor transmission rate (WVTR) of less than 0.02 g/100 sq.in. of bag area per 24 hr; type II has a WVTR between 0.02 and 0.08 g/100 sq.in./24 hr. The bags should meet type I requirements for water vapor transmission rate, flexibility, electrostatic discharge protection, mechanical

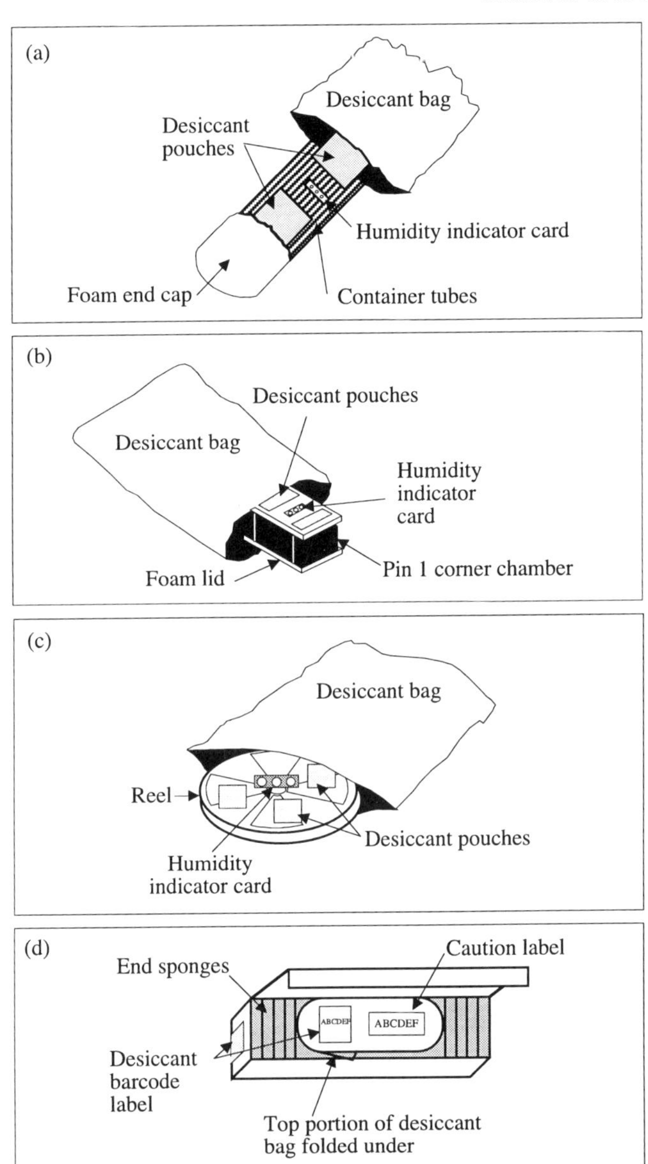


**Figure 5.12** *Examples of box and bag second level packaging [Intel 1993]*

**Table 5.5 Shipping methods and quantities [Texas Instruments 1992]**

| Package[a] | Tube | Tape/reel | Trays | Dry pack |
|---|---|---|---|---|
| 56-pin SSOP | 20 | 500 | N/A | No |
| 44-pin PLCC | 27 | 500 | N/A | No |
| 68-pin PLCC | 18/19[b] | 250 | N/A | Yes |
| 64-pin SQFP | N/A[c] | N/A | 50 | Yes |
| 80-pin SQFP | N/A | N/A | 50 | Yes |
| 120-pin SQFP | N/A | N/A | 50/84[d] | Yes |
| 80-pin SQFP | N/A | N/A | 50 | Yes |

[a] SSOP, skinny small-outline package; PLCC, plastic-leaded chip carrier; SQFP, small quad flatpack

[b] Eighteen packages can be packed in a single tube when pin is used as a tap; nineteen packages can be packed when plug is used as a tap

[c] N/A, not applicable

[d] depending on tray size

strength, and puncture resistance [IPC-SM-786A 1994]. Most of the air should be removed from the bag by manually compressing the outside of the bag or by using suction equipment. However, the bag should not be completely evacuated, since this would reduce the effectiveness of the desiccant and could also damage the enclosed components.

Desiccant is a moisture-absorbing material that is enclosed in the moisture-barrier bag to absorb moisture from the air inside the bag and maintain its relative humidity below 20%. A number of different materials are used as desiccants. Figure 5.13 shows desiccant moisture absorption rates as a function of desiccant type.

Desiccant material should meet MIL-D-3464 type II [IPC-SM-786A 1994]. It must be dustless and noncorrosive. The amount of desiccant needed per bag to maintain an interior relative humidity below 20% is a function of bag surface area, bag water vapor transmission rate, and storage time. The quantity of desiccant is measured in desiccant units. A desiccant unit is the amount of desiccant that at 25°C will absorb a minimum of 2.85 grams of water vapor at 20% relative humidity and 5.7 grams at 40% relative humidity, as defined in MIL-D-3464. The IPC-SM-786A standard also recommends a general

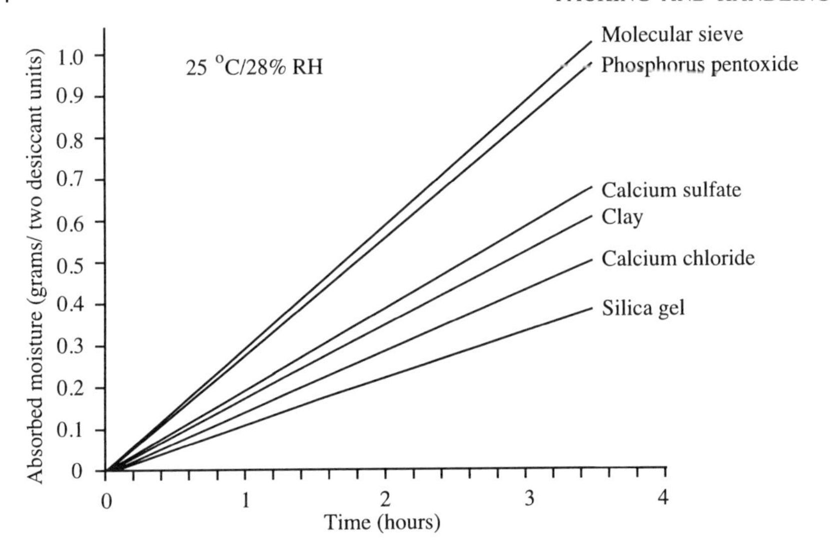


**Figure 5.13** *Desiccant moisture absorption rates as function of desiccant type [Intel 1993]*

formula for calculating the desiccant units required per moisture-barrier bag:

$$U = \frac{0.304\ T_{maxstor}\ (WVTR)\ A_{bag}}{2.85} \tag{5.2}$$

where $U$ is the amount of desiccant in desiccant units, $T_{maxstor}$ is the maximum storage time in months, $WVTR$ is the water vapor transmission rate in grams/100 sq. in./24 hr, and $A_{bag}$ is the surface area of the bag.

Kalidas [1991] has proposed modifications to Equation 5.2 to adjust desiccant units for storage conditions that may differ from those by which desiccant units are defined, and for the moisture capacity of the desiccant:

$$U = \frac{0.304\ A_{bag}\ (WVTR)\ T_{maxstor}(DRF)}{O_{des}} \tag{5.3}$$

where $DRF$ is the desiccant reduction factor, shown as a function of storage temperature and relative humidity in Table 5.6, and $O_{des}$ is the minimum moisture capacity of the desiccant at maximum interior humidity, in grams/unit.

Dunnage, a cellulose or synthetic packing material, is often used for wrapping and cushioning containers inside moisture-barrier bags. Two types of dunnage/filler material are generally used Padpack and bubble pack. Padpack is a machine-processed, three-ply kraft paper sheet dunnage system. Bubble pack is made of thin polyethylene plastic sheets with air pockets trapped between the

**Table 5.6 Desiccant reduction factor (DRF) [Kalidas 1991]**

| Storage temperature (°C) | Relative humidity (%) | | | |
|---|---|---|---|---|
| | 90 | 80 | 70 | 60 |
| 40 | 1.00 | 0.89 | 0.78 | 0.67 |
| 35 | 0.76 | 0.68 | 0.59 | 0.51 |
| 30 | 0.58 | 0.51 | 0.45 | 0.38 |
| 25 | 0.43 | 0.38 | 0.33 | 0.29 |
| 20 | 0.32 | 0.28 | 0.25 | 0.21 |

layers in hemispherical shapes; these layers can be either dissipative or conductive. Both padpack and bubble pack materials are recyclable. Additional desiccant is enclosed in the dry-pack to account for moisture that may be present in the dunnage. The total desiccant units, needed to meet dry-pack requirements with dunnage can be calculated as

$$T_{desiccant} = U + D \tag{5.4}$$

where $U$ is the amount of desiccant obtained from Equation 5.2 in desiccant units, and $D$ is the additional amount of desiccant required for dunnage. As an example, additional desiccant should be added at the rate of eight units per pound for cellulose dunnage and one-half unit per pound for synthetic dunnage.

A humidity indicator card (HIC) is a means of measuring the relative humidity of the air surrounding the card. A card is inserted in the moisture-barrier bag before sealing to help determine the effectiveness of the dry pack. An illustration of a humidity indicator card is given in Figure 5.14. The card has at least three color dots, providing indications from 10% to 30% relative humidity. A pink dot means the card has been exposed to at least the given amount of humidity; a blue dot means it has not been exposed to that much humidity. If the 30% dot starts to change from blue to pink, the relative humidity inside the moisture-barrier bag is higher that 20%, and the desiccant should be replaced. A humidity indicator card must be dry before it can be packaged in a dry pack; if it is not, it should be baked dry or put in a desiccant bag to dry. The chemical reaction that causes the change in color is reversible, so cards are reusable.

A warning label is mounted on each dry-packed moisture-barrier bag. The label adhesive and markings are water-resistant. The labels include a moisture-

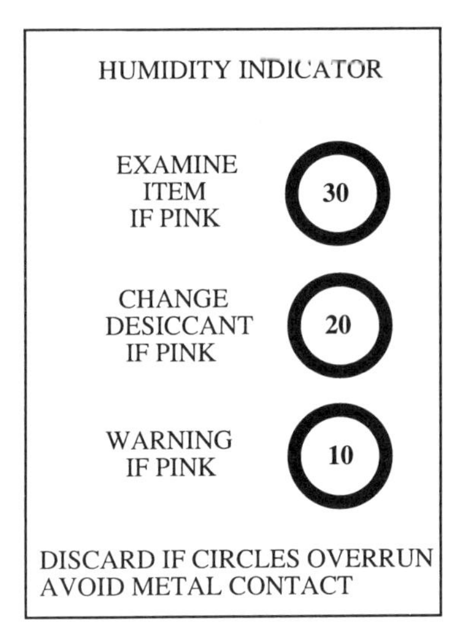


**Figure 5.14** *An illustration of a humidity indicator card [IPC-SM-786A 1994]*

sensitivity caution symbol, instructional notes, and a bag seal date. An example of the standardized warning label specified by JEP 113, Symbol Labels for Moisture-Sensitive Devices, is shown in Figure 5.15. The instructional notes may vary slightly amongst manufacturers.

### 5.2.3 Third Level

The third level of PEM packaging is the outer box, often made of corrugated cardboard. Bags with magazines, trays, or tape reels are stacked into the outer carton, which is then taped shut. Bar-code labels, usually placed on the outer box, indicate the presence of plastic-encapsulated microcircuits and the seal date. An example of a desiccant bar-code label is shown in Figure 5.16 [Intel 1993].

## 5.3 HANDLING OF DRY PACKAGES

Dry plastic-encapsulated microcircuits have a finite floor life, after which they absorb moisture above a safe level. Component floor life can be defined as the time period that begins after moisture- sensitive devices are removed from the moisture-barrier bag and the controlled storage environment, and ends when they have absorbed enough moisture to be susceptible to damage during reflow. Floor life characteristics are dependent upon many factors. The moisture-barrier bag should not be opened until the components can be board-mounted. Before the bag is opened, seal integrity should be verified. Often, the shelf-life of dry-packed

**CAUTION**

This Bag Contains

**MOISTURE SENSITIVE DEVICES**

1. Shelf life in sealed bag: 12 months minimum at $< 40\,^{\circ}$C and $< 90\%$ Relative Humidity (RH).

2. Upon opening this bag, devices to be subjected to I.R., V.P.R. or equivalent process must be :
   a) Mounted within _____ hours/days at factory conditions of $\leq 30\,^{\circ}$C/60% RH, or
   b) Stored at $\leq 20\%$ RH.

3. Devices require baking, before mounting, if:
   a) Humidity Indicator Card is $> 20\%$ when read at $23\,^{\circ}\text{C} \pm 5\,^{\circ}\text{C}$, or if
   b) 2a or 2b are not met.

4. If baking is required, devices may be baked for,
   a) 192 hours at $40\,^{\circ}\text{C} + 5\,^{\circ}\text{C}/-0\,^{\circ}\text{C}$ and $< 5\%$ RH for low temperature device container, or
   b) 24 hours at $125\,^{\circ}\text{C} \pm 5\,^{\circ}\text{C}$ for high temperature device containers.

Bag Seal Date:_______________________________

(if blank, see bar code label)

241187-31

**Figure 5.15** *An example of the standardized warning label specified by JEP 113 Symbol Labels for Moisture Sensitive Devices [IPC-SM-786A 1994]*

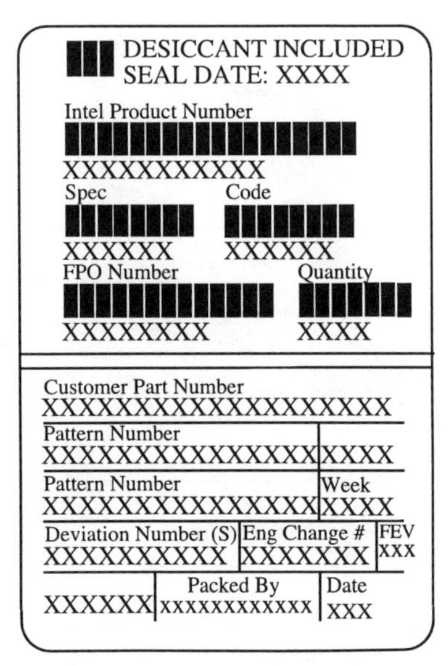


**Figure 5.16** *An example of a desiccant bar-code label [Intel 1993]*

components is less than twelve months from the seal date (the shelf life is stated on the bag warning label). If leakage is found or the shelf-life is exceeded, the plastic-encapsulated microcircuits should be dried before they are used for any reflow soldering.

The length of time moisture-sensitive PEMs may be exposed to the manufacturing floor environment after being removed from the moisture-barrier bag depends on the moisture content of the components and the floor environment temperature and relative humidity. Figure 5.17 shows typical component weight-gain versus time for various manufacturing floor environments. For example, plastic-leaded chip carriers absorb moisture beyond the safe moisture content of 0.11% by weight in approximately 130 hr. in a floor environment of 30°C/60% RH, as shown in Figure 5.3.

***Classification.*** Exterior configuration is not the sole indicator of moisture/reflow sensitivity. The same number of days of ambient exposure can result in different amounts of moisture being absorbed by the package, depending upon the molding compound material and package design. The same amount of moisture can cause different amounts of damage in different packages. Packages with the same

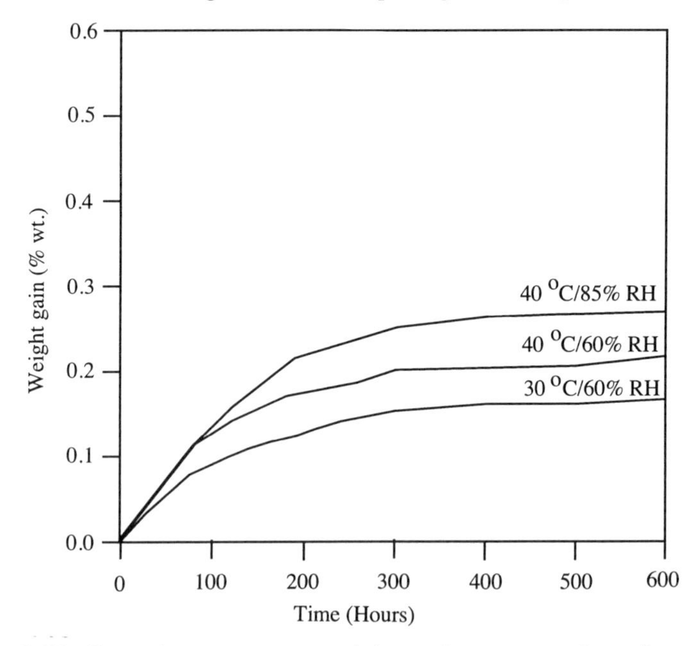


**Figure 5.17** *Typical component weight gain versus time for various manufacturing floor environments*

**Table 5.7 Classification and floor life of desiccant packed components [IPC-SM-786A 1994]**

| Level | Storage floor life (out of bag) at board assembly site | [a]Preconditioning moisture | |
|---|---|---|---|
| | | °C/% RH | Time (hrs) |
| 1 | unlimited at ≤ 85% RH | 85/85 | 168 |
| 2 | 1 year at 30 °C/60% RH | 85/60 | 168 |
| 3 | 1 week at ≤ 30 °C/60% RH | 30/60 | 168 + [b]MET |
| 4 | 72 hr at ≤ 30 °C/60% RH | 30/60 | 72 + MET |
| 5 | as specified on label range: 24 to 71 hr at ≤ 30 °C/60% RH | 30/60 | time on label + MET |
| 6 | mandatory bake<br>bake before use and once baked must be reflowed within the time limit specified on the label | 30/60 | time on label |

[a]Preconditioning moisture - Preconditioning moisture soak to simulate total exposure time conditions

[b]MET - Manufacturer's Exposure Time, that is the compensation factor that accounts for the time after bake that the component manufacturer requires to process components prior to bag seal. It also includes a default amount of time to account for shipping and handling

exterior configuration (leadcount, thickness, and so on) may not have the same interior construction, and absorption rate and sensitivity may differ between packages that otherwise look identical. Consequently, the percentage weight gain moisture content is not useful for establishing moisture sensitivity. Different package types, materials, lead counts, and die-attach areas will reach different levels of moisture absorption before reflow damage occurs.

Packages are now classified into six levels based upon their storage floor life (out of bag) at the board assembly site. Table 5.7 shows the test conditions and the storage floor life of packages at the six levels [IPC-SM-786A 1994]. The packages are first subjected to the appropriate test conditions based on the level they are being tested for. They are then subjected to a prescribed reflow simulation, and are evaluated for moisture/reflow induced damage. They are classified into the lowest level at which they pass the acceptance criteria. The classification technique is discussed in IPC-TM-650 Test Method 2.6.20A: Assessment of Plastic Encapsulated Electronic Components for Susceptibility to Moisture/Reflow Induced Damage.

***Board-level rework of moisture-sensitive components.*** If packages must be removed from the board, localized heating should be used and the maximum temperature excursion should not exceed 200°C. The component temperature should be measured at the top center of the package body; if the temperature exceeds 200°C, the board may require a bake-out prior to rework. If the part is to be reused, it should be baked dry before remount; new replacements should be kept dry as well. Localized replacement reflow heating avoids resubjecting the entire board to reflow temperature profiles. The impact of nearest-neighbor heating should be evaluated as part of the preconditioning flow for qualification. If the packages are determined to be nonmoisture sensitive after assembly, then controlling the board-level moisture content is not necessary for rework.

If the moisture-barrier bag is opened and closed several times, the cumulative floor exposure time should not exceed the stated floor life; the time may be extended by storing the plastic-encapsulated microcircuits in an environment with humidity less than 20%. The floor life differs from manufacturer to manufacturer. Intel claims that only their plastic-leaded chip carriers and plastic quad flatpacks with fewer than 100 leads may be in a floor humidity environment of less than 20% for extended periods of time. Hitachi [1991] contends that its plastic-packaged devices can be exposed to the floor environment for more than one week of cumulative time outside the moisture- barrier bag before reflow soldering.

Once the moisture-barrier bag is opened, component use or handled options may be pursued in the following order of preference:

- The components may be mounted within a specified time before moisture is reabsorbed beyond the safe moisture content level.
- The components may be stored outside the moisture-barrier bag in an environment of less than 20% relative humidity until future use.
- The components may be resealed in the moisture-barrier bag with new fresh desiccant added and reseal and exposure time noted; the total exposure time before mounting must be kept within the allowed time for the exposure conditions.

- The components may be resealed in the moisture-barrier bag using the original desiccant; this method does not allow the floor life of the devices to be extended beyond the time indicated by the seal date;

For lead protection, it is recommended that components not be removed from their original shipping containers until mounting; contact with the leads can cause positional or coplanarity problems. It is also recommended that automatic or semiautomatic handling equipment be used in lieu of manual handling, especially for surface-mounted packages.

Precautions should be taken against electrostatic discharge, especially if the relative humidity is below 20%, as it is during baking; air ionizers can be installed in baking ovens to remove static electricity. Water leakage can washout the water-soluble antistatic coating in shipping containers, or cause the coating to peel off and loose its effectiveness. Storing plastic-encapsulated microcircuits in antistatic-coated containers beyond six months should be avoided because the coating may warp over time. Antistatic-coated containers should not be re-used.

## 5.4 ENVIRONMENTAL CONSIDERATIONS

Protecting delicate leads during handling and transit is the main function of the tubes and trays used in packing. However, this protection comes at a price. Before "green" manufacturing became fashionable, packing paraphernalia was treated as regular, disposable "brown" trash; landfill environmental concerns were never raised. As a result, packing materials are regularly piled and dumped without much consideration of the ecological implications. With sparse land for landfill sites, European countries were the first to raise the possibility of recyclability in electronic manufacturing. Nontoxic, lead-free packaging was one practical outcome of that initial push.

Environmental pressure has forced recyclability to become a prime concern. Since the current packing materials are made mostly of thermoplastics, recycling is relatively trouble-free. Used polyethylene tubes or polivynyl chloride trash can be ground, pelletized, and remolded. The end products can be resold as packing grade with the "Made from recyclable materials" label or similar logo. Several industries have already sprung up to address this market niche. Used packing materials are collected from the OEMs or assembly houses, where parts are unloaded from their packing carriers to be mounted. Cleaning, grinding, remelting, and molding the materials can be a profitable enterprise.

## 5.5 SUMMARY

Handling costs for moisture-sensitive plastic-encapsulated microcircuits may constitute as much as 10% of the total cost. Poor handling can lead to loss of product yield during assembly of plastic-encapsulated microcircuits on printed wiring boards. Package cracking during surface mounting is directly associated with improper packing and/or handling of components, but can be avoided.

## REFERENCES

Alger, C., Huffman, W.A., Gordon, S.F., Prough, S., Sandkuhle, and Yee, K. Moisture Effects on Susceptibility to Package Cracking in Plastic Surface Mount Components. *Intel Corporation* (15 December 1987).

Alger, C.L., Pope, D.E., Rehm, P.M., and Subramaniam, N. Solderability Requirements for Plastic Surface Mount Packages. *Proceedings of the 7th IEEE-CHMT, IEMTS* (1990).

Altimari, S., Goldwater S., Boysan P., and Foehringer R. Role of Design Factors for Improving Moisture Performance of Plastic Packages. *Proceedings of the 42nd Electronic Components & Technology Conference*, (May 1992) 945-950.

ANSI/IPC-SM-786. Recommended Procedures for Handling of Moisture Sensitive Integrated Circuit Packages *Institute for Interconnecting and Packaging Electronic Circuits*, 7380 N. Lincoln Avenue, Lincolnwood, IL (1990).

Clifton, L., and Pope, E. Package Cracking in Plastic Surface Mount Components as a Function of Package Moisture Content and Geometry. *Proceedings of 5th IEEE-CHMT International Electronics Manufacturing Technology Symposium,* (October 1988) 89-92.

Hitachi. *Surface Mount Package Users Manual* (March 1991).

Intel Corporation. Recommended Procedures for Handling of Moisture Sensitive Plastic Packages (1988).

Intel Corporation Packaging Handbook. Recommended Procedures for Handling of Moisture Sensitive Plastic Packages (1993).

IPC-SM-786A Procedures for Characterizing and Handling of Moisture. Reflow Sensitive ICs. Institute for Interconnecting and Packaging Electronic Circuits (1994).

Kalidas, N. *Dry-Pack Procedure for Moisture-Sensitive Plastic Surface-Mount Devices*. Texas Instruments Inc. Semiconductor Group (May 1991).

Linker, F., Levit, B., and Tan, P. Ensuring Lead Integrity. *Advanced Packaging*, (Spring 1993) 20-23.

MIL-B-81705B. Military specification. Barrier Materials, Flexible, Electrostatic-free, Heat sealable. U.S. Department of Defense, Washington, DC (1989).

Nguyen, L.T., Chen, K.L., and Lee, P., Leadframe Designs for Minimum Molding-Induced Warpage, *Proceedings of the 44th Electronic Components and Technology Conference* (1993).

Nippon Electronic Company. Plastic SOP Dimensions. *Packing and Handling Manual* (1986).

Nippon Electronic Company. Plastic DIP Dimensions. *Packing and Handling Manual* (1986).

Nippon Electronic Company. Plastic SOJ/PLCC Dimensions. *Packing and Handling Manual* (1988).

Nippon Electronic Company. Plastic QFP Dimensions. *Packing and Handling Manual* (1988).

Pope, E. Moisture Barrier Bag Characteristics for PSMC Protection. *Technical Proceedings, SEMICON-East* (September 1988) 59-69.

Switky A. Package Cracking and Moisture Absorption in Plastic Surface Mount Components. *National Semiconductor Corporation* (1988).

Texas Instruments. *FIFO Surface Mount Package Information* (1992).

# 6

# FAILURE MECHANISMS, SITES, AND MODES

When a product's performance is no longer acceptable, it is said to have failed. Failure occurs when some failure mechanism — a chemical, electrical, physical, mechanical, or thermal process acting at some site in a structure — induces a failure mode, such as an electrical open, a short, or parametric drift. Typical failure mechanisms, corresponding sites in plastic packages, and failure modes are summarized in Table 6.1. Common failure sites for the various failure mechanisms in plastic-encapsulated microelectronic packages are schematically illustrated in Figure 6.1.

Interactions can occur simultaneously between various failure mechanisms when different types of loads are applied. For example, a thermal load can trigger mechanical failure due to a thermal expansion mismatch between adjacent

**Table 6.1 Failure sites, failure modes, failure mechanisms and environmental loads on a plastic-encapsulated microelectronic device**

| Failure site | Failure mode | Failure mechanism | Environmental load | Critical interactions and remarks |
|---|---|---|---|---|
| Die edge, corner or surface scratch due machining, dicing or handling | spalling crack, vertical crack, horizontal crack, electrical open | crack initiation, crack propagation | temperature gradients and changes | encapsulant shrinkage, modulus of elasticity of encapsulant, CTE mismatch among die, die-attach and encapsulant |
| Metallization traces, edges | corrosion, increase in resistance, electrical short, or open, notching, electrical parameter drift, metallization shift, intermittence, intermetallics | electromigration, oxidation, electrochemical reaction, interdiffusion, thermal mismatch with encapsulant | current density, humidity, voltage bias | often passivation cracking precedes; residual stresses in the metallization |
| Die passivation defect, stress concentration on the passivation from bearing of sharp edge of filler particle | transistor instability; corrosion of metallization, electrical open, shift in parametrics | overstress, fracture, oxidation, electrochemical reaction | cyclic temperature, temperature below glass transition temperature, humidity | encapsulant shrinkage, sharp edges of filler, mismatch in CTE of chip, passivation and encapsulant |

**Table 6.1 (cont.)**

| Failure site | Failure mode | Failure mechanism | Environmental load | Critical interactions and remarks |
|---|---|---|---|---|
| Die-attach void, crack, contamination site | delamination of die, nonuniform transfer of stresses to the die pad from the die; eventual loss of electrical function | crack initiation and propagation | cyclic temperature | moisture in die attach [Harada et al. 1992]; viscosity of die attach, die-attach thickness, CTE mismatch between die pad and die |
| Bonding wire | breakage | axial fatigue | cyclic temperature | filler particle bearing, mismatch in CTE of wire and encapsulant |
| Ball bond | electrical open, increased junction resistance, shift in electrical parametrics; bond lift, cratering, intermittence | shear fatigue, axial overstress, Kirkendall voiding, corrosion, diffusion and interdiffusion, intermetallics at the base of the bond, neck in the wire | absolute temperature, humidity, contamination | lack of interdiffusion barrier layer |
| Stitch-bond heel, base, neck | electrical open, increased junction resistance, shift in electrical parametrics, bond lift, cratering, intermittence | fatigue, Kirkendall voiding, corrosion, diffusion and interdiffusion | absolute temperature, humidity, contamination | lack of interdiffusion barrier layer |
| Bond pad | substrate cracking, bond-pad lift-off, eventual loss of electrical function, shift in electrical parametrics, corrosion | overstress, corrosion | humidity, electrical bias | CTE mismatch between bond pad and substrate, lack of passivation layer |

**Table 6.1 (cont.)**

| Failure site | Failure mode | Failure mechanism | Environmental load | Critical interactions and remarks |
|---|---|---|---|---|
| Die and encapsulant interface | eventual electrical open | deadhesion, delamination | humidity, contamination, temperature cycling about glass transition temperature | residual stresses, formation of moisture layer at the interface, loss of adhesion, CTE mismatch |
| Passivation and encapsulant interface | corrosion of metallization, shift in electrical parameters | | humidity, contamination, temperature cycling about glass transition temperature of encapsulant | lead design, residual stresses, loss of adhesion, CTE mismatch between encapsulant and lead, lead pitch |
| Bond wire and encapsulant interface | Corrosion of die bond pad, increase in junction resistance, electrical open | deadhesion, shear fatigue | humidity, contamination, temperature cycling about glass transition temperature of encapsulant | CTE mismatch between encapsulant and wire |
| Encapsulant | eventual loss of electrical function, loss of mechanical integrity | thermal fatigue cracking, depolymerization | temperature cycling about glass transition temperature of encapsulant | delamination of die pad and encapsulant, corner radius |
| Leads | reduced solderability, increased electrical resistance | dewetting | contamination, solder temperature | lead finish porosity |
| Die and encapsulant interface, encapsulant cracking at die corner and emerging either at the leads, base or top of the device | popcorning: eventual loss of electrical function, electrical open | vaporization of the entrapped moisture, doping of the device bottom, dynamic fracture of the encapsulant | temperature above the boiling point of entrapped moisture, rate of change of temperature | adhesion strength of encapsulant and die-pad, entrapment of vaporized moisture |

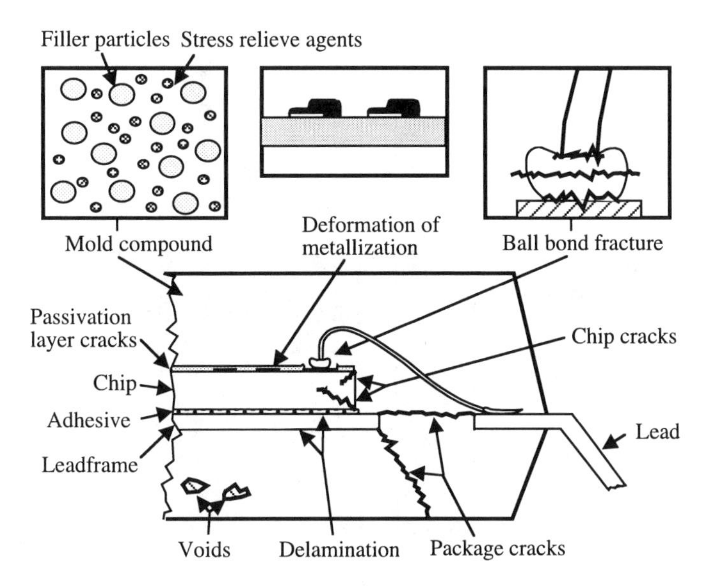


**Figure 6.1** *Typical failure mechanisms, sites, and modes in plastic-encapsulated devices*

materials in a structure. Other interactions include stress-assisted corrosion, stress-corrosion cracking, field-induced metal migration, passivation and dielectric cracking, hygrothermal-induced package cracking, and temperature-induced acceleration of chemical reactions. In such cases, the combined effect of the failure mechanisms is not necessarily the sum of the individual effects.

In this chapter, the emphasis is on understanding the failure mechanisms to avoid failures.

## 6.1 CLASSIFICATION OF FAILURE MECHANISMS

Failure mechanisms have been classified by Dasgupta and Pecht [1991], based on the rate of damage accumulation. Figure 6.2 shows the classification of failure mechanisms. The classification is especially useful in reliability analysis studies, where time to failure is a critical parameter.

Failure mechanisms fall into two broad categories: overstress and wearout. Overstress failures are often instantaneous and catastrophic. Wearout failures accumulate damage incrementally over a long period, often leading first to performance degradation and then to device failure. Further classification of failure mechanisms is based on the type of load that triggers the mechanism — mechanical, thermal, electrical, radiation, or chemical.

Mechanical loads include physical shock, vibration (e.g., underneath an

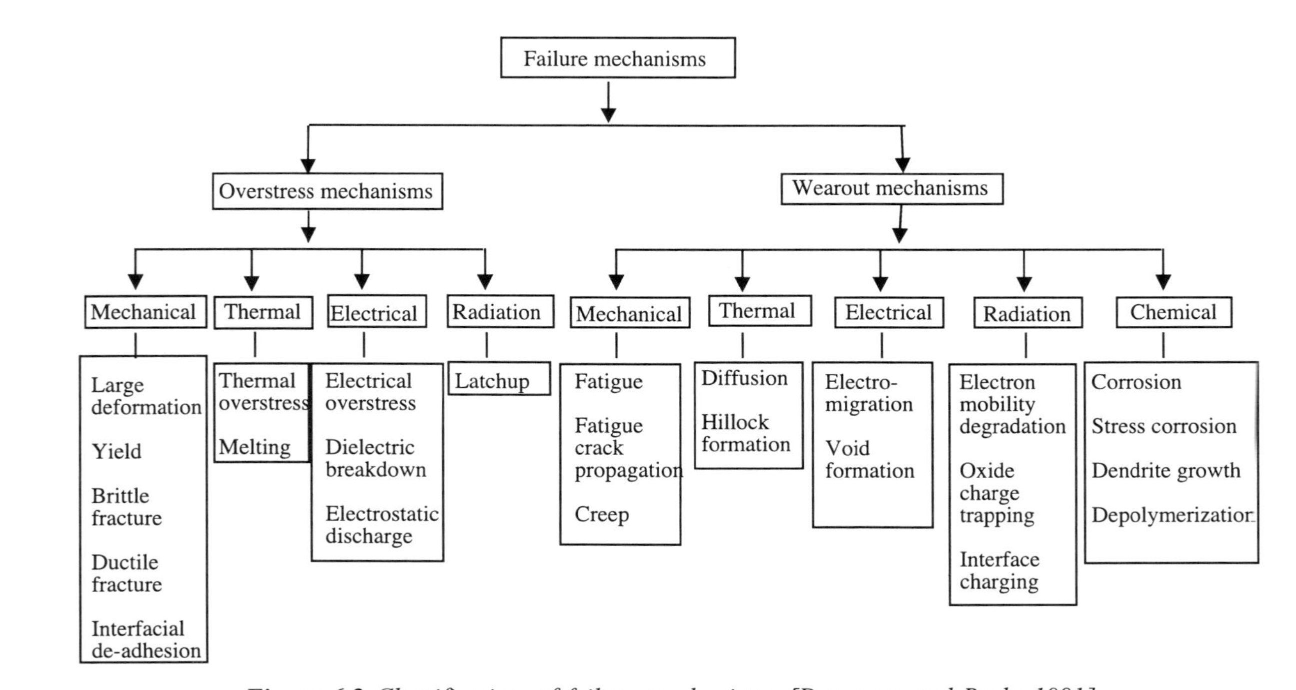


**Figure 6.2** *Classification of failure mechanisms [Dasgupta and Pecht 1991]*

automotive engine hood), loads exerted by filler particles on the silicon die (because of encapsulant shrinkage upon curing), and inertial forces (e.g., in the fuse of a cannon shell being fired). Structural and material responses to these loads may include elastic deformation, plastic deformation, buckling, brittle or ductile fracture, interfacial separation, fatigue crack initiation, fatigue crack propagation, creep, and creep rupture.

Thermal loads include exothermic die-attach curing, heating prior to wirebonding, encapsulation, postmold exothermic curing, rework on neighboring components, dipping in molten solder, vapor-phase soldering, and reflow soldering. External thermal loads lead to changes in dimensions because of thermal expansion, and can change such physical properties as the creep flow rate or cause burning of a flammable material. Mismatches in coefficients of thermal expansion often cause local stresses that can lead to failure of the package structure.

Electrical loads include, for instance, a sudden current surge through the package (e.g., in the ignition system of a car engine during start-up), fluctuation in the line current due to a defective power supply or a sudden jolt of transferred electricity (e.g., from improper grounding procedures), electrostatic discharge, electric overstress, applied voltage, and input current. These external loads may produce dielectric breakdown, surface breakdown of voltage, dissipation of electric power as heat energy, or electromigration. They may also increase electrolytic corrosion, current leakage due to formation of dendrites, and thermally induced degradation.

Radiation loads are principally either alpha particles found in packaging materials from trace radioactive elements, such as uranium and thorium, or cosmic rays that can upset memory devices, degrade performance, and depolymerize the encapsulant. The problem is more acute in memory devices that are highly sensitive to alpha particles, in which even a minor alpha flux through the device can flip the binary state of a memory bit. Due to the stochastic nature of this failure mode, preventive measures usually are built in. A polyimide die coating of proper thickness is a simple measure that can block damaging rays. Compounding ultrapure fillers with low alpha residues in the resin is another way to minimize erratic device behavior due to ionizing charges.

Chemical loads include chemically severe environments that result in corrosion, oxidation, and ionic surface dendritic growth. Moisture in a humid environment can be a major load on a plastic-encapsulated package because of the permeability of moisture through encapsulants. Moisture absorbed by the plastic can leach catalyst residues and polymerization by-products from the encapsulant, then ingress to the die metallization bond pads, semiconductor, and various interfaces and activate failure mechanisms that degrade the package. For instance, reactive flux residues coating the package after assembly can migrate through the encapsulation to reach the die surface.

## 6.2 ANALYSIS OF FAILURES

Specific failure mechanisms that occur at known failure sites and result in damage or failure of a plastic-encapsulated device are considered in this section.

### 6.2.1 Die Fracture

Surface scratching and cracks in dies may form at the thermal processing, die scribing, and dicing stages. If, after the manufacture of the die, a preexisting initial crack is equal to or greater than the critical crack size (depending on the magnitude of applied stress), the die can catastrophically fracture in a brittle manner. For example, microcracks nucleated at the top surface from surface scratching of the wafer may propagate to an active transistor and cause device failure. Edge cracks developed from die separation damage are most likely to propagate at the corner of the die, due to high longitudinal stresses common in that location when the die leadframe structure is temperature-cycled. Voids in the attachment material may also affect the die fracture.

Figure 6.3 illustrates the extreme adverse effect of high stresses on a silicon die. Given the appropriate combination of factors, die cracking occurs as shown in Figure 6.3a. The crack runs diagonally across the die surface of this device (decapped for failure analysis) after it has been subjected to extended thermal shock cycles. Figure 6.3b shows the back of the die after removal of the die attach. Die cracking results from a coarse wafer backgrind, high plunger forces during die attach, and high molding stresses.

Chiang and Shukla [1984] summarized die-cracking statistics for samples that had edge or center voids and were subjected to ten cycles of thermal shock. The devices with center voids showed no cracks, while devices with edge voids had nearly a 50% failure rate due to die cracks. Rigid attachments, such as glasses, cannot absorb the dimensional changes caused by thermal mismatches between the die and the substrate material; this leads to cracks in the die and eventual device failure, mostly during thermal cycling. Although cracks can originate from edge damage during sawing, fine control of saw speed and saw blade quality has eliminated most of the defects. Laser scribing is currently used. Die cracks may still originate from flaws on the backside due to backgrinding operations that thin wafers for thin package-outline requirements.

(a)

(b)

**Figure 6.3** *Cracking of the die due to high molding stresses: (a) crack running transversely across the surface of a decapped package; (b) back of a die after removal of the die attach, showing the transverse crack [Courtesy of National Semiconductor 1994]*

Depending on the grit size used, grinding can either reduce or increase the chance of cracking; typically the size of flaws approaches the grit size. Thus, all coarse flaws should be eliminated by either fine grinding or backside etching.

During die attach, each die is ejected from the backing tape by a transfer plunger. Damage to the die may be introduced at this stage, since the plunger tip and the loading speed govern the size and depth of the indentation.

The thermal stresses due to the CTE mismatches among the various materials in the plastic package can be expressed by

$$\sigma = c_5 \int_{T_1}^{T_2} \left[ \frac{\alpha_p - \alpha_i}{\frac{1}{E_p} - \frac{1}{E_i}} \right] dT \tag{6.1}$$

where $c_5$ is a geometry-dependent constant, $\alpha$ is the coefficient of thermal expansion, $E$ is the modulus of elasticity, and subscripts $p$ and $i$ denote the encapsulant material and the material in contact with the encapsulant material (i.e., the chip or leadframe material) [Manzione 1990]. If the thermally induced stress given by Equation 6.1 is greater than the allowable stress, the package fails. An estimate of allowable stress can be obtained from fracture mechanics principles [Broek 1991].

### 6.2.2 Loss of Chip Passivation Integrity

The chip passivation layer is usually made of brittle glass films, such as silica ($SiO_2$), phosphosilicate glass (PSG), or silicon nitride ($Si_3N_4$). The aluminum bond pad under the passivation film breaks easily from the bonding pressure and cannot sustain the passivation film. Phosphorus, added to the silicon dioxide passivation layer for ion gettering, stress-relaxes the layer and avoids cracking. An optimum phosphorous content is 1.8% by weight [Merrett 1985]; a high concentration (> 8%) accelerates corrosion failures of the die due to the formation of corrosive phosphoric acid with ingressed moisture. Corrosion of the die metallization can result in various electrical failures, including device leakage, shorts, or opens. Plasma-enhanced chemical vapor-deposited (PECVD) silicon nitride has better moisture resistance than phosphorus-doped silicon dioxide. Two examples of passivation cracking are shown in Figure 6.4.

Invariably, a larger distribution of cracks is found near the edges of the device, rather than in the die center. As mapped out with piezoresistive test chips, high shear stresses are always observed at the edges. The cracks are not easily seen under the optical microscope; however, once exposed to a mild etchant, the

(a)

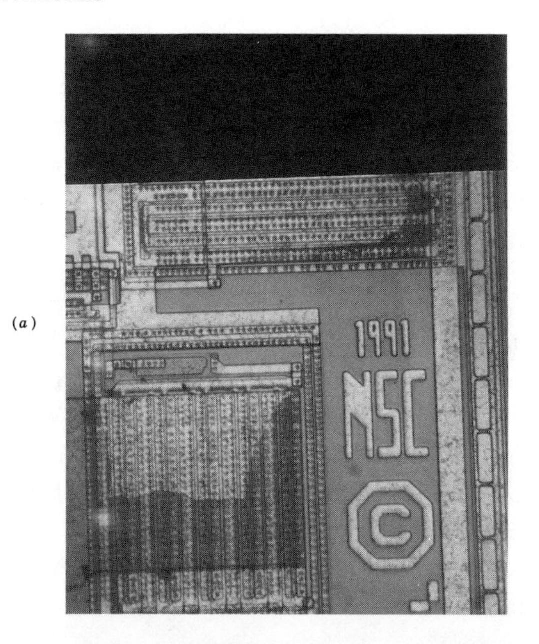


(b)

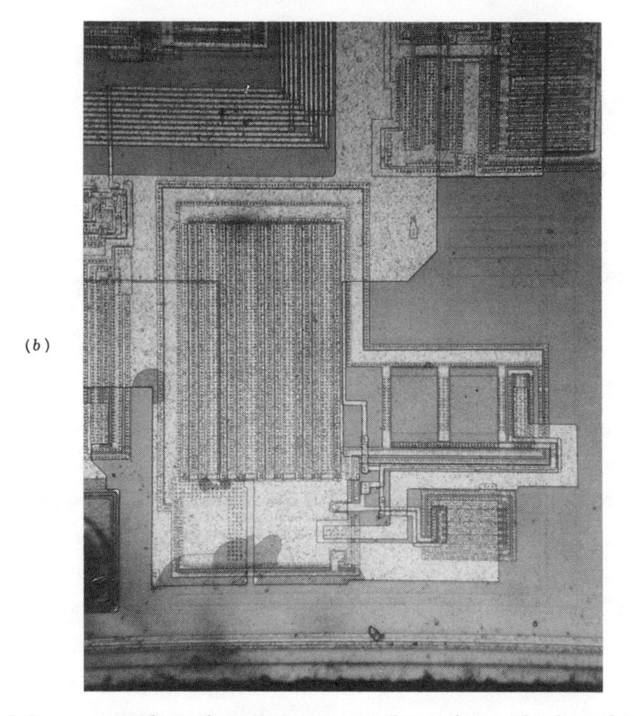

**Figure 6.4** *Two examples of passivation cracking due to high molding stresses; the dark areas are caused by metal dissolved by the etching solution permeating through the passivation cracks [Courtesy of National Semiconductor 1994]*

affected areas become more visible. The solution diffuses through the microcracks in the passivation, attacking the underlying metal and forming dark areas wherever the metal has been removed.

Cracking of passivation occurs due to shearing stresses imposed by the molding compound [Okikawa 1983]. Passivation layer cracking is often associated with ball-bond liftoff and shearing because of the close proximity of these structures to passivation layer damage on the edges of the chip.

Polyimide film can also be used as the passivation layer. The film is relatively soft and sufficiently ductile to withstand bonding pressure and high temperatures; although the film distorts under stress, it does not crack. Inayoshi et al. [1979] evaluated passivation cracking and showed that it can be avoided by using a ductile passivation. Polyimide also offers good adhesion with the molding compound, enhancing the interfacial integrity critical to the long-term reliability of the package during thermal excursions [Chen et al. 1993]. Another advantage of a polyimide coating, especially for memory devices, is the protection it offers against alpha particles.

### 6.2.3 Die Metallization Corrosion

Several factors affect corrosion of the aluminum metallization: system acidity, metal composition, encapsulation material, passivation glass, ionic contamination, temperature, relative humidity, applied voltage, and moisture resistance of the package [Gallace et al. 1978].

The most severe condition for the plastic package and a main cause of corrosion is the interaction of moisture with ionic contaminants and ingredients in the molding compound. Moisture can reach the wirebond by diffusion through the encapsulant or through such defects in the package as separation between the leadframe and the encapsulant where the leads enter the package body. It may react with chlorine, in the form of chloride ions present in many manufacturing processes imposed on the package. Bromides and antimony trioxide, both used as flame retardants, can be released from the molding compound when high temperatures occur in the presence of moisture. Figure 6.5 shows corrosion of the die metallization after temperature cycling tests and temperature-humidity-bias tests [Condra 1993].

Galvanic corrosion requires some voltage potential the interface of the dissimilar metals gold and aluminum sets up an approximately three-volt potential. In the presence of water and an ionic contaminant, galvanic corrosion can easily take place. Although most of the die metallization is covered by a passivation layer, defect sites can be present in the passivation, allowing

(a)

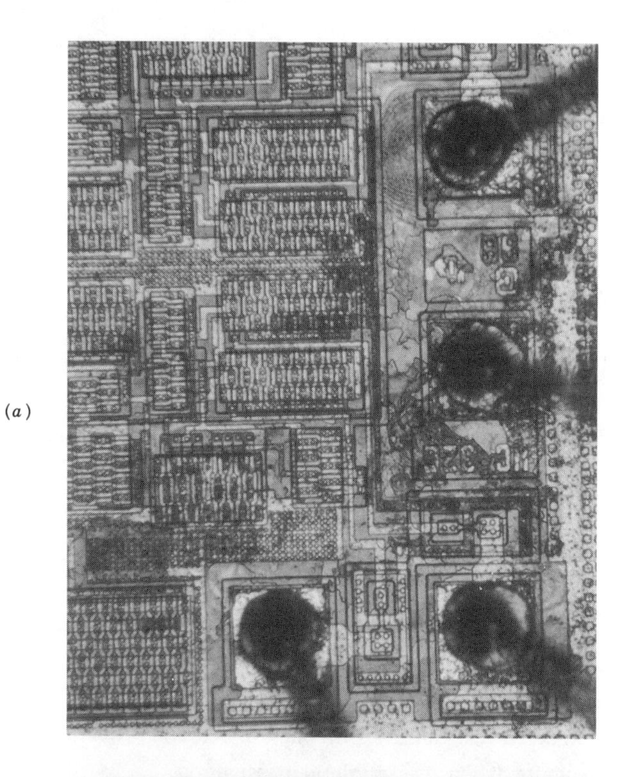

(b)

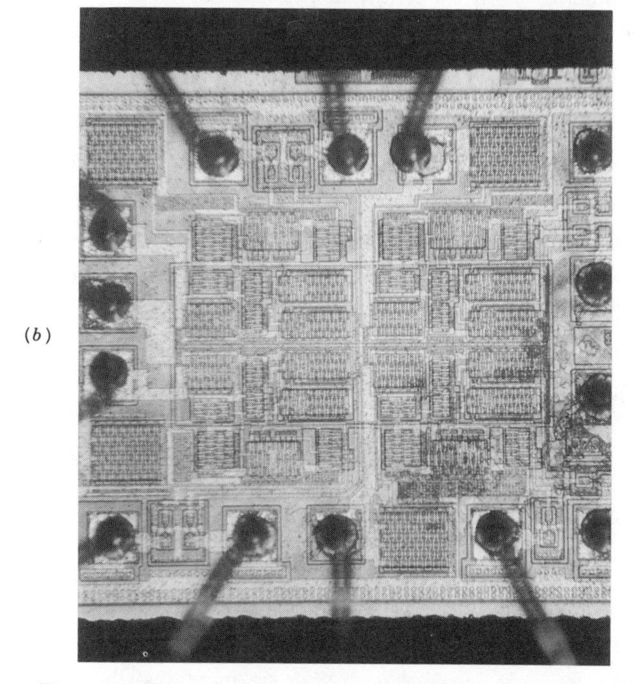

**Figure 6.5** *Corrosion of die metallization [Condra 1993]*

corrosion of the metal features. Obviously, because there is no passivation in the bond-pad area, this region is also vulnerable to any corrosive electrolyte. Complete coverage of the bond pad by the wire is usually impossible since the bond pad is square and the corresponding ball bond is circular. Unless high bonding force is used to flatten the ball to obtain greater coverage or round bond pads are designed, exposed aluminum will remain a fact of package assembly.

Techniques to passivate bond pads for reliability without hermeticity are expensive and are not typically employed for commercial packages. Round bond pads have been tried in the past, but were abandoned because the process control vision systems of current wirebonders recognize only square features. Measures to avoid and control corrosion-related failures include reducing ionic impurities in the encapsulant; using impurity-ion catchers or ion scavengers in the encapsulant; increasing the encapsulant-to-leadframe adhesion strength; using fillers in the encapsulant to elongate the path for moisture diffusion; using a low water-uptake encapsulant, and vapor-depositing inorganic or organic coatings for hermeticity before encapsulation [Nguyen 1991a, Nguyen and Jackson 1993]. Good passivation coverage (using phosphorus glass or silicon-nitride) is necessary for moisture resistance. Silicone rubber and silicone epoxy are also effective encapsulants when used without passivation.

### 6.2.4 Metallization Deformation

The aluminum metallization on the chip can be plastically deformed by shrinkage forces imposed by the molding compound, as shown in Figure 6.6. Deformation is highest at the edges of the chip, where the shear stresses are greatest. In this case, the encapsulant is constrained from moving over the chip by its adhesion to the chip surface adjacent to the metal trace; any loss of adhesion between the encapsulant and the chip increases the potential for metallization deformation.

The stress profile on the die can be mapped out with piezoresistive strain gauges embedded in the silicon, which provide a good correlation between measurements and finite element analysis [Nguyen et al. 1991]. For a given die and package configuration, the average in-plane shear stress is about one-tenth the magnitude of the corresponding principal stresses. Near the die edges, this scaling factor approaches one-third. The principal stresses in the die center increase with die size, while the reverse applies to in-plane shear stresses. Edge stresses are sensitive to die dimensions, increasing with die size.

The amount of free die-paddle space around the die also affects the stress level. Generally, stresses increase as free space increases; thus, tailoring a die pad to closely fit die size optimizes the stress level. Orientation and die aspect

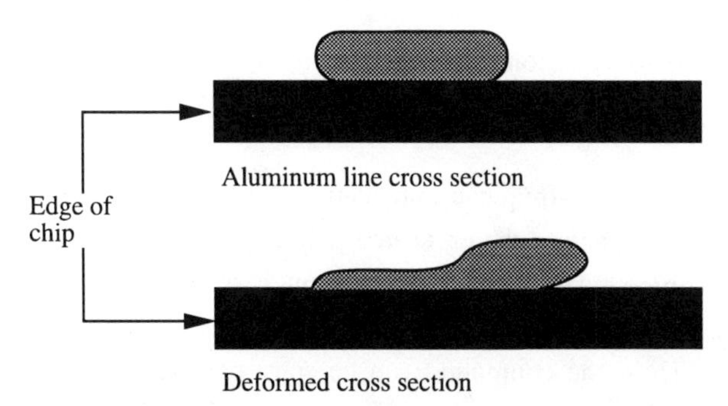


**Figure 6.6** *Schematic illustration of aluminum metallization deformation due to shear stress from mold compound*

ratio also influence edge stresses much more than center stresses. For example, an asymmetrical die with its axis perpendicular to that of the package encounters lower stresses than one aligned parallel to the axis.

The amount of stress imposed on a die depends on the thickness and volume of the surrounding molding compound. Thin packages experience much less stress than thick packages, and are the current trend in the packaging industry. The in-plane packaging stresses responsible for metal deformation are caused by lateral compression of the die. Eliminating such compressive stresses can be accomplished through a glob-top coating or by depositing a thin trail of silicone gel alongside the die for buffering. A good working knowledge of the die stress profile can help in the design of future products that minimize stress-induced damage.

### 6.2.5 Wire Sweep

Wire sweep usually denotes visible wire deformation, typically a lateral movement in the direction of the compound flow through the cavity. Under this lateral deformation, ball bonds and sometimes stitch bonds can develop kinks at the attachment point as the wire is pulled off the axis of the bond. The kink is most often noted on the side of the chip nearest the gate through which the molding compound enters the cavity.

Reliability concerns with wire sweep are device shorting and current leakage. Shorting can be from wire to wire, from wire to leadfinger, or from wire to die edge. Failure can be immediate or may not show up until the package experiences stress excursions.

Wire sweep can occur from any one of a number of causes: high resin viscosity, high flow velocity, unbalanced flows in the cavities, void transport, late packing, and filler collision [Nguyen 1988b, Nguyen and Lim 1990, Nguyen et al. 1992b].

***Viscosity.*** During radio-frequency preheating, the pellets of mold compound reach up to 100°C. Compacting the preheated pellets with a plunger in the molding pot causes the compound to heat further through a "fountain flow effect." The material near the wall is hotter than the rest and has a lower viscosity than the colder core. Thus, the compound can be convected to the cavities near the pot while the core is relatively cold. The ensuing high viscosity may cause wire sweep. Figure 6.7 shows radiographs of wire sweep, due to high resin viscosity and flow rate, in both a decapped part and in situ.

***High flow velocity.*** For cavities located far away from the pot, sweep can be induced by a different mechanism. At this stage, sufficient heat has been convected to the compound to lower viscosity to a level within the processing window. The majority of the cavities are already partially filled, resulting in a reduced flow path. As a result, for the same filling pressure, the velocity of the flow front in the end cavities can be quite high, raising the potential for wire sweep.

The influence of flow velocity on wire deformation is illustrated in Figure 6.8. In this case, a 14-lead device bonded with 25-μm (1.0-mil) wire was exposed to three fronts at different velocities [Nguyen and Lim 1990]. A clear silicone fluid with a viscosity similar to that of standard epoxy compounds was used. Shown is a polar plot of nondimensional sweep (sweep/pad pitch) as a function of wire location. Higher deformation was recorded with higher flow rates. The two distinctive lobes in the sweep profile were due to the die layout and ensuing wire length. For instance, wires 3 to 4 and 10 to 12 were nearest the die, resulting in the shortest wire lengths and best structural stability; these wires exhibited less sweep than the others.

***Unbalanced flows in the cavities.*** Cavity gates, if improperly designed, may produce flow fronts that are not balanced, above and below the die-paddle assembly, resulting in a "racetrack effect." One front precedes the other and blocks the vent line before complete filling, trapping air in the package. This flow imbalance creates a rolling bending moment that deflects the leadframe and affects the stability of the wirebonds.

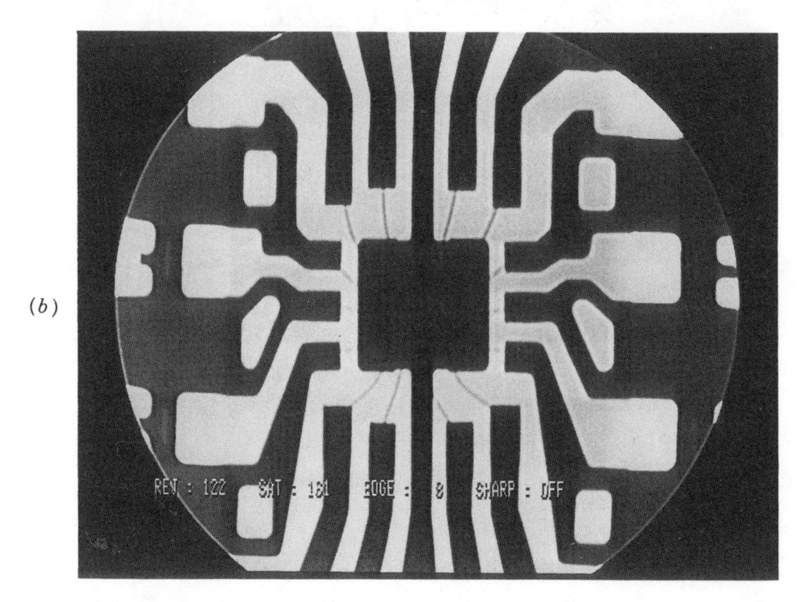


**Figure 6.7** *Radiographs showing wire sweep, due to high resin viscosity and flow rate, (a) in a decapped part and (b) in situ (bottom) [Courtesy of National Semiconductor 1994]*

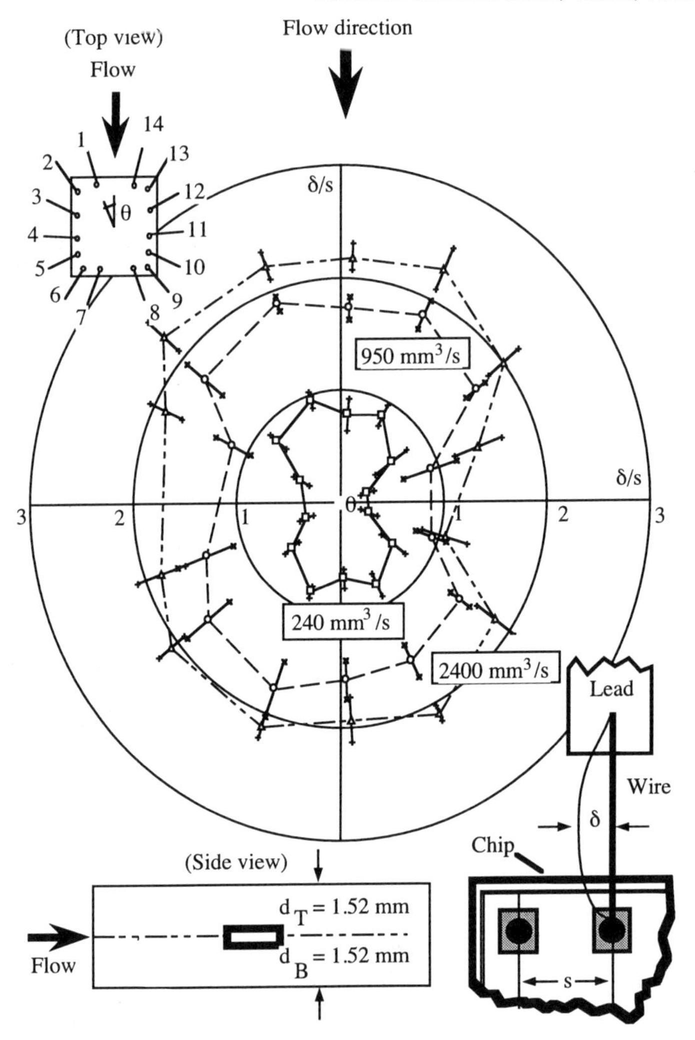


Mold transfer rates: 200 - 300 $mm^3$ /s

Diameter of bond wire: $2.54 \times 10^{-2}$ mm

**Figure 6.8** *Polar plot of the extent of wire sweep measured in a 14-lead package as a function of flow rate [Nguyen and Lim 1990]*

***Void transport.*** Voids have been reported as the cause of wire sweep [Nguyen and Lim 1990]. The change in interfacial tension as the void is convected across the die surface by the flow front can move the wirebonds. With a rapid succession of small voids, the wires can oscillate from their equilibrium position; such motion can be captured permanently by the rapidly curing resin. Molding voids tend to be more frequent with conventional molding than with multi-plunger (gang pot) molding, since gang pot presses provide a shorter flow path for the mold compound and better control of transfer speed and pressure [Nguyen et al. 1992a].

***Late packing.*** The filling profile programmed into standard molding presses usually requires filling 90 to 95% of the mold first under "velocity control." The profile is then switched to "pressure control," during which constant pressure is maintained to ensure that a certain density is obtained. The switchover must be made while the compound still has a low viscosity; a highly viscous gel could slide across the die and collapse the wirebonds [Nguyen 1993].

***Filler collision.*** Collision of fillers with fragile wirebonds is another cause of wire sweep. Wirebonds are usually between 23 to 38 μm (0.9 to 1.5 mils), depending on the current-carrying requirements of the device. Fillers, on the other hand, can have multimodal distributions for better packing. Sizes can range from a submicron up to 100 μm. By comparison, the thickness of gates in typical molds is around 200 to 250 μm (8 to 10 mils). As the highly filled front flows across the die, collision with any large filler particles can result in plastic deformation of wirebonds.

Molding pressure is a good indicator of wire sweep, since it depends on the product of velocity and viscosity, which dictates the amount of drag on the wire [Nguyen 1988b, Lim and Nguyen 1991]. Nguyen analyzed wire sweep and proposed a network flow model to depict the filling characteristics of the transfer molding. Wire sweep can be estimated by the following empirical equation:

$$D_{wire} = K_w \frac{\eta V_{mc}}{\ln\left(\eta / V_{mc}\right)} \tag{6.2}$$

where $D_{wire}$ is the wire displacement, $v_{mc}$ is the velocity of the molding compound, $\eta$ is the viscosity of the molding compound, and $K_w$ is the proportional coefficient determined from experiments [Manzione 1990]. If the estimated wire sweep is larger than the maximum allowable wire sweep, a low-viscosity material must be selected.

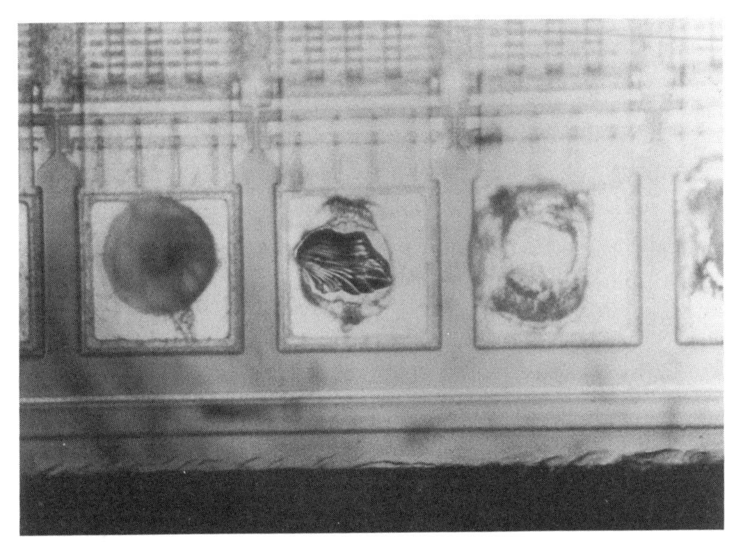

**Figure 6.9** *Pad cratering due to improper bonding conditions [Courtesy of National Semiconductor 1994]*

### 6.2.6 Cratering of Wirebond Pad

Cratering is a failure mechanism that occurs during manufacturing, usually as a result of improper bonder setup. Bonding parameters that affect the quality of a bond include temperature, ultrasonic power, force, and bond time [Gee et al. 1991]. Figure 6.9 shows cratering caused by improper bonding conditions. In this case, high ultrasonic power combined with low bonding forces generated sufficient stress to damage the pad metallurgy and crack the underlying silicon.

In ultrasonic bonding, even though the metal flow is equal in all directions, stacking faults in the silicon occur perpendicular to the direction of bonding tool motion, verifying that ultrasonic energy can introduce defects into single-crystal silicon [Winchell 1976, Winchell and Berg 1978]. Cratering is common in ultrasonic bonding, but not in thermocompression bonding. Similarly, Koyama et al. [1988] found that force and temperature cannot cause silicon nodule cratering without ultrasonic energy.

Contamination of the bond pad can cause improper setup of the bonding parameters, including the required bonding energy, since contaminated bond pads require more ultrasonic energy and higher temperature to make strong bonds. Moreover, too high or too low a static bonding force can lead to cratering in wedge bonds [Kale 1979]. Kale suggests that the optimum bonding force results in more efficient energy transfer, thus lowering the total energy requirement for the bond. Winchell [1976] notes the effect of bond pad thickness on the phenomenon of cratering; the bond pad serves as a cushion protecting the underlying silicon, silicon dioxide, polysilicon, or gallium arsenide from the stresses of the bonding process.

The tendency to crater predominates in thin metallizations. One percent silicon is sometimes added to the aluminum metallization in order to prevent spiking in shallow junctions. Micrometer-size silicon nodules in aluminum bond pads act as stress raisers and crack the underlying glass during thermosonic gold ball bonding. Koch et al. [1986], Ching and Schroen [1988], and Koyama et al. [1988] have generally confirmed the silicon-nodule cratering effect. After bonding is complete, these nodules decrease in the bonding region and damage appears in the insulation. Corrective action includes bonding at higher temperatures (250°C), using lower ultrasonic power, reducing molding stress, and removing a fracture-prone phosphorus glass layer from under the bond pad. To avoid cratering, Ching uses a hard titanium-tungsten underlayer for the pad, a modified bonding profile, and rapid ball touchdown.

A specially designed sensor has been used to study substrate cracking during bonding [Gee et al. 1991]. In this case, leakage currents in the p-n junction underneath the bond pads are correlated with bond pad damage before and after fatigue life testing by thermal shock. Results indicate that higher bond forces, power, and temperatures and longer bonding times produce higher bond strengths, mainly due to the increase in ball diameters. If ball diameters get too large, however, ball placement becomes a problem, because inaccuracy can generate cracks in the brittle passivation lip encircling the pads. Although high bond strengths are obtained with long bond times and high power, these can damage the silicon. A point of diminishing return must thus be identified for a given device technology (e.g., bond-pad metallization structure) and set of machine parameters.

Bond-pad hardness also affects cratering. A softer bond pad inhibits cratering by absorbing the energy from the bonding process. A harder bond pad, on the other hand, readily transmits the energy from the bonding process to the substrate. The best bonds are formed when the hardness of the wire and of the pad are reasonably matched [Ravi and Philoski 1971, Hirota et al. 1985].

Ching [1988] developed a theoretical model for contacting surfaces, based on the Hertzian contact theory relating a sphere and a flat surface, used to study the stresses induced at the surfaces during bonding. The assumptions are that both surfaces are elastic, that the intermetallic region possesses the same properties as the gold ball, and that the contact is made between the sphere of the intermetallic and the flat semiconductor bond pad. The ultrasonic effects are neglected.

Manufacturing variables include the time to touchdown after bond formation, the moisture content, and the hardness of the gold bond.

The Hertz model, as applied to cratering, starts with a circle being pressed against a bond pad. The radius of curvature establishes the contact area. The initial force creates stresses many times the yield strength or fracture strength of the bond pad or wire, implying that the stress is transferred to the underlying silicon, initiating the crater. The metal has a yield strength far lower than that of the brittle silicon, allowing the deformation of both the wire and the bond-pad metal. As the wire flattens, the stress drops rapidly below the metal yield strength. The Hertz model predicts the initiation of cratering at the center of the bond-pad area, contrary to experimental observations that the worst damage occurs along the perimeter of the bond pad. The conclusion drawn from this model is that to reduce bond-pad cracking, the contact area at touchdown should be maximized, while the force exerted on the bond pad should be minimized. The Hertz model is also pessimistic, since it assumes elastic properties, whereas bonding results in significant plastic deformation of the metals.

Clatterbaugh et al. [1984] studied gold-aluminum intermetallics using the ball-shear test, and measured the variation in the probability of cratering versus time at 250°C. The probability of cratering peaks at approximately 20% after 4 hr The increase is attributed to the increased volume of the original aluminum metallization due to the formation of intermetallic phases, including $Au_2Al$, which cause high stress under the bond. Continued heat treatment at 250°C, however, decreases the probability of cratering after 35 hr; this is due to the reduction in stress under the bond as the intermetallic phases recrystallize and the small-volume formation of other intermetallics.

### 6.2.7 Loss of Wirebond

Impurities often cause a loss of surface bondability and premature bond failure during the operational life of the device. Impurities and interdiffusion in wirebonds can cause Kirkendall voids [Horsting 1972]. With gold-aluminum bonds, the interdiffusion front moves through the gold, forming intermetallic alloy phases. Due to the lower solubility of the intermetallic, impurities swept ahead of the front precipitate there and act as sinks for the vacancies produced during the diffusion process.

Plating baths are a major source of such impurities as potassium-gold-cyanide, buffers, lactates, citrates, carbonates, and phosphates. Other impurities, like thallium, lead, or arsenic, are added to the bath to increase plating speed and reduce grain size.

Increasing the plating current density increases the co-deposited impurity level exponentially. Variations in plating process parameters, such as bath temperature and bath concentration, also cause bondability problems.

Thermocompression gold bonding is highly steady-state temperature-dependent in the presence of organic films covering the bond pad before bonding [Jellison 1975]. Gold does not form a stable oxide film at room temperature, and gold-gold bonds are very difficult to obtain when deformation is low. This problem has been attributed to the inability of gold to destroy contaminant surface films by absorption during solid-phase welding at moderately elevated temperatures (approximately 192°C). Jellison [1975] investigated the temperature dependency of both conventional thermocompression bonding (using a heated substrate during the bonding process) and pulsed thermocompression bonding (using a heated capillary and unheated substrates, and subjecting the bonding interface to a very small temperature cycle). The bond strength of gold bonds with contaminated surfaces increases for all bonding temperatures after the surfaces are cleaned with ultraviolet light. The length of substrate exposure to high bonding temperature dictates the bond strength; shorter applications result in stronger bonds. However, the shear force at failure increases with pulsed thermocompression bonding temperature, both with and without contaminated surfaces.

Postheat treatment of gold-gold bonds increases the bond strength, while preheat treatments decrease the strength. Nickel degrades bondability and reduces reliability. It may be introduced into the gold plating by accident or by grain-boundary diffusion upward from the thin underplating, thermally induced during die attachment or heat treatment [Hall and Morabito 1978, Loo and Su 1986]. Nickel spreads on the surface through surface diffusion, oxidizes, and renders the surface unbondable. Copper from plating bath contamination and leadframes can follow the same route as nickel and oxidize on the surface, resulting in decreased bondability. And chromium, often used in microelectronic components to promote adhesion between substrates and vacuum-deposited gold films, can rapidly diffuse to the surface through grain boundaries and subsequently oxidize.

Hydrogen bubbles in the gold film contribute to bondability degradation. Gas entrapment decreases film bondability by increasing ultrasonic energy absorption. The gas bubbles rupture during bonding, preventing the surfaces from coming into intimate contact. Bubbles coalesce and grow through structural rearrangement [Joshi and Sanwald 1973]. Hydrogen concentration is a function of plating current density, bath impurity, low bath concentration, and low bath agitation. During the bonding of gold-plated copper wire to gold films, the lowest bondability occurs for plating currents in the range of $1.6 \times 10^{20}$ to $2.7 \times 10^{20}$ A/m$^2$, which corresponds to the onset of hydrogen evolution at the cathode and dendritic-like surface morphology [Huettner and Sanwald 1972]. Gold films containing hydrogen are harder than pure films, but hardness decreases rapidly during high-temperature annealing at 350°C for 8 hr [Joshi et al. 1973]. However, free hydrogen inside the package has been found to inhibit the formation of gold-aluminum intermetallic compounds by filling the vacancies in aluminum and its grain boundaries [Shih and Ficalora 1978].

### 6.2.8 Wirebond Fracture and Liftoff

Wire can break at the heel of a wedge bond due to the reduced cross-section of the wire. Cracks in the heel of a ball bond can arise as a result of excessive flexing of the wire during loop formation, especially when the level of the second bond is significantly lower than the first; repeated flexing and pulling of the wire occurs as the device heats and cools during temperature cycling, due to thermal mismatches [Harman 1974]. Fatigue failure by crack propagation can occur at the heel [Harris et al. 1980]. A wire can also break at the neck of a wirebond, leading to an electrical open. Thallium, a major source of wirebond neck failures, forms a low-melting eutectic with gold and can be transferred to gold wires from gold-plated leadframes during crescent-bond break-off [McDonald and Palonberg 1971, McDonald and Riach 1973]. Thallium diffuses rapidly during bond formation and concentrates over grain boundaries above the neck of the ball, where it forms a eutectic. During plastic encapsulation or temperature cycling, the neck breaks and the device fails [James 1980]. Ball-bond fracture causes bond liftoff. The fracture can be a result of either tensile or shear forces induced by thermal stress or the flow of encapsulant during molding. The latter, also known as wire wash, happens only sporadically; when it does, however, it signals a molding compound that has expired or been improperly conditioned [Nguyen et al. 1992a]. Generally, bond strength is more a function of temperature cycling than of steady-state temperatures between -55°C and 125°C, although bond strength decreases as a function of temperature above 150°C for gold-aluminum bonds [Newsome 1976] and above 300°C for gold-copper bonds [Hall 1975].

### 6.2.9 Wirebond and Bond-Pad Corrosion

Wires inside packages are susceptible to corrosion in the presence of moisture and contaminants. Chlorine residues from the molding compound, introduced by the capillary action of the water along the wire, can concentrate around wirebonds. The aluminum reacts with the dissolved chlorine to produce $AlCl_3$, a process accelerated by the elevated temperatures occurring during burn-in and high-temperature storage. Corrosion of gold wirebonds on aluminum pads results from the liberation of bromine from brominated flame retardants, a process accelerated by high temperatures. Bromide ions react with the aluminum in the gold-aluminum intermetallic phase ($Au_4Al$), forming $AlBr_3$ and precipitating gold. The aluminum bromide hydrolyzes to $Al_2O_3$, with the liberated bromide ions providing the driving force until all the intermetallic $Au_4Al$ is consumed.

Bond corrosion may not result directly in failure, but it increases the electrical resistance of the interconnect until the device becomes nonfunctional. Often, the molding compound exerts a compressive force on the die surface and the adjacent wirebonds, and interconnection problems are not revealed until extensive corrosion has occurred. In mild cases, subtle changes take place in parametric functionality; in more severe cases, however, complete metallization attack occurs, resulting in open-circuit failures.

Corrosion can completely open one end of the wire (and occasionally both ends), permitting the wire to move within the package and cause intermittent electrical short circuits. Since the wire is held together by the molding compound, contact is usually maintained, unless delamination occurs at the encapsulant-wire interface or the compound expands and moves the wire away from the connecting pads during a thermal excursion.

Figure 6.10 shows the effect of bromine-induced intermetallic failure. Exposure to 175°C for an extended period fosters sufficient intermetallic formation to cause complete bond lifting after decapping.

Residues of the porous intermetallic phases can be seen under the ball bond. As the intermetallic phases grow, bond resistance increases, as shown in Figure 6.11a. A dramatic rise in low output voltage of a molded package is recorded after approximately 200 hr, while a part similarly decapped (and thus without any pad contact with the molding compound) behaves normally. Continuous testing of the affected junction returns the pad resistance to normal values, since the extra current flow fuses the pad junction.

### 6.2.10 Leadframe Corrosion

While leads generally have a nickel plating under the primary finishes of nickel, tin, and tin-lead to protect the base metal (usually Alloy-42 or copper) from corrosion, the assembly process for leadframes generates high residual stresses and high surface contamination. Any cracks or open voids in this plating can initiate corrosion in the presence of moisture and contaminants. During assembly and handling, the leadfingers are bent; cracks can develop and easily expose any

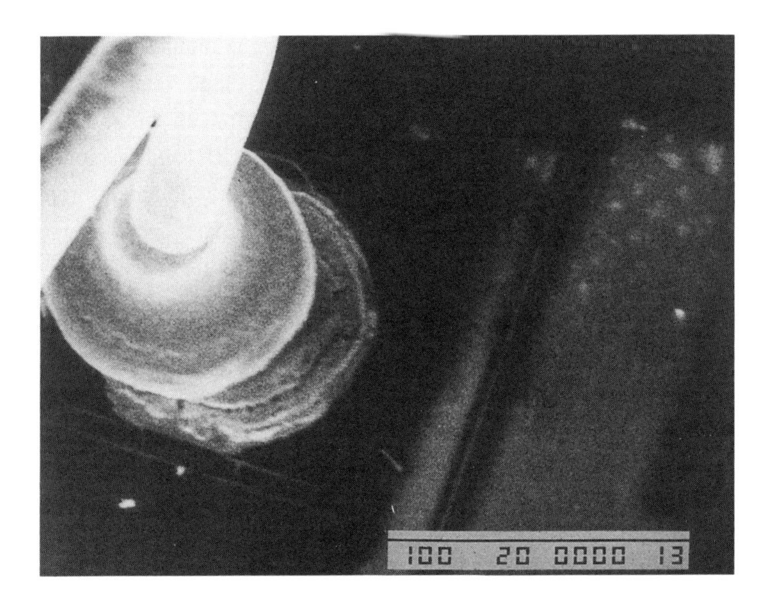


**Figure 6.10** *Effect of mold compound outgassing during high-temperature storage-life testing on bond-pad integrity [Courtesy of National Semiconductor 1993b]*

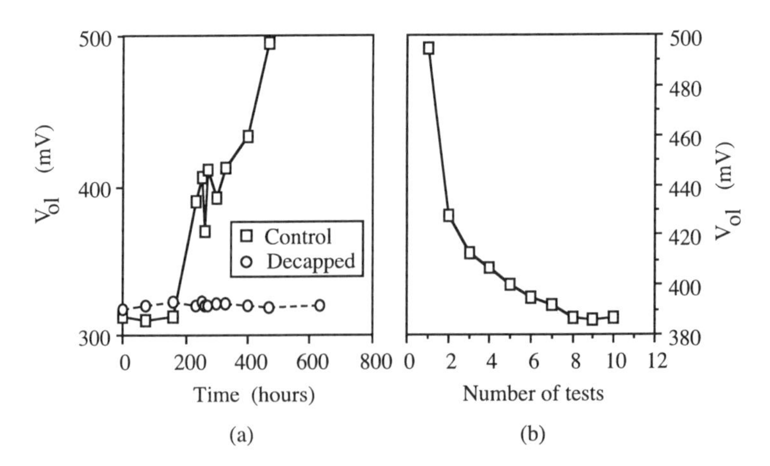


**Figure 6.11** *(a) Increase in low output voltage due to the formation of intermetallics from outgassing; (b) Rewelding of junction under repeated testing [Nguyen 1991a]*

corrodible surface to the external environment. Stress-corrosion-induced cracking can also occur, especially in Alloy-42 leads.

The rate of galvanic corrosion can be high, because the lead finish is often a cathodic metal with respect to the lead base material. Although a leadframe can undergo corrosion anywhere on its external surface, the most sensitive area is at the interface between the molding compound and the leadframe.

### 6.2.11 Leadframe Deadhesion and Delamination

Moisture ingress, either through the bulk encapsulant or along the interface between the leadframe and molding compound, can accelerate delamination in plastic packages [Kim 1992]. Experiments with moisture sensors reveal that when the adhesion between the molding compound and the leadframe is good, the main path of ingress into the package is through the bulk of the encapsulant [Nguyen 1988a, Nguyen 1990]. However, when this adhesion is degraded by improper assembly procedures — for example, oxidation from bonding temperatures, leadframe warpage from insufficient stress relief, or excessive trim and form forces [Nguyen 1991], — delamination and microcracks are introduced at the package outline, and water vapor can diffuse readily along this path.

At each interface, moisture can hydrolyze the epoxy, degrading the interfacial chemical bonds. However, different molding compounds respond differently to moisture exposure. Low-stress epoxy compounds, for instance, with silicone modifiers added for stress reduction, tend to be more susceptible to changes induced by moisture than molding compounds without silicone. A low glass transition temperature also adversely affects moisture absorption. The effect of package moisture absorption on the adhesion of three enhanced-epoxy mold compounds to copper leadframes is shown in Figure 6.12 [Kim 1992]. Figure 6.13 shows the percentage moisture absorbed by an epoxy resin as a function of glass transition temperature (after 168 hr. at 85°C/85% RH) [Kim 1992].

Degradation in bond strength can be detected by a gradual decrease in signal intensity during analysis with acoustic microscopy. Scanning acoustic tomographs conducted on plastic devices suggested that moisture absorbed by the encapsulant tended to migrate to the various interfaces in the package [Nguyen et al. 1993b].

Surface cleanliness is a crucial requirement for good adhesion. Oxidized surfaces, such as copper-alloy leadframes exposed to high temperatures, often lead to delamination [Kim 1991, Yoshioka et al. 1989]. The presence of a nitrogen or forming-gas shroud helps to avoid oxidation and is recommended during high-temperature processes.

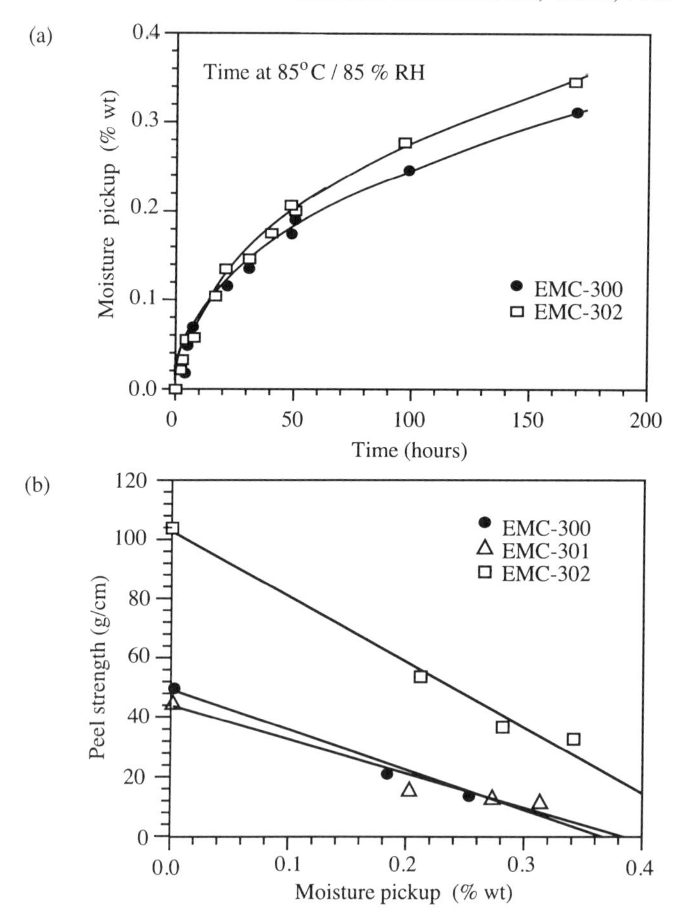


**Figure 6.12** *Package moisture absorption and its effect on the adhesion of three enhanced epoxy mold compounds (EMC-300, 301, and 302) to copper leadframes [Kim 1992]*

Low-affinity surface finishes, such as silver-spot plating, enhance interfacial adhesion. Traditionally, silver plating of the die pad is employed for bias control and to prevent oxidation on leads. Unfortunately, silver plating adheres poorly to molding compounds [Nguyen et al. 1991, Nguyen 1993b]. Newer leadframe designs use spot plating to minimize both the amount of precious metal used and the leadframe coverage, which is susceptible to delamination.

Lubricants and adhesion promoters in molding compounds have been shown to encourage delamination in plastic packages, and must be delicately balanced [Kim 1992]. Lubricants facilitate removal of the molded parts from the mold cavities at the risk of greater interfacial delamination. On the other hand, adhesion promoters ensure good interfacial adhesion between the compound and

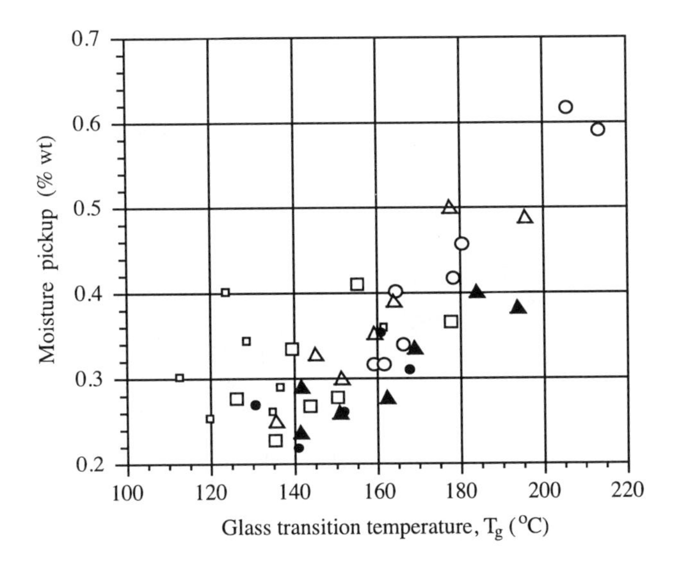


**Figure 6.13** *Moisture absorption as a function of glass transition temperature of epoxy encapsulant after 168 hr at 85°C/85% RH [Kim 1992]*

the component, but parts may be hard to pry out of the mold cavity. Kim's test results on the peel strength of epoxy resins with a variety of lubricants are shown in Figure 6.14. The effects of adhesion promoters and catalysts on glass transition temperatures and peel strength are indicated in Figures 6.15a and in Figure 6.15(b), respectively.

### 6.2.12 Encapsulant Rupture

Resistance to crack initiation in a material is strongly dependent on geometry, material properties, and loading. Kitano et al. [1991] developed a method for measuring the critical stress intensity factor for an encapsulant ruptured by brittle fracture. The specimen employed was cut from a plastic package and loaded to cause fracture, as shown in Figure 6.16 and 6.17. Typically, the crack initiated at an angle of approximately 135° clockwise from the chip bottom-encapsulant interface. The critical value of the stress intensity factor is inferred from the maximum load at fracture.

The relation between the stress intensity factor distribution and the applied load was obtained from elastic stress analysis. The specimen geometry was modeled using finite element methods. The encapsulant stress distribution around the corner was calculated at discrete points.

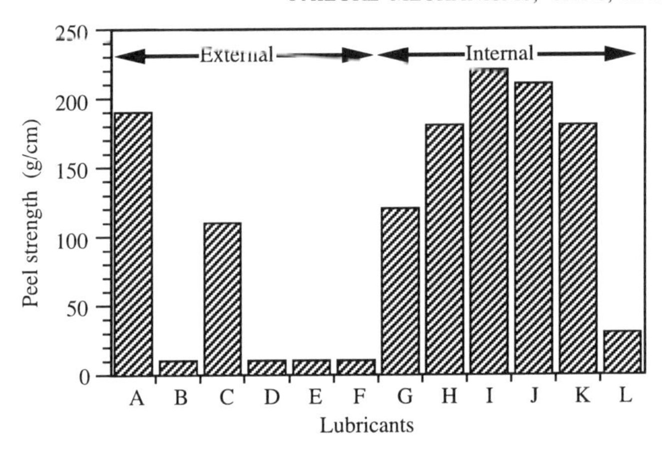


**Figure 6.14** *Effect of various lubricants on peel strength [Kim 1992]*

Assuming a power law distribution of the stress,

$$\sigma(r) = K_{stress} r^{-\lambda} \tag{6.3}$$

where $\sigma(r)$ is a stress component at distance $r$ from the corner, $K_{stress}$ is the stress intensity factor, and $\lambda$ is the exponent. The value of the exponent was determined from the slope of the straight line obtained by plotting log $\sigma$ versus log $r$. The assumption of the power law stress distribution was verified by plotting the stress intensity factor for the circumferential stress component as a function of the angle, measured from the base of the die; the plot shows that the stress intensity factor peaks at an angle of 135°. Because crack initiation takes place at the same angle, the failure criteria or critical stress intensity factor was computed using a correspondence technique — that is, for 1 N load in the finite element analysis, the value of $K_{stress}$ along the angle axis at $\theta = 135°$ is $x$; then the value of $K_{stress}$ at the fracture load is $x * f_{load}$, where $f_{load}$ is a scaling factor for the load.

### 6.2.13 Encapsulant Fatigue Fracture

Cyclic stresses in a package occur due to mismatches in thermal expansion caused by temperature changes and temperature gradient. An estimate of package life subjected to thermal cycles is obtained using Paris' law [Paris et al. 1961] for brittle crack propagation.

The stress intensity factor for a change in temperature is calculated using finite element methods, and the relationship is expressed as a polynomial, $f(a)$. Employing Paris' law, [Paris et al. 1961] the number of temperature cycles, $N_{cycles}$, for propagating a crack from length $a_0$ to $a_c$ is given by

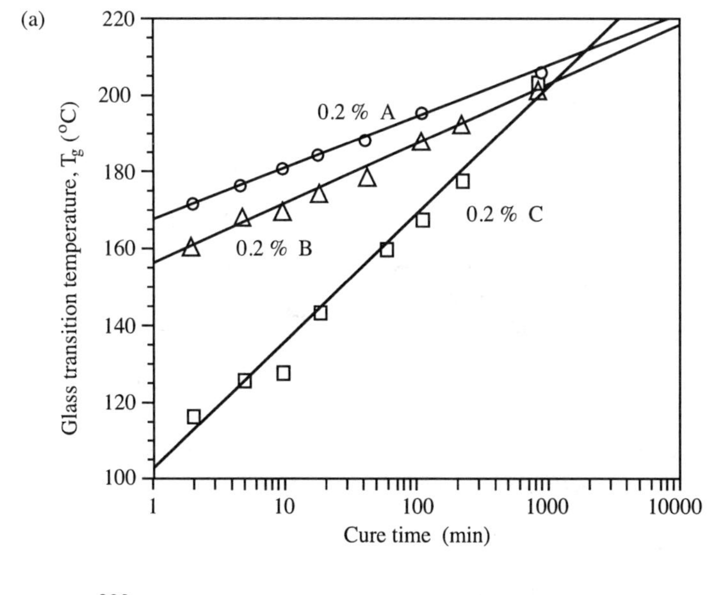


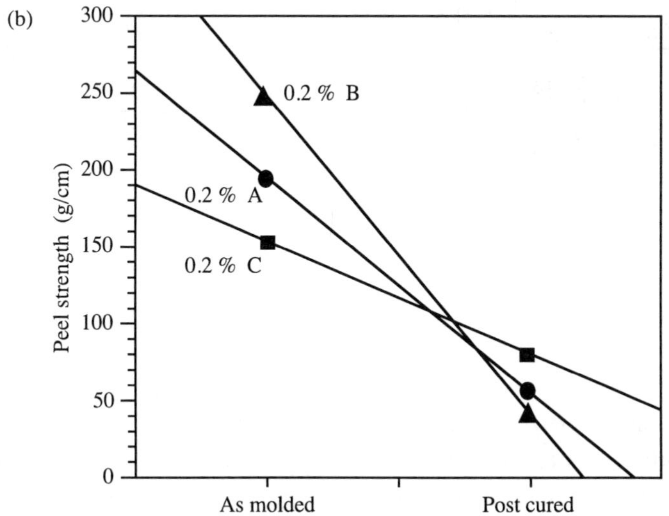


**Figure 6.15** *(a) Effect of cure time on glass transition temperature for three different catalysts; (b) peel strength as a function of catalyst used, after both molding and curing [Kim 1992]*

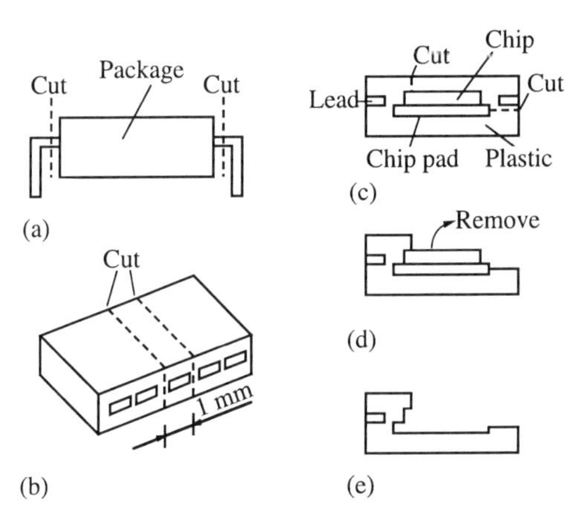


**Figure 6.16** *Specimen cut out from a plastic package for obtaining critical stress intensity factor for the encapsulant [Kitano et al. 1991]*

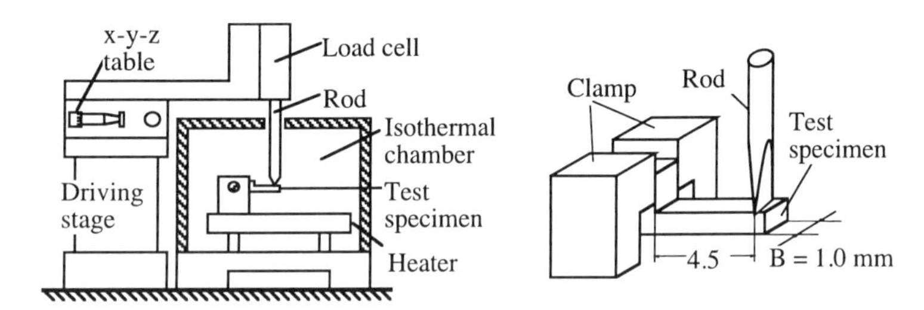


**Figure 6.17** *Loading to fracture of the specimen [Kitano et al. 1991]*

$$N_{cycles} = \int_{a_0}^{a_c} \frac{da}{c_3 \, f\,(a)^m} \tag{6.4}$$

where the material constants $c_3$ and $m$ have been determined for the encapsulant materials using single-edge-notch specimens [Nishimura et al. 1987]. The initial crack length, $a_0$, is taken to be some small value dependent on manufacturing defects, but the total number of cycles to failure is relatively insensitive to initial crack length, provided the crack is small.

### 6.2.14 Package Cracking ("Popcorning")

Package cracking results from internal stresses generated by the reflow solder temperature profile during the assembly of plastic-encapsulated devices on circuit cards. This phenomenon is discussed in Chapter 4, because the failure phenomenon is associated with improper assembly procedures.

In brief, during reflow soldering, the assembly temperature is rapidly raised above the glass transition temperature of the molding compound (approximately 140 to 160°C) to a temperature of approximately 220°C or more, which is required to melt the solder. At these temperatures, the thermal mismatch between the molding compound and the materials adhering to it, such as the leadframe and silicon die, can be large enough to make the interfaces susceptible to delamination. Moisture absorbed by the molding compound can vaporize into steam, and the resulting volume change can produce a pressure greater than the adhesion strength of the interface, causing delamination. Delamination may occur between the molding compound and the leadframe, or between the compound and the die surface. In extreme cases, the package may crack, ball bonds may shear from the bond pads, or the silicon may crater beneath the ball bonds. The cracks often propagate from the die paddle to the bottom surface of the package, where they are difficult to see during visual inspection of soldered boards. Occasionally, depending on package dimensions, the cracks may propagate to the top of the package or along the plane of the leads to the package sides, where they become more visible [Ito et al. 1987]. Proper handling and use after unpacking the assembly from the desiccant will deter this failure mechanism.

### 6.2.15 Electrical Overstress and Electrostatic Discharge

Electrostatic discharge (ESD) is the transfer of electrical charge between two bodies at different potentials, either through direct contact or through an induced electrical field. Electrostatic discharge can occur during manufacturing, handling, or service. It is neither unique nor more prevalent in plastic-encapsulated microcircuits than in hermetic packages. Its failure modes can include gate oxide breakdown and conductive path formation, with or without shorts. The failure site is often at sharp edges, such as corners of bond pads and metallizations, where the electric field strength is the greatest.

Electrostatic charges are generated by frictional sliding contact between dissimilar materials, or by induction charging of a conductor from an external charged insulator. Electrostatic discharge arises primarily from several sources: human contact, leading to static electricity discharge from the human body through the device; discharge of a statically charged device due to sliding against a carrier surface; and discharge of an induced charge in the device conductor from an external, electrostatically charged insulator — for example, a coffee cup carried across a carpet. Thus, the cause is electrical abuse.

Electrostatic discharge is frequently associated with electrical overstress (EOS). Electrical overstress is the damage to an electrical circuit due to thermal overstress caused by excessive electrical power dissipation during a transient electrical pulse. Like electrostatic discharge, electrical overstress failures are often caused by misapplication of the device in a circuit, allowing surges in a power line. Failure modes due to electrical overstress include junction spiking, latchup (common in MOS and CMOS devices), melted metallization, metal electromigration, and open bond wires.

Electrical damage from electrostatic discharge and electrical overstress have increased in importance due to the progressive decrease in integrated circuit dimensions. Metal oxide semiconductor (MOS) chips are more susceptible to electrostatic discharge damage because of small geometries and thin gate oxide dielectrics.

### 6.2.16 Soft Errors

Soft errors are temporary upsets in the state of a memory bit due to electron-hole pairs induced by alpha-particle radiation. Soft errors first showed up in 16k dynamic random access memory devices (DRAMs); the mechanism was discovered by May and Woods [1978]. These errors did not show a repeat pattern, and were erased with each refresh clock cycle.

The charge difference in memory cells, which distinguishes a "0" state from a "1", is called the critical charge, which for a 64K DRAM is about 1 to 2 x $10^6$ electrons. An alpha particle produces 1.4 x $10^6$ electrons and an equal number of holes as it penetrates 25 μm (1 mil) into a silicon chip. The holes are electrically rejected by the memory cell, and the electrons collected can change the state of the memory cell until the next refresh cycle restores the proper state. As the dimensions of the memory cells are reduced, the critical charge decreases, compounding the radiation of alpha particles. Like electrostatic discharge, soft errors are not unique to plastic-encapsulated microcircuits, and must also be addressed with hermetic packages.

### 6.2.17 Solder Joint Fatigue

Solder joint fatigue is the same for PEMs on circuit cards as it is for hermetic-package assemblies. The drivers of solder joint fatigue are primarily the global mismatch in the CTEs of the component and the circuit card, and the local mismatch in the CTEs of the solder and the lead, and the solder and the circuit card. The global mismatch can be further increased under operating conditions when there are temperature differences between the component and the circuit card. It may be advantageous to use a PEM over a hermetic part when the board material coefficient of thermal expansion more closely matches the PEM than the hermetic (metal or ceramic) package; this is the case with most of the common organic board materials, such as FR-4, cyanate ester, and bismaleimide triazine. Excellent overviews and models of the solder joint failure mechanism can be found in Frear et al. [1994].

## 6.3 FAILURE ACCELERATORS

The major accelerators of failure mechanisms in plastic-encapsulated integrated circuits are moisture, temperature, solvents, lubricants, contaminants, general environmental stresses, and residual stresses.

### 6.3.1 Moisture

Because the hygrothermal behavior of the epoxy network is complex, the diffusion kinetics can exhibit dual-mode absorption rather than the Fickian behavior of simpler systems [Nguyen and Kovac 1987, Liutkus et al. 1988, Nguyen 1988a, 1990]. When water permeates an epoxy-based encapsulant, swelling occurs that may lead to microcracking. Baking the encapsulant removes the moisture from the bulk, but the cracks do not self-heal, as they do with thermoplastics. When the encapsulant is again exposed to hot, humid conditions, moisture is reabsorbed, and the cumulative microdamage remains. The absorption and resorption coefficients are similarly affected [Belton 1989].

One-dimensional models have been used to compute the moisture concentration at the die-encapsulant interface [Shirley and Hong 1991, Kitano et al. 1988]. The one-dimensional diffusion of moisture is often valid, since the encapsulant layer adjacent to the die is relatively thin, for the size of the PEM. Ignoring the temperature gradient in the encapsulant and evaluating molding compound properties at the mean temperature of the die and case, the moisture concentration at the die-encapsulant interface is [Shirley and Hong 1991]

$$O_{conc}(x,t) = O_0 - \sum_{n=0}^{\infty} \left\{ \frac{4(-1)^n O_0}{\pi(2n+1)} - F_n \right\} \exp\left[ -\frac{1}{4l^2}(2n+1)^2\pi^2 Dt \right] \cos\left[ \frac{1}{2l}(2n+1)\pi x \right] \quad (6.5)$$

where $F_n$ are Fourier coefficients given by

$$F_n = \frac{2}{l}\int_0^l O_0(x) \cos\left[\frac{1}{2l}(2n+1)\pi x\right] dx \quad (6.6)$$

The history effect — that is, the variance of the ambient humidity due to time —

is input into the solution through the term $O_0(x)$. Figure 6.18 shows moisture absorption and desorption curves for a plastic package as a function of exposure time in hours [Kitano et al. 1988]. The level of moisture saturation is defined as the ratio of the actual absorbed moisture mass to the saturated absorbed moisture mass. Absorption curves correspond to the number of hours at 85°C/85% RH; desorption corresponds to 80°C.

The moisture diffusion constant for molding compounds $O_{diff}$ is expressed as

$$O_{diff} = c_1 \exp\left(-\frac{E_a}{kT}\right) \tag{6.7}$$

where $c_1$ is a constant, $E_a$ is an activation energy (eV), $k$ is Boltzmann's constant (8.617 x $10^{-5}$ eV/K), $T$ is the absolute temperature of the compound in (K), and the units of $O_{diff}$ are cm$^2$/sec. Kitano et al. [1988] found that $c_1$=0.472, and $E_a$=0.5 $eV$.

An assumption associated with the development of Equation 6.5 is that the moisture flux at the die-encapsulant interface is equal to zero:

$$\frac{\partial O(x=0,t)}{\partial x} = 0 \tag{6.8}$$

At the surface exposed to the ambient, a steady-state moisture concentration is given by the ambient temperature, $T_a$, the relative humidity, $RH$, and the encapsulant moisture saturation coefficient as

$$O(x=h,t) = RH\ P_{sat}(T_a)\ O_{sol} \tag{6.9}$$

where the $P_{sat}(T_a)$ is the saturation vapor pressure at the ambient temperature. The moisture solubility coefficient depends on the temperature of the molding compound:

$$O_{sol} = c_2 \times 10^{-4} \exp\left(\frac{E_a}{kT_m}\right) \tag{6.10}$$

where $c_2$ is a constant (often set to 4.96 x $10^{-4}$), $E_a$ is the activation energy (often set to 0.40 eV), $T_m$ is the mold compound temperature (K), and $O_{sol}$ is in moles/MPa-cm$^3$ [Kitano et al. 1988]. The initial condition is the moisture concentration, as a function of distance from the die-encapsulant interface:

$$O_{conc}(x,t=0) = O_0(x) \tag{6.11}$$

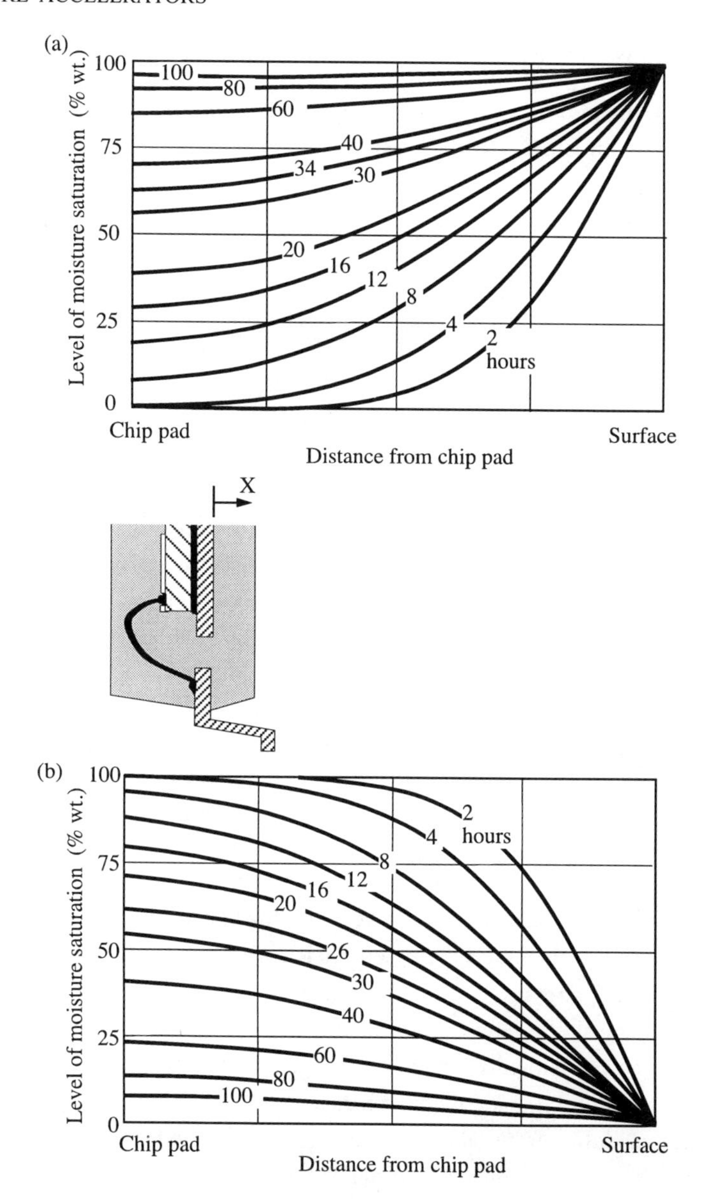


**Figure 6.18** *(a) Moisture absorption in 85°C/85% RH and (b) desorption in 80°C for a plastic package [Kitano et al. 1988]*

### 6.3.2 Encapsulant Swelling

Swelling due to moisture absorption is one of the reasons for plastic package deformation. In high filled materials, moisture can either condense in microcracks (it is believed they are on the interface of the filler particle - polymer) or dissolve into a polymer. As a first approximation, a simple additive elongation equation can be used:

$$\frac{\Delta l_e}{l_e} = \frac{1}{3}(1-\nu_{filler})\frac{\Delta m}{M_c}\frac{\rho_C}{\rho_W} \qquad (6.12)$$

where $\nu_{filler}$ is the volume part of the filler, $\Delta m$ is the quantity of absorbed water, $M_c$ is the mass of composite, $\rho_c$ the specific density of composite, and $\rho_w$ the specific density of water.

For typical epoxy novolac composites the press cooker test (120°C/100% RH/ 100 hr) leads to a moisture uptake of 0.9 to 1%. According to Equation 6.12 the elongation is 0.27 to 0.33%. Direct measurements with epoxy novolac premixes gives elongations between 0.22 to 0.3%. This means that most of the moisture in encapsulating composites is absorbed in the polymer.

An elongation of 0.2 to 0.3% is equal to an increase in temperature of 90 to 110°C from sorption isotherm studies. Therefore, the combination of moisture absorption and increased temperature may cause deformation to a package.

### 6.3.3 Temperature

Aside from moisture, temperature is a key stress factor, especially in terms of the glass transition temperature, the coefficients of thermal expansion, and thermomechanical stresses.

The glass transition temperature is the point at which the modulus of a polymer decreases markedly, though such a reduction is gradual over a small range of temperature. As a polymer approaches the glass transition temperature, the flexural modulus drops, the coefficient of thermal expansion increases, the ionic/molecular mobility increases, and the adhesion strength decreases. An improperly cured compound generally has a high coefficient of thermal expansion, poor adhesion strength, and a lower glass transition temperature because of the reduced cross-link density of the encapsulant.

Migration paths of contaminants can vary, depending on the history of the integrated circuit package. Outgassing by-products of the die-attach curing operation are deposited directly onto the leadframe or the die surface. Once encapsulated by the molding compound, polymerization residues (e.g., catalyst fragments or initiator radicals) can be entrained by the diffusing water. Similarly, external contaminants on the leads or package surface can also be dragged into the package by absorbed water. A solid transport medium is not required, especially with corrosion induced by outgassing products [Nguyen 1991a]. In this case, exposure to fumes released by the molding compound at high temperatures is sufficient to degrade a bonding interface. These fumes originate from the decomposition of silicone modifiers and the resin blend, and can attack any exposed aluminum on the bond pads to form porous intermetallics highly detrimental to device performance. Ionic contaminants can be found in corrosion by-products. Contamination due to saliva results in a corrosion product containing aluminum, potassium, chlorine, sodium, calcium, and magnesium in trace amounts. The presence of zinc may be due to perspiration contamination, since zinc is a common element in many antiperspirants.

### 6.3.4 Combined Environmental Stress Conditions

The performance of plastic and hermetic microcircuits has been compared during temperature cycling and coupled temperature, humidity, and bias tests for avionic environments [Condra 1993]. Plastic and hermetic integrated circuits, as discrete packages and as assembled on circuit cards, were temperature-cycled within the range of -55 and +85°C through 1000 cycles, and 26 parametric values were observed in terms failure rates or shifts. The circuit card assemblies were also tested for 650 hr. at 85°C and 85% relative humidity. The integrated circuit versions of the cards were coated with either polyurethane or parylene, and along with unassembled integrated circuits were tested for 1000 hr. at 85°C, 85% relative humidity with intermittent bias.

Figures 6.19 a and b show a diagram of the plan for the temperature cycling and temperature-humidity-bias tests, respectively. No differences in plastic and hermetic pass-fail results or in parametric outputs were observed with temperature cycling, or with temperature cycling followed by temperature-humidity-bias tests. Life estimates for the microcircuits were estimated to be more than twenty years.

### 6.3.5 Exposure to Contaminants and Solvents

Solvent exposure occurs during cleaning operations, which may use acetone, isopropyl alcohol, methylene chloride, or even kerosene. How quickly such solvents permeate the encapsulant, what type of chemical or physical damage they do to the epoxy network, and what influence they have on performance are

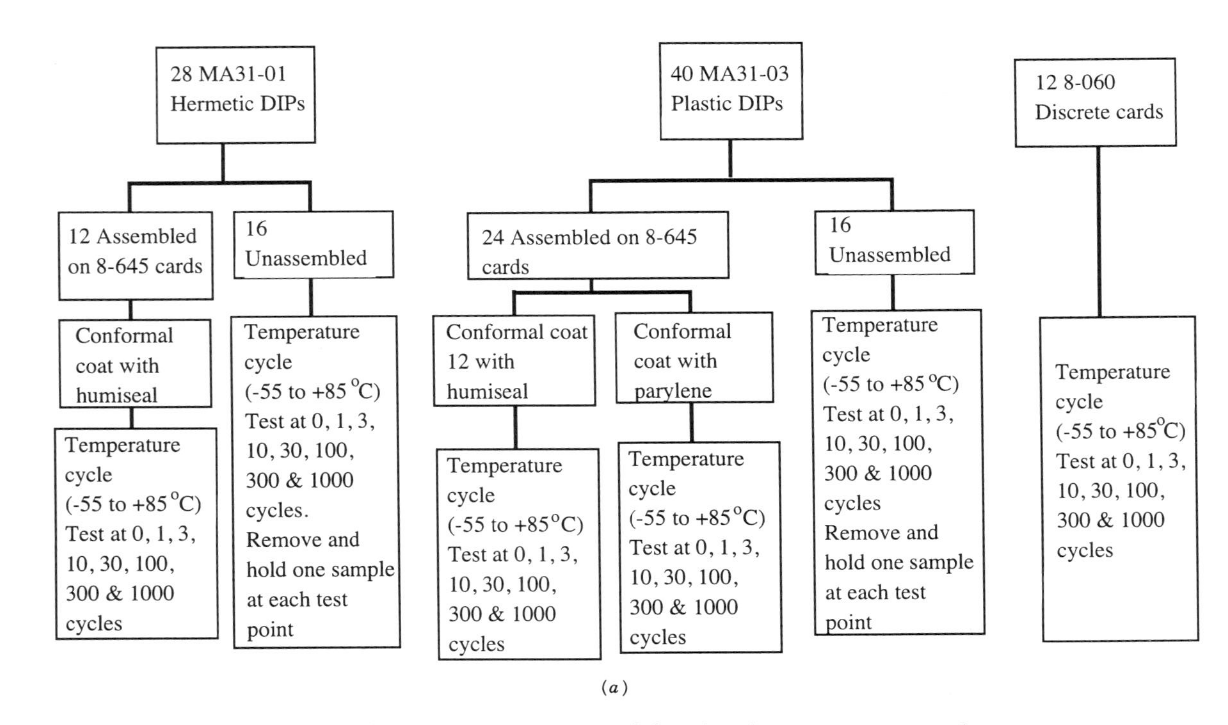


**Figure 6.19** *(a) Diagram of the plan for temperature cycling*

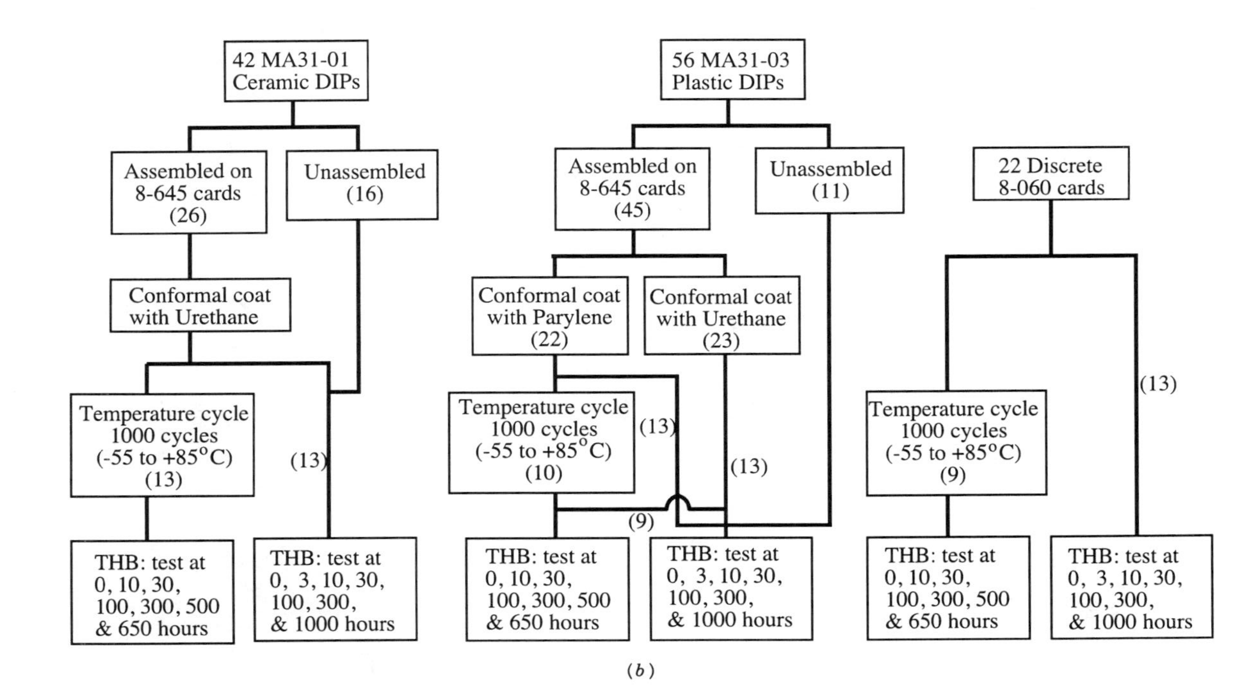


**Figure 6.19** *(b) Diagram of the plan for temperature-humidity-bias tests*

still unanswered questions. For applications in which exposure to solvents (such as jet fuel or hydraulic oil) is routinely unavoidable during preventive maintenance or field conditions, more reliability field data need to be collected.

Contaminants provide sites for failure mechanism initiation and propagation. Sources of contaminants are atmospheric pollutants, moisture, solder flux residues, ionic impurities in plastic materials, corrosive elements produced by thermal degradation, and outgassing by-products in die-attach adhesives (usually epoxies).

### 6.3.6 Residual Stresses

Residual stresses are generated in a package immediately after die attach. Depending on the nature of the die-attach (e.g., polyimide, silicone, or silicone-modified epoxy), various stress levels can be achieved [Nguyen et al. 1991]. New molding compounds are being synthesized for lower coefficients of thermal expansion, lower stiffness, and higher glass transition temperatures. However, trying to optimize the stress level requires more than just selecting the lowest-stress die attach or molding compound materials. The best combination for a package configuration will involve trade-offs [Chen et al. 1993]. Naturally, molding imparts stresses that are quite high, since the encapsulant shrinks more than other package materials. Stress-test chips are being employed to characterize assembly stresses [Nguyen 1990, Beaty et al. 1983]. The leadframe stamping process leaves residual stresses and sharp burrs that act as stress concentrators. The magnitude of the residual stresses depends on the package design, which includes both the paddle and the leadframe. If the residual stresses in the package are high, the amount of absorbed moisture required for package cracking will be lowered.

### 6.3.7 General Environmental Stress

Depolymerization is characterized by the breaking of polymeric bonds, which often turns the solid polymer into a gummy liquid comprising monomers, dimers, and other lower-molecular-weight species. Elevated temperatures and a closed environment usually accelerate depolymerization.

Exposure to seemingly harmless conditions, such as outdoor sunlight, may lead to gradual depolymerization. The ultraviolet rays in sunlight and atmospheric ozone can be powerful agents that depolymerize epoxy by scissioning the molecular chains.

Depolymerization can be prevented either by avoiding conditions that induce reversion or by using polymers tested for reversion resistance. Products that need to perform under hot and humid conditions require reversion-resistant polymers. Prediction of a material's behavior from chemical structure classification alone is unreliable. In addition, due to the proprietary nature of epoxies, the exact formulation is often not provided by compound manufacturers to integrated circuit suppliers [Nguyen et al. 1993a].

To insure that polymer reversion will not occur, compound suppliers usually do extensive experimentation on the ingredients, using various stoichiometric ratios to produce a mix with the optimum properties. Currently, there are no realistic industrial specifications or tests that address the reversion resistance of polymers.

### 6.3.8 Manufacturing and Assembly Loads

Failure may result from manufacturing and assembly conditions, including high and low temperature, temperature change, handling loads, and loads on wirebond and die-paddle assemblies due to flowing encapsulant. Popcorning of surface-mount plastic packages is one example. The design team must identify manufacturing and assembly stresses and account for them in developing appropriate guidelines.

## 6.4 MODELS FOR FAILURE MECHANISMS

Potential failure mechanisms in plastic-encapsulated integrated circuits include large deformation, corrosion, diffusion, interdiffusion, interfacial delamination, deadhesion, compound depolymerization, brittle fracture, ductile fracture, fatigue crack initiation, fatigue crack propagation, electrostatic discharge, electrical overstress, and radiation. This section discusses typical failure mechanisms and reviews the models available for describing them.

### 6.4.1 Corrosion

In a plastic-encapsulated package, corrosion is the process of chemical or electrochemical degradation of metallic interconnect elements. A time-dependent wearout failure process, the rate of corrosion depends on the component materials, the availability of an electrolyte, the concentration of ionic contaminants, the geometry of the metallic elements, and the local electric field. Common forms of corrosion are uniform chemical, galvanic, and pitting [Pecht 1991]. Generally, all three types are accelerated by elevated temperatures.

Uniform corrosion is a heterogeneous chemical reaction at a metal-liquid electrolyte interface, that involves the oxidation of the metal. Typically, this process occurs on chemically uniform metal surfaces, such as bare bond pads and unpassivated metallization. Ideally, chemical corrosion occurs evenly over the surface. If the corrosion product does not dissolve readily in the corrodent, the process is eventually self-limiting, because of the increasing thickness of the corrosion product. The rate of corrosion will depend on the stability of this layer.

Galvanic corrosion occurs where two metals of different electrode potential are exposed to corrosive environments, such as the interface between the aluminum metallization and the gold bond wire. Corrosion occurs at bond pads and can proceed along the passivation-metallization interface. Its rate is proportional to the biasing field. In plastic packages, it is most severe in those structures in which a small aluminum anode is in contact with a gold wire cathode of much larger area. The conductivity of the corrosion medium affects both the rate and distribution of galvanic attack.

In pitting corrosion, the attack on the metal is highly localized at specific areas that eventually develop into pits. Chemically active metals like aluminum, as well as alloys that depend on aluminum-rich passive oxide films for resistance to corrosion, are susceptible. Corrosion usually appears at well-defined boundaries on the surface, but can grow in any direction. It is usually the result of localized, autocatalytic corrosion cell action.

Levels of water-extractable ionics of molding compounds vary; good molding compounds have less than 25 parts per million (ppm) for halides (chlorides and bromides), 20 ppm for calcium, 10 ppm for potassium and sodium, and 3 ppm for tin. High impurity levels are sufficient to initiate corrosion under typical environmental conditions. Indeed, the conductivity of water extracts and the concentration of water-leachable ionics are generally good predictors of reliability [Nguyen and Jackson 1993] but good die passivation and ion getter can negate the problem.

The device failure modes associated with corrosion include opens in lead wires; shorting due to flaking, peeling, dendritic growth, and conductor migration; and parametric degradation due to electrical leaks.

The mechanism for electrolytic corrosion of aluminum (Al) in a moist environment starts with the adsorption of chloride ions ($Cl^-$) on the aluminum surface, subject to an electric field. The chloride ions ($Cl^-$) compete with the hydroxyl ions ($OH^-$) and water molecules for open bonding sites on the hydrated aluminum hydroxide surface. Aluminum hydroxide reacts with the adsorbed chloride ions to form a basic hydroxychloride aluminum salt. This reaction is represented as [Paulson and Lorigan 1976, Iannuzzi 1982]

$$Al(OH)_3 + Cl^- \rightarrow Al(OH)_2Cl + OH^- \quad (6.13)$$

Once the surface oxide is dissolved, the underlying aluminum reacts with the chloride ions:

$$Al + 4Cl^- \rightarrow Al(Cl)_4^- + 3e^- \quad (6.14)$$

The $AlCl_{4-}$ subsequently reacts with the available water:

$$2Al(Cl)_4^- + 6H_2O \rightarrow 2Al(OH)_3 + 6H^+ + 8Cl^- \quad (6.15)$$

This reaction liberates chloride ions, which are then free to continue the corrosion process by the reactions given in Equations 6.12 through 6.14. Gold on aluminum produces a galvanic couple, which accelerates the corrosion process by providing the driving force for the aluminum oxidation reaction.

A corrosion mechanism at high temperatures has been proposed by Ritz et al. [1987] for gold-plated aluminum bonds in a bromine environment — formed, for instance, by the flame retardants added to the epoxy encapsulant. At temperatures higher than the molding temperature of 175°C, the brominated flame retardants decompose, releasing in the process a potpourri of highly corrosive by-products, such as methyl bromide and hydrogen bromide:

$$\begin{aligned} CH_3Br &\rightarrow CH_3 + Br^- \\ 4HBr + 2O &\rightarrow 4Br^- + 2H_2O \end{aligned} \quad (6.16)$$

The bromide ion ($Br^-$) reacts with the aluminum in the $Au_4Al$ intermetallic phase, forming aluminum bromide ($AlBr_3$) and gold ($Au$). The aluminum bromide oxidizes, sustaining the reaction until all the $Au_4Al$ intermetallic is consumed:

$$\begin{aligned} Au_4Al + 3Br^- &\rightarrow 4Au + AlBr_3 \\ 2AlBr_3 + 3O &\rightarrow Al_2O_3 + 6Br^- \end{aligned} \quad (6.17)$$

The time to corrosion failure is the sum of the induction time and the time to failure due to wearout corrosion attack. In the absence of paths facilitating moisture permeation, such as cracks and voids, induction time is determined by diffusion of the moisture to the encapsulant-metal interface. With sufficient contaminants leached to the die surface, at least three monolayers of condensed water molecules [Cvijanovich 1980] are needed to start the flow of electrolytic current.

Mechanical stresses can influence corrosion development, especially if they cause defects to form in the passivation layers, or cracks between the chip and the plastic package. There also exists evidence that mechanical stresses open up sites for corrosion to occur. That is, the volume of aluminum corrosion products is three times more than aluminum, suggesting that a certain free volume (or a gap) above the metallization is necessary to develop corrosion. On the other hand, corrosion will be suppressed if mechanical stresses place the encapsulating material in compression with respect to the chip surface.

Finally, moisture absorption in an encapsulating material causes package swelling. In the case of low residual compressive stresses, swelling can result in delamination and water accumulation in the gap.

### 6.4.2 Diffusion

Diffusion is the ability of a molecular, atomic, or ionic species to migrate into the bulk phase of a second material time-dependently. From an atomic or molecular perspective, diffusion in solids is the migration of atoms or molecules of the diffusing species from one lattice site of the solid to another. The atoms must have sufficient energy to break bonds from the previous lattice site and form new bonds at the next sites. One-dimensional diffusion is described by the following governing differential equation:

$$\frac{\partial O_{conc}}{\partial t} = O_{diff} \frac{\partial^2 O_{conc}}{\partial x^2} \tag{6.18}$$

where $O_{conc}$ is the moisture concentration at time $t$ and $O_{diff}$ is the moisture diffusion constant of the diffusion medium [Crank 1975]. Often, failure mechanisms such as corrosion, electromigration, and outgassing are driven by diffusion.

Interdiffusion occurs when two different bulk materials are in intimate contact at a surface. The molecules of each material migrate into the other by diffusion. The diffusion rate of one solid into another often follows a parabolic relationship:

$$x = k_D \sqrt{t} \tag{6.19}$$

$$D_0 = c_9 \, e^{\left(-\frac{E_a}{kT_T}\right)} \tag{6.20}$$

where $D_0$ is the diffusion coefficient, $x$ is the intermetallic layer thickness (cm), $t$ is the time (seconds), $c_9$ is a constant (cm$^2$/s), $E_a$ is the activation energy (eV), $k$ is Boltzmann's constant (8.617 x $10^{-5}$ eV/K), and $T_T$ is the steady-state temperature (K) [Kidson 1961].

Differences in the diffusion rates of two materials bonded together cause one material to suffer a depletion of molecules at the interface, leading, for example, to Kirkendall voiding [Tu 1985, Ramsey 1973]. This interdiffusion provides a mechanism for interfacial adhesion in limited quantities; however, when interdiffusion is excessive, gross voiding can occur. The intermetallics and the voids reduce the mechanical strength of the bond. A common example is the leaching of gold into aluminum at the wirebond interface, leading to "purple plague" [Philosky 1970a, b].

Interdiffusion occurs at the interface of two metals. Sites of interdiffusion in plastic packages include wirebonds and the chip-to-metallization interface. Interdiffusion at the wirebond site occurs because the diffusion rates of the interface materials are often vastly different, leading to voids concentrated at the interface, or Kirkendall voiding.

The diffusion rate is a function of lattice defects. Metallizations and poor bonds with many grain boundary defects, vacancies, dislocations, and free surfaces have much higher rates of diffusion than bulk metal, and are susceptible to Kirkendall voiding. Vacancies in the lattice can pile up and condense to form voids. Interdiffusion can be enhanced by such impurities as nickel, iron, cobalt, boron, chlorine, bromine, and antimony trioxide. The impurities become concentrated ahead of the intermetallic diffusion front, precipitate, and act as sinks for vacancies produced by the diffusion reaction, resulting in Kirkendall-like voids. The diffused metal forms intermetallic compounds. Locations for interdiffusion failures include gold-aluminum, gold-copper, and copper-aluminum interfaces at the wirebonds.

When gold wire is connected to an aluminum bond pad, the interdiffusion voids often form on the gold side, along the interface. High temperature helps the diffusion process and facilitates the formation of voids and intermetallics; in the presence of halogenated species, this process is accelerated at a lower temperature than normal [Khan and Fatemi 1986, Ritz et al. 1987, Gallo 1990]. Mechanical stress arises due to the volumetric expansion resulting from an increase in gold under the bonds, the thermal mismatch between the intermetallic compound and the parent materials, and the hardness of each phase.

Five intermetallic compounds can be formed at the gold-aluminum interface: $Au_5Al_2$, $Au_2Al$, $AuAl_2$, $AuAl$, $Au_4Al$. Comparison of the relative growth rates for each of the intermetallic compounds shows that $Au_5Al_2$ grows much faster than other phases of intermetallic [Philosky 1970 a, b]. $AuAl_2$ is purple and, as a result, the formation of gold-aluminum intermetallics is called "purple plague." The brittle and porous intermetallic compounds begin to form during the wirebonding process of thermocompression or thermosonic bonding. An excess of such intermetallics may lead to premature bond failures. Thermal cycling is detrimental to gold-aluminum bonds with excessive intermetallics [Philosky 1970a,b, 1971]. The presence of hydrogen inside the package reduces the rate of intermetallic formation [Shih and Ficalora 1978], but presents a potential problem because it embrittles any iron, nickel, and palladium inside the package. Hydrogen also combines with oxygen to produce water, which can ultimately lead to corrosion [Harman 1989]. Palladium added to gold thick films ultrasonically bonded to aluminum results in a stable ternary compound of gold-aluminum-palladium, which slows down the rate of gold-aluminum diffusion and results in longer bond life [Horowitz et al. 1979, Hund and Plunkett 1985]. Thinner metallization also minimizes voiding by restricting the availability of one of the intermetallic elements. Intermetallic failures can be largely prevented if the ratio of the width of the aluminum wedge bond to the thickness of the gold film is greater than four and the storage temperature is less than 350°C [Kashiwabara and Hattori 1969, Philosky 1970 a, b].

Copper-aluminum systems also form intermetallics. The intermetallic compounds that develop at elevated temperatures include $CuAl$, $CuAl_2$, $Cu_9Al_4$. At lower temperatures, the intermetallics occur in pairs [Gershinski et al. 1977, Campisano et al. 1978], including $CuAl$-$CuAl_2$ and $CuAl_2$-$Cu_9Al_4$. Olsen and James [1984] aged copper-aluminum bonds in air, vacuum, and nitrogen, and found that intermetallic formation occurs under all aging conditions. Studies on copper-aluminum bonds in the temperature ranges of 400 to 535°C [Funamizu and Watanabe et al. 1971] and 300 to 500°C [Wallach and Davies 1971] show a significant decrease in bond strength and ductility after aging, directly related to the total thickness of the intermetallics; the critical thickness of the intermetallic range for most studies has been around 16 μm. The rate of decrease in bond strength versus aging time increases with the increase in aging temperature [Hall et al. 1975]. Pitt and Needes [1982] study thermosonic bonds of goldwire to thick-film copper. Their findings suggest that the copper-gold bond is not suitable for use in microelectronic devices operating at temperatures above 300°C. However, at lower temperatures, no intermetallic formation is observed in gold-copper bonds [Johnson 1973]. The use of a thin nickel film approximately 5,000 ± 2,500 Å in thickness as an intermetallic barrier between copper and gold prevents the degradation of the wirebond.

### 6.4.3 Delamination and Deadhesion

Delamination occurs at the interface of adhering materials. The work required to separate the interfaces includes the work necessary for deadhesion as well as the work required to deform, elastically or inelastically, the separating bulk phases for crack extension. The total fracture energy, $G$, can be written as

$$G = W_a + W_p \tag{6.21}$$

where $W_a$ is the reversible work of adhesion and $W_p$ is the irreversible work of deformation in the two phases. $W_a$ can be defined as

$$W_a = \gamma_\alpha + \gamma_\beta + \gamma_{\alpha\beta} \tag{6.22}$$

where subscripts $\alpha$ and $\beta$ indicate the surface tensions of the two distinct phases, and the subscript $\alpha\beta$ refers to the interfacial surface tension between them. Thus, the total adhesive strength of a joint depends not only on its interfacial properties, but also on the mechanical properties of the bulk phases. Delamination between the plastic encapsulant and the material surfaces that are in contact with it is thus a function of package materials [Kim 1991].

In experimental terms, interfacial bonding strength is often characterized in terms of the electron binding energy between pairs of materials, and is a unique property of that pair. Thus, the interfacial fracture toughness must be measured for commonly used material pairs in packaging applications.

### 6.4.4 Brittle Fracture

Brittle fracture can occur in materials that exhibit low levels of yielding and inelasticity. In otherwise ductile materials, brittle fracture can occur due to formation of brittle intermetallics. Under excessive stress levels, sudden catastrophic crack propagation can initiate from a microscale flaw. For all loads on a structure, a singular stress field exists in the vicinity of a sharp crack tip.

Fracture mechanics can be used to estimate failure at defect sites with singular stress fields. The failure criterion is based on the critical value of the stress intensity factor, $K_I$, that characterizes the severity of the crack-tip stress field as a function of crack type, crack size, body geometry, and applied stress. The failure criterion can be written as

$$K_{\max} < K_{Ic} \tag{6.23}$$

where $K_{max}$ is the maximum stress intensity factor at the potential failure site in the structure [Broek 1991]. Failure occurs when $K_{max}$ equals or exceeds the critical stress intensity factor, $K_{Ic}$. Often called the fracture toughness of the material, this material property characterizes the ability of a material to resist fracture.

Mechanical failure of a silicon die is often due to brittle fracture that may or may not lead to instantaneous electrical failure of the device. Although the crack may initiate from the back of the die, bifurcation and growth of the crack tip may lead to transverse propagation. Thus, the crack can propagate through the entire die without traversing any sensitive structures. The crack may initiate from backgrinding or backlapping steps, aggravated further by stresses generated by the die-attach or molding processes. During die-attach assembly, the machine plunger pushing the die away from the adhesive backing tape can impart microscopic damage to the back of the die; microcracks introduced at this stage would be magnified by packaging stresses [van Kessel et al. 1983, 1987].

### 6.4.5 Ductile Fracture

When plasticity around the crack tip is appreciable, the J-integral is often used to provide a measure of the energy release rate of an advancing crack. The value of the J-integral, the crack-driving force for fracture in ductile materials, is often calculated using an engineering approach [Kumar et al. 1981, Kanninen and Popelar 1985]. The linear elastic component of the J-integral is related to the stress intensity factor, $K_{stress}$

$$J = \begin{cases} \dfrac{K_{stress}^2}{E} & \textit{plane stress} \\ \dfrac{(1-\nu^2)K_{stress}^2}{E} & \textit{plane strain} \end{cases} \tag{6.24}$$

where $\nu$ is Poisson's ratio. The plastic component of the J-integral can be obtained using the pure bending plane strain solution [Shih and Needleman 1984]. The presence of initial crack-like defects is not important for ductile fracture, because the crack propagation time often dominates the time to failure.

Epoxy molding compounds do not usually fail by brittle rupture. Although the material is highly loaded with silica fillers, the polymer is still viscoelastic. The crack tip typically exhibits a localized yielding zone. Depending on the formulation of the molding compound (for example, whether it includes either silicone modifiers or flexibilizers incorporated directly into the molecular network), crack-tip blunting may occur. The fracture toughness of the molding compound also depends on the composition of the formulation. Resin chemistry and filler technology (type, size, distribution, and interfacial adhesion treatment) both play an important role [Yamaguchi et al. 1991].

### 6.4.6 Fatigue Failure

When mechanical or thermal loads applied to an assembly cause repetitive cyclic stresses below the ultimate strength of the material, failure can occur due to accumulated fatigue damage. Fatigue failure is a wearout mechanism that initiates a crack at a point of discontinuity or defect.

The fatigue properties of a material are characterized by a stress-life (S-N) curve, which empirically relates the stress range to the cycles to failure by a power law function [Osgood 1982]. The number of cycles to failure is given by the relation

$$N_{ft} = b\,\sigma_{loc}^{c} \quad \textit{and} \quad N_{fs} = b'\,\tau_{M}^{c'} \tag{6.25}$$

where $N_{ft}$ and $N_{fs}$ are the cycles to failure due to tensile fatigue and shear fatigue, respectively; $b$ and $c$ are the tensile fatigue constants; $b'$ and $c'$ are the shear fatigue constants of the material; $\sigma_{loc}$ is the local principal stress; and $\tau_M$ is the local von Mise's stress.

### 6.4.7 Fatigue Crack Propagation

Fatigue crack propagation is a stable increase in crack length under cyclic stresses. The rate of increase in crack length per cycle in encapsulants can be described by Paris' law [Nishimura et al. 1987]:

$$\frac{da}{dN_{cycles}} = c_3\,(\Delta K)^{c_4} \tag{6.26}$$

where $a$ is the crack length, $N_{cycles}$ is the number of cycles, $\Delta K$ is the stress intensity factor range, and $c_3$ and $c_4$ are material constants. For encapsulants, $c_4$ is approximately 20 [Nishimura et al. 1987], much higher than the typical value for fatigue crack propagation in metals (typically from 2 to 8), indicating faster crack propagation in encapsulants.

### 6.4.8 Radiation

Radiation is known to cause degradation of devices employing metal oxide semiconductor dies through voltage shifts and degradation of circuit speed [Hughes et al. 1989]. Failure modes due to buried-oxide failure are latch-up in the CMOS circuit, single-event upset, and dose-rate upset sensitivities. Failure sites in the die due to radiation include buried oxide, lightly doped drain, oxide spacers, and lateral oxide-isolation structures.

A model has been proposed for the threshold-voltage shift ($\Delta V_{TH}$) due to radiation in MOS devices [Hughes 1989]. The model is based on the trapped oxide charge ($N_{OT}$) and the radiation-induced interface charge ($N_{IT}$). The threshold voltage shift is given as

$$\Delta V_{TH} = \Delta V_{OT} + \Delta V_{IT} \tag{6.27}$$

$$\Delta V_{OT} = \frac{q\,\Delta N_{OT}}{C_O} \quad and \quad \Delta V_{IT} = \frac{q\,\Delta N_{IT}}{C_O} \tag{6.28}$$

where $C_0$ is the gate oxide capacitance, $q$ represents the electronic charge, $\Delta V_{OT}$ is usually negative in value for silicon dioxide gate insulators, and $\Delta V_{IT}$ is usually positive for N-channel devices and negative for P-channel devices.

Circuit-speed reduction can be attributed to the mobility reduction induced by increased $N_{IT}$; the signal propagation delay, $T_D$, is inversely related to channel mobility, $\mu$. Then,

$$T_D \sim \frac{1}{\mu(V_A - V_{TH})} \tag{6.29}$$

where $V_A$ is the applied voltage.

### 6.4.9 Charge Instability

As it was shown in Section 5.5, the origin of PEMs' failures often remains undetermined. Charge instabilities in elements of PEM can cause some of the difficult-to-determine failures. This may be especially important for contemporarily ICs because the decrease of MOS gate width enhances the moisture effect on charge characteristics of the $Si\text{-}SiO_2$ interface.

Charge instability in PEMs can arise due to either increased lateral electrical conductivity along the interface between chip and encapsulating material, or to reduced electrical resistance of encapsulating material. High humidity and delamination of plastic are favorable to the first mechanism, whereas low humidity and high temperatures are favorable to the second. In either case, redistribution of charges and potential on the surface of the passivation film or on the interface between dielectric films will occur. The result is parasitic channel formation between elements of integrated circuit or a threshold rise in MOS elements.

The behavior of charges on the surface of MOS structures can be explained with the help of an equivalent diagram that is represented as a R-C-G chain, where C is the specific capacity of oxide film, R is the surface resistance (in a general case it is nonlinear and is determined by the amount of accumulated charges), and G is the volumetric conductivity of the oxide. The solution of the relevant telegraph equation gives the time dependance of the redistributed potential $V(x,t)$.

When the conductivity of the encapsulating polymer predominates, the potential distribution along the surface of MOS structures can be described by a system of differential equations (Laplace equation and continuity of the electric displacement and current). The result function $V(x, t)$ can be thus determined, so that when this value exceeds some limit, a failure may occur.

Failures due to a parasitic channel formation were common in 1970s. However, while measures have been adopted to eliminate these failures, more sophisticated mechanisms of charge instabilities have been revealed. For example, without an electrical field in the oxide, moisture will not affect charges on the interface Si-$SiO_2$. However, under a negative bias applied to the gate of a MOS structure, a positive effective charge in the interface will occur.

One of the reasons for this phenomenon is an increasing of hydrogen concentration in the oxide due to generation on the Al-$SiO_2$ interface. Hydrogen has a high diffusivity even if the temperature is low. Hydrogen in the oxide leads to a formation of a deep donor like level on the Si-$SiO_2$ interface. Moisture effects can be gradually ionized under a negative potential on the $SiO_2$ surface, causing the accumulation of positive charges. Thus the threshold voltage increases in the areas free from metallization and near the electrode edges, leading to parametric shifts, which can cause failure.

### 6.4.10 Leakage Current

Long-term storage of PEMs in high-humidity conditions can lead to leakage current. The increase of the current is often explained by high surface conductivity of the oxide film. But the value of $SiO_2$ surface conductivity even under condition of 100% RH does not exceed $10^{-11}$ at $1/\Omega$. At this condition the leakage current between adjacent pads should not exceed $10^{-10}$ A under a 10-V bias. In most cases such leakage current can be neglected.

After testing in moisture environments, the leakage current sometimes reaches values from $10^{-8}$ to $10^{-10}$. These values can be explained by the presence of delaminations (microgaps) either between the chip and the plastic or between adjacent leads or through-micropores in oxide film near the edge of the metallization. The micropores or gaps are gradually filled by water due to humidity diffusion into the package. The process is fairly slow, in that the characteristic time of filling is as follows:

$$\tau = \frac{h_{enc} t_{gap} \rho_w}{DS(P_a - P_c)} \tag{6.30}$$

where $t_{gap}$ is the gap thickness, $h_{enc}$ is the encapsulant thickness, $\rho_w$ the specific density of water, $S_c$ is the sorption coefficient, and $P_a$, $P_c$ are water steam pressure outside the package and inside the gap.

## 6.5 RANKING OF POTENTIAL FAILURES

Failure distributions differ from device to device and manufacturer to manufacturer. Table 6.2 presents a summary of failure analysis results in the form of a Pareto ranking of various field failures in very large-scale integration (VLSI) class devices for demanding commercial applications, such as automotive manufacturing and telecommunications [Ghate 1992]. The sources of failures include manufacturing fallout, qualification, reliability monitors, and customer returns. Failures due to electrical overstress and electrical discharge dominate, with electrical overstress due to misapplication and voltage overstress as primary causes. The major causes of failures were poor handling and assembly. Of the remaining failure mechanisms, wirebond failure was significantly, but corrosion of die metallization was a relatively small factor.

Vahle and Hanna [1990] reported a distribution of early field failures of plastic -encapsulated integrated circuits. Their data indicate that 36.3% of failures are unverified, 22.2% are of undetermined origin, 17% are due to wirebond failure, 8.3% result are from electrical overstress, 2.3% are from mask/etch defects, 2.1% are due to electrostatic damage, and 11.9% are attributed other causes.

Another study, by Bloomer [1989], presented failures in commercial integrated circuits components (Figure 6.20). The components analyzed were supplied by a large industrial user from 1984 to 1988. The pie chart includes failures during manufacturing, incoming component testing, final board testing, and field operation. Like other studies, this one found that the major causes of failure are electrical overstress, electrostatic discharge, and wirebond weaknesses.

Finally, Pecht [1991] collected electronics system and device failure data and noted some of the reliability problems associated with electronics systems. They also noted that reliability improvements related to inherent component stress considerations are generally ineffective in fault prevention, if the stress levels are within design limits. This contrasts Arrhenius models, in which increasingly lower stress (steady-state temperature) is believed (but is not physically proven) to provide increasingly improved reliability.

**Table 6.2 Pareto ranking of failure mechanisms of 3400 commercial VLSI-class devices from multiple sources, including manufacturing fallout, qualification, reliability monitors, and customer returns (an * indicates possible packaging or assembly-related failures) [Ghate 1992]**

| Failure mechanisms | % of failed integrated circuits |
|---|---|
| Electrical overstress and electrostatic discharge | 19.9 |
| Unresolved | 15.9 |
| Gold ball-bond failure at bond | *9.0 |
| Not verified | 6.0 |
| Gold ball-bond failure at stitch bond | *4.6 |
| Shear stress, chip surface | *3.5 |
| Corrosion, chip metallization/assembly | *3.2 |
| Dielectric fail, poly-metal, metal-metal | 3.0 |
| Oxide defect | 2.9 |
| Visible contamination | 2.7 |
| Metal short, metal open | *2.6 |
| Latch-up | 2.4 |
| Misprocessed, wafer fab-related | 2.4 |
| Chip damage, cracks/scratches | *2.4 |
| Misprogrammed | 2.0 |
| Oxide instability | 1.9 |
| Design of chip | 1.7 |
| Diffusion defect | 1.5 |
| Final test escape | 1.4 |
| Contact failure | 1.2 |
| Bond failure, nongold | *1.2 |
| Protective coating defect | 0.9 |
| Assembly, other | *0.9 |
| Polysilicon/silicide | 0.8 |
| External contamination | *0.7 |
| Others | 5.3 |

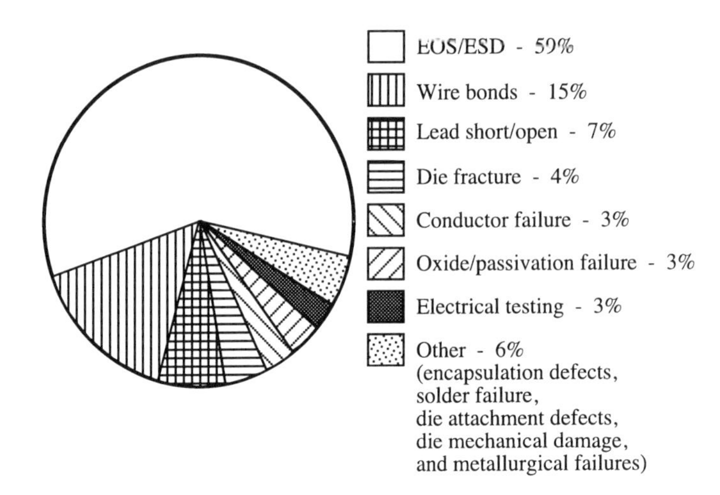


**Figure 6.20** *Distribution of failures in commercial integrated circuits components (PEMs) from a large industrial end-user [Bloomer et al. 1989]*

## 6.6 COMPARISON OF PLASTIC AND CERAMIC PACKAGE FAILURE MODES

Integrated circuit packages encapsulated with the first generation of epoxy molding compounds in the 1960s and 1970s exhibited rather poor performance when compared with their ceramic counterparts. Corrosion and thermally induced mechanical stress were two common failure modes. Ionic residues like chloride, potassium, and sodium exceeded several hundred parts per million in the early epoxies, compared with the less than 10 ppm obtained with the latest technology. The high level of contaminants, coupled with the poorer quality of the device passivation in the 60s and 70s, accounted for the low mean time to failure of plastic-encapsulated devices. Furthermore, the technology of compound stress modification was not recognized; silicone modifiers and stress absorbers were still in the conceptual stage. Today, with advances in molding compounds and device technology, plastic-encapsulated packages exhibit long-term reliability comparable to and, for many applications, better than hermetic packages [Taylor 1976, Taylor and Roberts 1978, Robock and Nguyen 1989, Nguyen and Jackson 1993 a, b].

Failure modes are similar in plastic and ceramic packages. Table 6.3 shows a pareto ranking of failures observed in hermetic packages by researchers at Texas Instruments. Comparing the pareto rankings in Tables 6.2 and 6.3 shows that electrical overstress, electrostatic discharge, corrosion, and thermomechanical

**Table 6.3 Pareto ranking of failure mechanisms of failed hermetic devices [Martin 1993]**

| Failure mechanisms | % of integrated circuits failed |
|---|---|
| Electrical overstress and electrostatic discharge | 16.4 |
| Hermeticity rejects | 6.6 |
| Not verified | 5.8 |
| Polysilicon/silicide | 5.1 |
| Pin rejects | 3.8 |
| Visible contamination | 3.2 |
| Oxide defect | 2.3 |
| Chip damage —- cracks/scratches | 2.1 |
| Metal open | 2.1 |
| Metal short | 1.9 |
| Contact failure | 1.7 |
| Passivation defect | 1.6 |
| Contamination on P.O. | 1.5 |
| Assembly —- Other | 1.4 |
| Bond failure —- nongold | 1.3 |
| External contamination | 1.0 |
| Dielectric fail, poly-metal | 0.8 |
| Waferfab-other | 0.8 |
| Misprocessed | 0.7 |
| Diffusion defect | 0.7 |
| Die-attach | 0.7 |
| Oxide instability | 0.6 |
| Package damage | 0.6 |
| Lead adhesion | 0.5 |
| Silicon defect | 0.4 |

stress are reliability concerns for both configurations. However, the relative importance of each mode varies between package types. For example, even though current epoxies are better purified and almost free of ionic residues, external contaminants entrained by absorbed moisture can still reach the die surface in contact with the epoxy. On passivated dies, corrosion often occurs at the exposed bond pads. In hermetic packages, corrosion is often due to organic contaminants from the die-attach process mixed with traces of moisture trapped within the hermetic confines of the package.

## 6.7 SUMMARY

With new package designs, materials, and architectures, many PEM failure mechanisms and modes have been eliminated, though new mechanisms can develop. A close look at various failure modes reveals that only a few failure mechanisms are responsible for most device failures. An understanding of these failure mechanisms, as applied to potential failure sites, can indicate the applications that a given design can tolerate for a desired cost and performance over time.

## REFERENCES

Beaty, R.E., Jaeger, R.C., Suhling, J.C., Johnson, R.W., and Butler, R.D. Evaluation of Piezoresistive Coefficient Variation in Silicon Stress Sensors Using a Four-Point Bending Test Fixture. *IEEE Transactions on Components, Hybrids, and Manufacturing Technology*, 15 (1983) 904-914.

Belton, D.J., Sullivan, E.A., and Molter, M.J. Moisture Transport Phenomena in Epoxies for Microelectronic Applications. *Proceedings of the ACS Symposium* 407 (1989) 286-320.

Bloomer, C., Franz, R.L., Johnson, M.J., Kent, S., Mepham, B., Smith, S., Sonnicksen, R.M., and Walker, L.S. Failure Mechanisms in Through-Hole Packages. *Electronic Materials Handbook*, 1, *Packaging*, ed. by M. L. Minges, ASM International, Materials Park, OH (1989) 969-981.

Broek, D. *Elementary Engineering Fracture Mechanics*, 4th ed., Boston: Kluwer Academic, Boston (1991).

Campisano, S.U., Costanzo, E., Scaccianoce, F., and Cristofollini, R. Growth Kinetics of the θ -Phase in AlCu Thin Film Bilayers. *Thin Solid Films* 52 (1978) 97-101.

Chen, A.S., Nguyen, L.T., and Gee, S.A. Effects of Material Interactions During Thermal Shock Testing on Integrated Circuits Package Reliability. *Proceedings of the IEEE Electronic Components and Technology Conference* (1993) 693-700.

Chiang, S.S., and Shukla, R.K. Failure Mechanism of Die Cracking Due to Imperfect Die-attachment. *Proceedings of the IEEE Electronic Components and Technology Conference* (1984) 195-202.

Ching, T.B., and Schroen, W.H. Bond Pad Structure Reliability. *Proceedings of the 26th Annual* International Reliability Physics Symposium (1988) 64-70.

Clatterbaugh, G.V., Weiner, J.A., and Charles, H.K. Gold-Aluminum Intermetallics: Ball Bond Shear Testing and Thin Film Reaction Couples. *IEEE Transactions on Components, Hybrids, and Manufacturing Technology*, CHMT-7 (1984) 349-356.

Condra, L. Private communications. ELDEC Corporation, Lynnwood, WA (1993).

Crank, J. *The Mathematics of Diffusion*, 2nd ed., Oxford University Press, New York (1975).

Cvijanovich, G.B. Conductivities and Electrolytic Properties of Adsorbed Layers of Water. *Proceedings of the NBS/RADC Workshop, Moisture Measurement Technology for Hermetic Semiconductor Devices, II* (1980) 149-164.

Dasgupta, A., and Pecht, M. Material Failure Mechanisms and Damage Models. *IEEE Transactions on Reliability* 40 (1991) 531-536.

Frear, D., Norgan, H., Burchett, S., and Lau, J. The Mechanics of Solder Alloy Interconnects, Van Nostrand Reinhold, New York (1994).

Funamizu, Y. and Watanabe, K. Interdiffusion in the Aluminum-Copper System. *Transactions of the Japan Inst. Met.* 12 (May 1971) 147-152.

Gallace, L.J., Khajezadeh, H.J., and Rose, A.S. Accelerated Reliability Evaluation of Trimetal Integrated Circuit Chips in Plastic Packages. *Proceedings of the 14th Annual* International Reliability Physics Symposium (1978) 224-228.

Gallo, A.A. Effect of Mold Compound Components on Moisture-Induced Degradation of Gold-Aluminum Bonds in Epoxy Encapsulated Devices. *Proceedings of the 28th Annual* International Reliability Physics Symposium (1990) 244-251.

Gee, S.A., Nguyen, L.T., and Akylas, V.R. Wire Bonder Characterization Using a P-N Junction-bond Pad Test Structure. *MEPPE FOCUS 91* (1991) 156-170.

Gershinskii, A.E., Formin, B.I., Cherepov, E.J., and Edelman, F.L. Investigation of Diffusion in the Cu-Al Thin Film System. *Thin Solid Films*, 42 (1977) 269-275.

Ghate, R.B. *Industrial Perspective on Reliability of VLSI Devices*, Texas Instruments (1992).

Hall, P.M. and Morabito, J.M. Diffusion Problems in Microelectronics Packaging. *Thin Solid Films* 53 (1978) 175-182.

Hall, P.M., Panousis, N.T., and Manzel, P.R. Strength of Gold Plated Copper Leads on Thin Film Circuits Under Accelerated Aging. *IEEE Transactions on Parts, Hybrids, and Packaging,* PHP-11, 3 (1975) 202-205.

Harman, G.G. Metallurgical Failure Modes of Wirebonds. *Proceedings of the 12th Annual International Reliability Physics Symposium* (1974) 131-141.

Harman, G.G. *Reliability and Yield Problems of Wire Bonding in Microelectronics*, ISHM Technical Monograph (1989).

Harris, D.O., Sire, R.A., Popelar, C.F., Kanninen, M.F., Davidson, D.L., Duncan L.B., Kallis, Hiatt, J. Microprobing. *Proceedings of the 18th Annual International Reliability Physics Symposium* (1980) 116-120.

Hirota, J., Machida, K., Okuda, T., Shimotomai, M., and Kawanaka, R. The Development of Copper Wire Bonding of Plastic Molded Semiconductor Packages. *35th Proceedings of the IEEE Electronic Components Conference* (1985) 116-121.

Horowitz, S.J., Felton, J.J., Gerry, D.J., Larry, J.R., and Rosenberg, R.M. Recent Developments in Gold Conductor Bonding Performance and Failure Mechanisms. *Solid State Technology* 22 (March 1979) 37-44.

Horsting, C.W. Purple Plague and Gold Purity. *Proceedings of the 10th Annual International Reliability Physics Symposium* (1972) 155-158.

Huettner, D.J. and Sanwald, R.C. The Effect of Cyanide Electrolysis Products on the Morphology and Ultrasonic Bondability of Gold. *Plating and Surface Finishing* (August 1972) 750-755.

Hughes, H.L. Radiation Effects in MOS VLSI Structures, *Ionizing Radiation Effects in MOS VLSI Devices*, ed. T.P. Ma and P.V. Dressendorser, Wiley, New York (1989).

Hund, T.D., and Plunkett, P.V. Improving Thermosonic Gold Ball Bond Reliability. *IEEE Transactions on Components, Hybrids, and Manufacturing Technology* CHMT-8 (1985) 446-456.

Iannuzzi, M. Bias Humidity Performance and Failure Mechanisms of Non-Hermetic Aluminum Integrated Circuits in an Environment Contaminated with Cl2. *Proceedings of the 20th Annual International Reliability Physics Symposium* (1982) 16-26.

Inayoshi, H., Nishi, K., Okikawa, S., and Wakashima, Y. Moisture-Induced Aluminum Corrosion and Stress on the Chip in Plastic-Encapsulated LSIs. *Proceedings of the 17th Annual International Reliability Physics Symposium* (1979) 113-117.

Ito, S., Kitayama, A., Tabata, H., and Suzuki, H. Development of Epoxy Encapsulants for Surface Mounted Devices. *Nitto Technology Reports* (1987) 78-82.

James, H.K. Resolution of the Gold Wire Grain Growth Failure Mechanism in Plastic-Encapsulated Microelectronic Devices. *IEEE Transactions on Components, Hybrids, and Manufacturing Technology* CHMT-3 (1980) 370-374.

Jellison, J.L. Susceptibility of Microwelds in Hybrid Microcircuits to Corrosion Degradation. *13th Annual Proceedings of the International Reliability Physics Symposium* (1975) 70-79.

Johnson, D.R. The Contribution of Plating Variables to Thermocompression Bond Quality. *Plating in the Electronic Industry Symposium* (1973) 272-282.

Joshi, K.C. and Sanwald, R.C. Annealing Behavior of Electrodeposited Gold Containing Entrapments. *Journal of Electronic Materials* 2, No. 4 (1973) 533-551.

Kale, V.S. Control of Semiconductor Failures Caused by Cratering of Bond Pads. *Proceedings of the International Microelectronics Symposium* (1979) 311-318.

Kanninen, M.F., and Popelar, C.H. *Advanced Fracture Mechanics*, Oxford University Press, New York (1985).

Kashiwabara, M., and Hattori, S. Formation of Al-Au Intermetallic Compounds and Resistance Increase for Ultrasonic Al Wire Bonding. *Review of the Electrical Communication Laboratory,* 17 (1969) 1001-1013.

van Kessel, C.G.M., Gee, S.A., and Murphy, J.J. The Quality of Die-attachment and Its Relationship to Stresses and Vertical Die-Cracking. *IEEE Transactions on Components, Hybrids, and Manufacturing Technology* CHMT-6 (1983) 414-420.

van Kessel, C.G.M., Gee, S.A., and Dale, J.R. Evaluating Fracture in Integrated Circuits With Acoustic Emission. *Acoustic Emission Testing*, 5, 2nd ed., ed. G. Harman, *American Society for Non-Destructive Testing*, 5 (1987) 370-388.

Khan, M.M., and Fatemi, H. Gold Aluminum Bond Failure Induced by Halogenated Additives in Epoxy Molding Compounds. *Proceedings of the International Symposium on Microelectronics (ISHM)* (1986) 420-427.

Kidson, G.V. Some Aspects of the Growth of Diffusion Layers in Binary Systems. *Journal of Nuclear Materials*, 3, No. 1 (1961) 21-29.

Kim, S.S. The Role of Plastic Package Adhesion in Integrated Circuits Performance. *Proceedings of the 41st IEEE Electronic Components and Technology Conference* (1991) 750-758.

Kim, S.S. Improving Plastic Package Reliability through Enhanced Mold Compound Adhesion. *IEEE International Reliability Physics Symposium Tutorial*, Topic 2 (1992) 2d.1-2d.17.

Kitano, M., Nishimura, A., Kawai, S., and Nishi, H. Analysis of Packaging During Reflow Soldering Process. *Proceedings of the 26th Annual International Reliability Physics Symposium.* (1988) 90-95.

Kitano, M., Nishimura, A., and Kawai, S. A Study of Package Cracking During the Reflow Soldering Process (1st & 2nd Reports, Strength Evaluation of the Plastic by Using Stress Singularity Theory), *Transactions of the Japan Society of Mechanical Engineers* 57, No. 90 (1991) 120-127.

Koch, T., Richling, W., Whitlock, J., and Hall, D. A Bond Failure Mechanism. *Proceedings of the 24th Annual International Reliability Physics Symposium* (1986) 55-60.

Koyama, H., Shiozaki, H., Okumura, I., Mizugashira, S., Higuchi, H., and Ajiki, T. A Bond Failure Wire Crater in a Surface Mount Device. *Proceedings of the 26th Annual International Reliability Physics Symposium* (1988) 59-63.

Kumar, V., German, M. D., and Shih, F. An Engineering Approach for Elastic-Plastic Fracture Analysis. *Electric Power Research Institute Report EPRI-NP-1931* (1981).

Lim, F.J., and Nguyen, L.T. Pressure and Temperature Measurement during the Filling Stage of the Transfer Molding Process. *MEPPE FOCUS '91* (1991) 246-270.

Liutkus, J., Nguyen, L., and Buchwalter, S. Transport Properties of Epoxy Encapsulants. *46th SPE ANTEC* (1988) 462-464.

Loo, M.C., and Su, K. Attach of Large Dice With Big Glass in Multilayer Packages. *Hybrid Circuits (UK)*, No. 11 (September 1986) 8-11.

Manzione, L.T. *Plastic Packaging of Microelectronic Devices*, Van Nostrand Reinhold, New York (1990).

Martin, S.R. Private communications. Texas Instruments (1993).

May, T.C., and Woods, M.H. A New Physical Mechanism for Soft Errors in Dynamic Memories. *Proceedings of the 16th Annual International Reliability Physics Symposium* (1978) 33-40.

McDonald, N. C., and Palmberg, P. W. Application of Auger Electron Spectroscopy for Semiconductor Technology. *IEDM* (1971) 42.

McDonald, N.C., and Riach, G.E. Thin Film Analysis for Process Evaluation. *Electronic Packaging and Production* (1973) 50-56.

Merrett, P.P. Plastic-encapsulated Device Reliability, *Plastics for Electronics*, ed. Martin T. Goosey, Elsevier Applied Science Publication Ltd., New York (1985).

Newsome, J.L., Oswald, R.G., and Rodrigues de Miranda, W. R. Metallurgical Aspects of Aluminum Wire Bonds to Gold Metallization. *14th Annual Proceedings of the IEEE Electronics Components and Technology Conference* (1976) 63-74.

Nguyen, L.T. Moisture Diffusion in Electronic Packages, II: Molded Configurations vs. Face Coatings. *46th SPE ANTEC* (1988a) 459-461.

Nguyen, L.T. Wirebond Behavior During Molding of Integrated Circuits. *Polymeric Engineering Science* 28, No. 4 (1988b) 926-943.

Nguyen, L.T. Surface Sensors for Moisture and Stress Studies, *New Characterization Techniques for Thin Polymer Films*, ed. H-M. Tong and L. T. Nguyen, Wiley, New York (1990b).

Nguyen, L.T. Reliability of Postmolded Integrated Circuits Packages. *SPE RETEC* (1991a) 182-204.

Nguyen, L.T. On Lead Finger Designs in Plastic Packages for Enhanced Pull Strength. *International Journal Microcircuits and Electronic Packaging* 15, 1 (1991b) 11-33.

Nguyen, L.T. Reactive Flow Simulation in Transfer Molding of Integrated Circuits Packages. *IEEE Electronic Components and Technology Conference* (1993).

Nguyen, L.T. and Jackson, J.A. Identifying Guidelines for Military Standardization of Plastic-Encapsulated Integrated Circuits Packages. *Solid State Technology* (1993) 39-45.

Nguyen, L.T. and Kovac, C.A. Moisture Diffusion in Electronic Packages. I. Transport within Face Coatings. *SAMPE Electronics Materials and Processes Conference* (1987) 574-589.

Nguyen, L.T. and Lim, F.J. Wire Sweep during Molding of Integrated Circuits. *IEEE Electronic Components and Technology Conference* (1990a) 777-785.

Nguyen, L.T., Gee, S.A., and Bogert, W. F. Effects of Configuration on Plastic Packages. *Journal Electronic Packaging*, Transactions of the ASME, 113 (December 1991b) 397-404.

Nguyen, L.T., Walberg, R. L., Chua, C. K., and Danker, A. S. Voids in Integrated Circuits Plastic Packages from Molding. *ASME/JSME Conference on Electronic Packaging* (1992a) 751-762.

Nguyen, L.T., Danker, A.S., Santhiran, N., and Shervin, C.R. Flow Modeling of Wire Sweep During Molding of Integrated Circuits. *ASME Winter Annual Meeting* (1992b) 27-38.

Nguyen, L.T., Lo, R.H.Y., and Belani, J.G. Molding Compound Trends in a Denser Packaging World, I: Technology Evolution. *IEEE International Electronic Manufacturing Technology Symposium* (1993a).

Nguyen, L.T., Lo, R.H.Y., Chen, A.S., Takiar, H., and Belani, J.G. Molding Compound Trends in a Denser Packaging World, II: Qualification Tests and Reliability Concerns. *Semiconductor Conference* (1993b) 243-254.

Nishimura, A., Tatemichi, A., Miura, H., and Sakamoto, T. Life Estimation for Integrated Circuits Plastic Packages Under Temperature Cycling Based on Fracture Mechanics. *IEEE Transactions on Components Hybrids, and Manufacturing Technology* 12, No. 4 (1987) 637-642.

Olsen, D.R. and James, K.L. Evaluation of the Potential Reliability Effects of Ambient Atmosphere on Aluminum-Copper Bonding in Semiconductor Products. *IEEE Transactions on Components Hybrids, and Manufacturing Technology* CHMT-7 (1984) 357-362.

Osgood, C.C. *Fatigue Design*, Pergamom Press, New York (1982).

Paris, P.C., Gomez, M.P., and Anderson, W.E. A Rational Analytical Theory of Fatigue. *The Trend in Engineering* 13 (1961) 9-14.

Paulson, W.M., and Lorigan, R.P. The Effect of Impurities on the Corrosion of Aluminum Metallization. *Proceedings of the 14th Annual International Reliability Physics Symposium* (1976) 42-47.

Pecht, M.G. *Handbook of Electronic Package Design*, Dekker, New York (1991).

Philosky, E. Purple Plague Revisited. *Proceedings of the 8th Annual International Reliability Physics Symposium* (1970a) 177-185.

Philosky, E. Intermetallic Formation in Gold-Aluminum Systems. *Solid State Electronics*, 13 (1970b) 1392-1399.

Philosky, E. Design Limits When Using Gold Aluminum Bonds. *Proceedings of the International Reliability Physics Symposium* (1971) 11-16.

Pitt, V.A., and Needes, C.R.S. Thermosonic Gold Wire Bonding to Copper Conductors. *IEEE Transactions on Components, Hybrids, and Manufacturing Technology* CHMT-5 (1982) 435-440.

Ramsey, T.H. Metallurgical Behavior of Gold Wire in Thermal Compression Bonding. *Solid State Technology* (October 1973) 43-47.

Ravi, K.V., and Philosky, E. The Structure and Mechanical Properties of Fine Diameter Aluminum-1% Si Wire. *Metallurgical Transactions* 2 (March 1971) 712-717.

Ritz, K.N., Stacy, W.T., and Broadbent, E. K. The Microstructure of Ball Bond Corrosion Failures. *Proceedings of the 25th Annual International Reliability Physics Symposium* (1987) 28-33.

Robock, P.V., and Nguyen, L.T. Plastic Packaging, *Microelectronics Packaging Handbook*, ed. R.R. Tummala and E.J. Rymaszewski, Van Nostrand Reinhold, New York (1989).

Shih, D.Y., and Ficalora, P.J. The Reduction of AuAl Intermetallic Formation and Electro-Migration in Hydrogen Environments. *Proceedings of the 16th Annual International Reliability Physics Symposium* (1978) 268-272.

Shih, C.F., and Needleman, A. Fully Plastic Crack Problems, Part I: Solutions by Penalty Method. *Journal of Applied Mechanics* 51 (March 1984) 48-56.

Shirley, G.C., and Hong, C.E.C. Optimal Acceleration of Cyclic THB Tests for Plastic-Packaged Devices. *Proceedings of the 29th Annual International Reliability Physics Symposium* (1991) 12-21.

Taylor, C.H. Just How Reliable Are Plastic-Encapsulated Semiconductors for Military Applications and How Can the Maximum Reliability Be Obtained? *Microelectronics and Reliability* 15 (1976) 131-134.

Taylor, C.H., and Roberts, B.C. Evaluation of a U.K. Specification for the Procurement of Plastic-Encapsulated Semiconductor Devices for Military Use. *Microelectronics and Reliability*, 18 (1978) 367-377.

Tu, K.N. Interdiffusion in Thin Films. *Annual Review of Material Science* 15 (1985) 147-176.

Vahle, R.W., and Hanna, R.J. *Proceedings of the International Congress on Transportation Electronics*, Society Automotive Engineering (October 1990) 225.

Wallach, E.R., and Davies, G.J. Mechanical Properties of Al-Cu Solid-Phase Welds. *Metals Technology* 4 (April 1971) 183-190.

Winchell, V.H. An Evaluation of Silicon Damage Resulting from Ultrasonic Wire Bonding. *Proceedings of the 14th Annual International Reliability Physics Symposium* (1976) 98-107.

Winchell, V.H., and Berg, H.M. Enhancing Ultrasonic Bond Development. *IEEE Transactions on Components, Hybrids, and Manufacturing Technology* CHMT-1 (1978) 211-219.

Yamaguchi, M., Nakamura, Y., Okubo, M., and Matsumoto, T. Strength and Fracture Toughness of Epoxy Resin Filled With Silica Particles. *Nitto Technology Reports* (1991) 74-81.

Yoshioka, O., Okabe, N., Nagayama, S., Yamaguchi, R., and Murakami, G. Improvement of Moisture Resistance in Plastic Encapsulants MOS-Integrated Circuits by Surface Finishing Copper Leadframe. *Proceedings of the 39th IEEE Electronic Components and Technology Conference* (1989) 464-471.

# 7

# QUALITY ASSURANCE

Quality is the degree of conformance of the product to the applicable product specifications, guidelines, and workmanship criteria. Poor quality is usually due to defective materials, out-of-control manufacturing processes, and improper handling. Poor quality has high penalties in terms of scheduled product delivery, manufacturer warranty expenses, repair costs, and the risk of losing customers.

Quality conformance for qualified products is accomplished through monitoring and control of critical parameters within the acceptable variabilities already established during qualification. Quality conformance therefore helps to increase product yield.

Quality conformance ensures that previously qualified parameters are maintained within specified tolerances through the monitoring, verification, and maintenance of materials and process parameters. Quality conformance is not intended to check the ability of the nominal product to meet usage criteria, but to ensure that all variabilities beyond a specified tolerance are identified and controlled, by eliminating any product that is out of tolerance. Parameter variability may be due to any one or a combination of the following factors:

- raw material property variability between lots and suppliers;
- variability in a manufacturing process parameter due to inaccuracies of process monitoring and control devices;
- human error and workmanship inadequacies; and
- unintended stresses (e.g., contaminants, particles, vibration) in the manufacturing environment.

Quality assurance requires the manufacturer to (1) identify potential defects using in-line process monitors and statistical process control; (2) conduct root cause analysis to determine the defects and failure mechanisms that would cause early-life failures of the product; (3) determine where in the process flow the defects arose, and implement process improvements to solve the process problems; (4) evaluate the economy of screening, and as a last resort, select screens that activate the failure mechanisms which will expose the potential defects; and (6) reduce or eliminate screening as warranted. The key ingredients (e.g., defect identification, process assessment, and screening) to accomplishing each of these steps are discussed in this Chapter.

## 7.1 DEFECTS IN PLASTIC-ENCAPSULATED MICROCIRCUITS

Defects in a plastic-encapsulated microcircuits can result from defective materials, design inconsistencies, faulty manufacturing, and improper handling. Defective materials that properties which do not meet design and manufacturing process requirements. Defects introduced by faulty manufacturing include die defects, die passivation defects, die-attach voids and poor adhesion, leadframe and die-paddle defects, wirebonding defects, and encapsulation defects, as well as defects that can be introduced during the assembly of plastic-encapsulated microcircuits on a printed wiring board, including bent lends, lack of lead coplanarity, solder joint defects, and damage to plastic-encapsulated microcircuits during the soldering process. Table 7.1 lists common latent defects in plastic-encapsulated microcircuits, along with their potential sources.

**Table 7.1 Plastic package defects and their potential sources in manufacturing**

| Defect | Potential sources |
|---|---|
| Fractured die | nonuniform die attach; improper setup of bonder; excessive force during bonding; dicing; electrical overstress in test |
| Corroded die | passivation cracks, pin holes, and delamination; inappropriate storage conditions; contamination |
| Passivation pin holes and voids | deposition parameters; viscosity-curing characteristics of spun-on passivation |
| Delaminated passivation | contamination on die |
| Deformed metallization | residual stresses in the encapsulant due to inadequate postcuring; improper die size to plastic thickness ratio |
| Wire sweep | encapsulant viscosity; flow velocity; voids and fillers in the encapsulant; poor wirebond geometry; delayed packing profile |
| Bond-pad cratering | improper bonder setup; insufficient pad metal thickness; wrong pad metal underlayer |
| Bond liftoff, shearing, and fracture | improper wirebonding parameters; contamination |
| Paddle shift | encapsulant viscosity and flow velocity; poor leadframe design |
| Package nonplanarity, warping, or bowing | die shift; die size; die-attach void or delamination; high molding stress |
| Pin holes in lead and paddle coating | improper coating deposition parameters; contamination; storage |
| Foreign inclusion | inadequate screening of encapsulant; improper molding process |
| Encapsulant voids | entrapping of air during raw material feeding; inadequate venting of the mold; high encapsulant viscosity; high moisture content in encapsulant |

**Table 7.1 (cont.)**

| Defect | Potential sources |
|---|---|
| Disbonded and delaminated regions with encapsulant | contamination or entrapped void |
| Package cracking popcorning | voids in the plastic encapsulant; excessive absorbed moisture; incorrect handling procedure; insufficient bake before reflow or vapor solder |
| Misaligned leads | improper handling or forming |
| Burrs on leadframe | improper etching; reversed blanking |
| Cracked lead | blanking parameters; defective metal sheet; improper trim and form |
| Poor solder wetting of land or lead | excessive solder temperature; contamination |
| Improper marking | high viscosity of the encapsulant and improper curing; surface contamination |

### 7.1.1 Die Defects

Die defects include metallization deformation, ionic contamination, radioactive contamination, sharp edges, spalling, and cracks. Deformation of aluminum metallization occurs because of encapsulant shrinkage upon cooling from glass transition temperature [Clatterbaugh and Charles 1989]. Damage often occurs at the die edges. Ionic contaminants may remain on the die during wafer processing because of poor cleaning operation. Alpha radiation from residual cleaning chemicals such as mineral acids can cause soft errors in memory dies [Hasnain and Ditali 1992, Ditali and Hasnain 1993]. The die may crack during die processing operations such as wafer scrubbing, slicing, and die separation. Surface scratches can occur during handling. Cracks initiated from the top surface of the die may propagate vertically to the bottom of the die causing die separation [Kessel 1983]. Edge cracks developed from die cutting damage can propagate from the corner of the die, and cause device failure by spalling cracks traversing across active circuit paths [Carlson et al. 1983].

### 7.1.2 Die Passivation Defects

Die passivation defects include pin holes, cracks, and delamination. Phosphosilicate glass passivation is inherently more susceptible to pin-hole defects than plasma-enhanced vapor-deposited silicon nitride passivation. Large differences in temperatures at which constituents (e.g., silicon oxides, metals, polysilicon, silox, and plastic encapsulant) are added, lead to a complex state of thermally induced tensile and compressive stresses in the die structure. For instance, polysilicon and aluminum metallization in the corners of the die are very susceptible to cracking due to high thermo-mechanical stresses.

### 7.1.3 Die-attach Defects

Die-attach defects include partial wetting and voids. These defects may lead to degradation in the performance of the device by decreasing heat conduction, causing a nonuniform temperature distribution (i.e., localized hot spots), and enhancing the susceptibility to delamination of the die from the die paddle, which in turn provides sites for the accumulation of contaminants and moisture. In severe cases, die cracking may occur.

In some cases where poor process control is maintained, a slight amount of die attach may be deposited on top of the die surface. With a conductive die- attach resin, unintentional layering of die attach will cause shorting due to the tight pad pitch on the silicon. Even with a nonconductive resin, the presence of the die attach can ruin the bondability of the device. This defect occurs due to "tailing" of the resin, whereby the material does not break off cleanly, leaving a trail of resin deposited on lead finger tips and the die. In this case, the viscosity of the resin either has not been formulated properly or has degraded due to long-term exposure to the external environment (e.g., extensive solvent evaporation). Remedies for this defect can be as simple as following the recommended shelf-life of the die attach to selecting the proper material for the equipment and application.

### 7.1.4 Leadframe and Die-paddle Defects

Leadframe and die-paddle defects include burrs, misaligned leads, cracked lead, unacceptable lead-to-bond clearance, slack lead, and pin holes in the protective coating. Burrs provide sites for stress concentration where cracks can initiate and cause delamination of the die-paddle and encapsulant interface. Misaligned leads, and noncoplanar leads on surface-mounted devices may cause solder joint problems during assembly of devices on circuit boards. Microscopic pin holes or contamination of the protective coating cause improper wetting of solder at the leads and possible corrosion of the leadframe base metal materials. Poor lead-forming processes can create cracks (called half-moons due to their characteristic shape) in the plastic above the plastic-leadframe interface.

### 7.1.5 Wirebonding Defects

Wirebonding defects include liftoff, shearing, and fracture of ball or stitch bonds, cratering of the bond pad on the die, and broken bond wires. Liftoff is the phenomenon caused by loss of contact between the bond and the bonding pad on the die. Liftoff and bond fracture can result during encapsulation, because of flow-induced tensile and shearing stresses. An improperly placed bond wire can reduce the contact area at the bond pad and cause wirebond liftoff.

Wirebonds can fracture or lift due to excessive Kirkendall voiding that may result from a poorly controlled temperature profile during the bonding process, especially in gold-aluminum metal systems. The presence of certain contaminants, such as bromine, also accelerate this mechanism.

Bond-pad cratering initiates from microcracks in the silicon and oxide layers under the bond pads, which may be introduced during wirebonding because of improper bonder setup or too thin a metal bond pad. Factors that influence microcracking are stresses at touchdown of the bonding head, the static force applied after touchdown, the ultrasonic energy, mechanical vibrations before or after bonding, and the hardness of the gold ball. Residual stresses occurring while cooling from cure temperatures can, if sufficiently large, shear wirebonds from die pads, sever the wire at the neck region of the bond, and induce cratering. Shear-induced cratering can result in damage to, or separation of, some portion of the silicon, glass, or other layers beneath the die-bonding pads. Complete separation of the ball from the bond pad is rare. Typically, cracks propagate through the oxide passivation layers and into the silicon itself, producing no visible damage but degrading device performance [Gallo 1990].

### 7.1.6 Encapsulation Defects

Encapsulation defects include nonuniform encapsulant thickness over the die, paddle shift, wire sweep, foreign inclusions, improper markings, voids, disbonded regions, and delaminations. Voids in the encapsulant provide space for accumulation of contaminants and moisture that can eventually degrade device performance. Voids indicate a lack of process control and may represent a design problem.

Package defects can arise due to incorrect design, handling, and/or encapsulation process parameters. Defects include loss of package planarity, extraneous metal, and popcorning of surface mount devices during the assembly process. As the package size increases, susceptibility to occurrence of these defects increases, particularly in the high-pin-count plastic leaded chip carriers and plastic quad flatpacks [ANSI/IPC-SM-786 1990].

Mishandling of the raw materials can introduce contaminants, such as sodium, while cleaning operations can introduce chlorine or other chemically active elements. In the presence of moisture, the contaminants can migrate to the die surface, shifting threshold values, corroding metallization, or deteriorating wirebond integrity.

The flame retardants required in many plastic packages often contain halogens such as bromine and chlorine. If the plastic curing process is well controlled, these potentially corrosive elements are chemically bound to the molecular chains. However, improper curing or operating the devices above the glass transition temperature may break the chemical bonds and the halogens become mobile. Once these mobile halogens reach the die surface, they can corrode the metallization.

Silica is often used as an economical filler in encapsulants. Again, if the process is not well controlled or the quality of the raw materials is not monitored, the fillers can agglomerate, forming nonuniformities in the encapsulant. These regions have higher density, irregular surface, and can form stress concentration sites at the die. Stress concentration sites can cause memory upset and can cause passivation fracture.

Paddle shift leads to an asymmetric package that causes undue stress concentration in the package under thermo-mechanical loads. The displacement of the die paddle away from the leadframe will stretch the wirebonds. This may lead to greater stresses on the ball and stitch bonds and induce overstress failure.

Wirebond sweep is caused by the viscous drag on the wirebond by the encapsulant during the molding process. Visible wire deformation occurs along the direction of encapsulant flow in the mold cavity. As a result of the drag force and any nonuniform encapsulant flow profile, broken wirebonds, flattened wire loops, and wire kinks can form.

Defects leading to delamination can occur at interfaces between encapsulant and metal such as the encapsulant and the leadframe, as well as at the die attach. Defective surfaces can cause rapid moisture ingress, provide a stress concentration site, and cause increase in thermal resistance. Failures that can be induced by delaminations include metal smear, pattern shift, open channels, passivation cracking, and fracture of the encapsulant [Vroonhoven 1993].

### 7.1.7 Solder-Joint Defects

Especially when using fine-pitch, surface-mounted, plastic-encapsulated microcircuits, the quality of the printed wiring assemblies depends on the board design, selection of components, handling and assembly techniques, solder paste selection and application procedures, and solder mask application process. The detailed parameters and potential problems due to these processing variables have been catalogued for surface-mounted, printed-wiring assemblies [Adams 1988], and the interested reader may wish to consult the books devoted to soldering [Pecht 1993, Manko 1992].

Key defects that are of concern in surface mounted PEMs include lack of device lead coplanarity, dewet land, dewet lead, nonwet land, nonwet lead, spattering of solder, presence of icicles, voids, holes, excess solder, wickers, cold solder, and rosin solder (entrapped flux) [Millard 1989].

## 7.2 SCREENING

The quality conformance task that addresses defect detection is called screening—an audit process to ensure that the product's materials and manufacturing conform to the control limits of the production processes. Screening involves both the early detection of product parameters that are out of tolerance and the precipitation of defects. Defect detection is most effective when conducted at the time the defect is created. Thus, in order to be proactive, screening should be part of the in-line manufacturing processes associated with quality control. Screening at individual process stages ensures that defects detected can be attributed to a specific manufacturing step, thereby facilitating immediate corrective action and minimizing troubleshooting and rework costs. In some cases, a defective part may be eliminated immediately, preventing additional costs from accruing on a product of poor quality.

### 7.2.1 Screen Selection

If screening must be conducted, then the preferred screens are nonstress, followed by stress screens. Nonstress screens are essentially non-contact-type tests and are used to detect defects rather than precipitate failures by applying loads. Nonstress screens include visual inspection, radiography, acoustical microscopy, electrical functional tests, and electrical parametric tests.

Stress screens expose defects by subjecting the product to electrical, mechanical, or thermal loads. The applied loads are not necessarily representative of service loads, and are often applied at an accelerated level to reduce the time-to-failure for a weak product. A stress screen may activate more than one failure mechanism at more than one site in a product.

Stress screens are further categorized as wearout screens and overstress screens based on the failure mechanism that causes failures in weak products. Wearout screens activate fatigue, diffusion, and tribological wear mechanism, whereas overstress screens cause the stress level at the defect site to exceed the local strength, leading to catastrophic failures. Wearout screens include temperature cycling and vibration. These screens lead to damage accumulation at the defect sites that eventually causes failures in weak products. A fraction of the useful life of the product is also consumed by the screens due to damage accumulation. Ensuring that the remaining useful life of the screened product meets the requirement is critical. The screen parameters should be such that failures occur only at the defect sites with minimal consumption of the useful life.

Overstress screens include bond pull test and thermal shock. An overstress screen is preferred over a wearout screen, because an overstress screen precipitates failure instantaneously and does not cause damage accumulation in a defect-free product. However, overstress screens must be very carefully implemented or they can cause product yield problems.

A defective product that will fail early in service by wearout failure mechanism may be detected by employing an overstress screen. For example, consider a defective product that has a crack in the solder joint which fails early in service due to stable crack propagation driven by environmental temperature cycle. An overstress screen constituting a mechanical shock applied to the leads can detect the lead with the cracked solder joint if the load is within the design limits, but high enough to cause unstable crack propagation in the solder joint if the crack length is larger than an allowable value.

Screens must be selected and tailored for the specific defects or failure mechanism(s) at the specific defect site(s) in order to be effective. The defects and the potential defect sites(s) depend on the product processing technology. Table 7.2 lists common screens and defects exposed by each screen.

Step-stress analysis is a common technique used to establish screening stress levels. In this procedure, progressively stronger stresses are imposed on a sample of the product. Failure analysis is conducted on the failed product to determine the cause of each failure. If the cause is due to a latent defect, and not due to overstress, the stress level is increased to the next higher level. The process is continued until overstress failures are observed. The stress level that causes overstress failures determines the upper limit of the stress and the stress for the screen is decided based on the defects that are required to be precipitated.

Another technique for establishing screen stress level is called "error seeding," whereby known measurable defects are introduced into the product. Thus, the failure mechanisms and the stresses that stimulate the defects are known. The stress level is progressively increased until all seeded defects in the sample are precipitated. The stress level is thus fixed to screen out the known defects. While simple in principle, error seeding is quite difficult and expensive in practice.

In addition to the two methods mentioned above, modeling of the failure site with the potential failure mechanism can be effectively used in screen selection. The method involves the development of models that incorporate possible manufacturing variabilities or flaws. The stresses required to precipitate the latent defects are then computed. The quantitative model can also provide a means to assess the damage introduced by the screen in defect-free products. These models often involve computer finite element analysis.

Regardless of the screens employed, the design life of the product must not be compromised. The life used up by screening can be calculated by continuously running the screen over and over again until failure occurs; the percentage life used from a single screen can then be calculated. One can also run an accelerated life test on the previously screened products and apply the appropriate acceleration model to compute the remaining service life at use conditions.

**Table 7.2 Screens and the defects they can expose**

| Screen | Defect exposed | Limitations |
|---|---|---|
| Visual inspection | package nonplanarity; surface defect, such as improper marking, passivation cracks, contamination, foreign material, burrs on leads and paddle; misaligned leads; cratering; improper bonding; lifted or broken wires; lifted or chipped die; improper die mounting; dimensional inaccuracies; encapsulant flash; corroded die; improper solder joint; board warping; metallization voids; bridging conductive paths; localized corrosion; bond intermetallics; chipped die and improper metallization | labor intensive; probability of escape increases with increasing complexity and magnification; automated inspection is often preferred to reduce human subjectivity and human error |
| Acoustic microscopy | encapsulant voids and inclusions; paddle shift; wirebond sweep; cracked die; interface cracks, delaminations, and disbonded areas | subject to operator interpretation; only scanning laser acoustic microscopy is production oriented |
| Radiography | broken lead connection; lead geometry defects in sections that cannot be seen with naked eyes; missing component part; extraneous metal; cracked lead; lead-to-bond clearance after encapsulation; slack lead; internal lead clearance; voiding configuration of die attach; contamination; passivation cracks, pin holes, and scratch | not production oriented (more of a sampling technique) functional test broken wirebond; cracked bonds; corroded metallization; metallization shift leading to open, short or degraded electrical characteristics; open channels; electrical shorts; cracked or open solder joint; circuit board net continuity; shifted electrical characteristics of traces and interconnect structures |

**Table 7.2 (cont.)**

| | | |
|---|---|---|
| Functional test | broken wirebond; cracked bonds; corroded metallization; metallization shift leading to open, short, or degraded electrical characteristics; open channels; electrical shorts; cracked or open solder joint; circuit board net continuity; shifted electrical characteristics of traces and interconnect structures | may require testing at hot and cold temperatures; test limit dependent, often must have gross failures for detection |
| Burn-in | bond liftoff; metal migration; reduced dielectric integrity; contact spiking cracked die; shorted or cracked interconnect structures; mobile contaminants particulates; bond intermetallics | generally requires functional testing; not the most efficient way to precipitate defects; many of these defects may require extremely high temperatures which can cause other unwanted problems to arise |
| Temperature cycling | cracked die; passivation cracking; encapsulant cracking; delamination of interfaces; metal smear; and wirebond liftoff | damage accumulation technique which can use up useful life |
| Thermal shock | cracked die; encapsulant fracture; delamination of interfaces | highly accelerated stress; can cause unwanted problems to arise (best as sampling technique as opposed to a screen) |
| Elevated temperature storage | premature corrosion; metallization defects; surface instabilities; surface contaminations; and wirebond intermetallics | low acceleration (see burn-in) |
| Vibration | weak solder joints; loose connectors and screws, and components that are not secured properly | not production oriented |
| Moisture resistance | contaminants in package, on the leads, or on the die; corrosion; poor lead plating | can use up useful life (best as sampling technique as opposed to a screen) |

### 7.2.2 Screen Duration

In stress screening, one may find multimodal distribution caused by a combination of early failures (due to defects) and wearout failures (due to physicochemical processes). The distribution of time-to-failure can vary in many ways. By plotting failure density along the y coordinate and time-to-failure, $T_f$, on the x coordinate, distributions such as monomodal, bimodal or multimodal can be obtained. The failure density is expressed by the relation

$$f(t) = \frac{1}{N}\frac{d[N - \bar{N}(t)]}{dt} \tag{7.1}$$

where $f(t)$ is the failure density at time $t$, $N$ is the device population at time $t=0$, and $\bar{N}(t)$ is the number of devices at time $t$.

A multimode distribution of failure density can arise when there are mixed products, designs, vendors, or manufacturing lines. Figure 7.1 illustrates a time-to-failure distribution characteristic of a mixed population of products from different vendors. The combination of the individual time-to-failure distributions results in a distribution with multiple peaks. The duration for the application of a screening stress, for a known failure distribution, should satisfy the condition that the remaining life be greater than or equal to the expected life of the product.

Figure 7.2 shows a failure distribution pattern with two peak subpopulation of devices, i.e., the area under the first two peaks, and a main population, i.e., the area under the largest peak, separated by a region of useful life which has low failure probability. A stress screen can be employed that consumes time $t_S$ at operating conditions, exposing the defective product. The useful life of the remaining devices is shortened by a time equal to $t_S$.

If the two peaks are close to each other, as shown in Figure 7.3, screening would consume a significant portion of the design life, $t_D$. In some cases, if the failure mechanisms associated with each peak can be activated independently, then screening is realistic, as long as the removal of the mechanism causing the earlier failures does not seriously reduce the useful life of the remaining devices. Screening should be conducted to precipitate the failure mechanism that causes failures of the first group so that the useful life of the remaining screened product is not reduced.

### 7.2.3 Screening Within the Process Flow

Screening is least costly at the lowest possible level of assembly or manufacture. The labor and material resources required to troubleshoot and repair a failed item

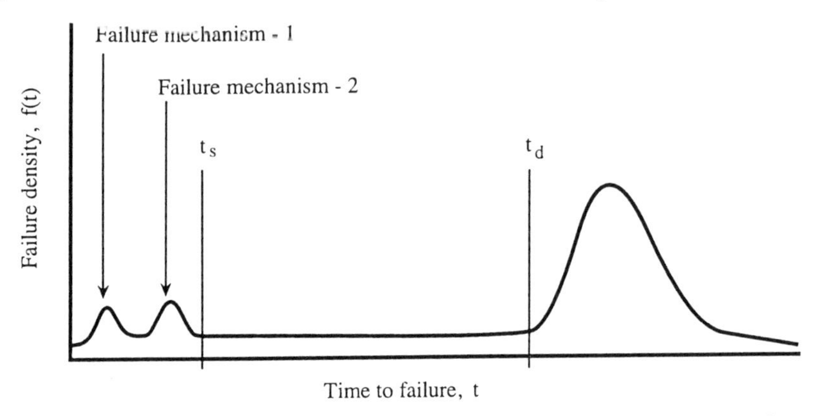


**Figure 7.1** *A time-to-failure distribution characteristic of a mixed population of products from different vendors*

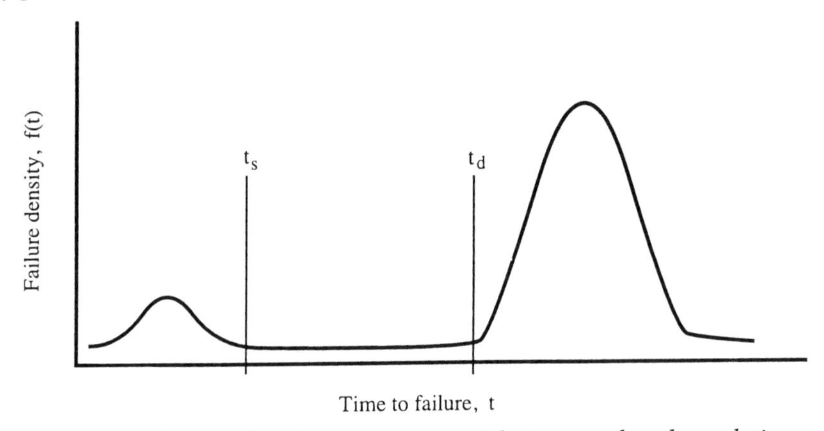


**Figure 7.2** *A failure distribution pattern with two weak subpopulation of devices*

increase as the product progresses in the manufacturing process. Process controls should be conducted at each critical manufacturing process during which defects are introduced into the product.

Screens (usually not 100% but only a sample) for product quality are often applied at various stages in the manufacturing cycle: raw materials acceptance testing, subassemblies (i.e., internal to the manufacturing process), and after the final product has been produced. Screen stations can either be on-line or off-line with respect to the product manufacturing process flow, depending on the type of screen.

Screens should weed out defects occurring between the previous screen station and the present one. For example, a wirebond pull screen after the bonding process precipitates poorly bonded interconnects. It is cost-effective to conduct

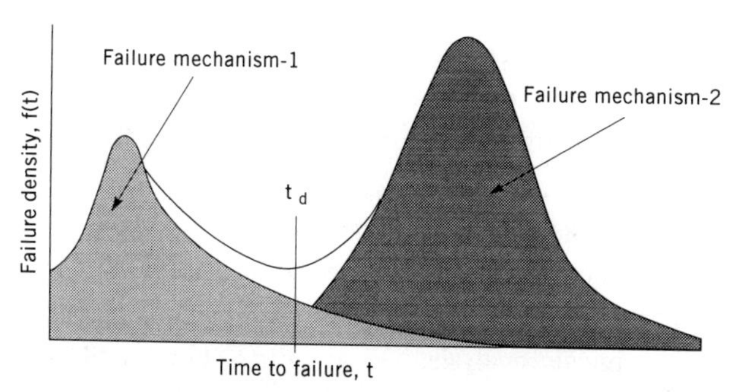


**Figure 7.3** *Failure distribution pattern with two peak subpopulation of devices close to each other*

the screen immediately after the die bonding process, but before more cost is added to the product by way of other manufacturing operations. The feedback from the screen may be used for controlling the wirebonding process. When more than one screen is required, the screens can be applied either in sequence or simultaneously depending on the failure mechanisms and the nature of the screens.

### 7.2.4 Root Cause Analysis

Failures that arise as a result of screening must be analyzed to determine their root cause. Fault tree analysis may help eliminate certain possible causes. Using simulation and controlled experiments, the most likely causes of the defects and their contributing factors are established. The major factors are in turn traced back to possible root causes in defective material, design, manufacture, handling, or testing. Manufacturing defects may be caused by process parameters that are out of the tolerance limits, unstable, not immune to noise, and by contaminants present in the environment. Once the root cause has been identified, lots of old and new material must be generated, processed through the stress, and the results statistically compared.

If almost all the products fail in a properly designed screen test, it shows that the design is incorrect. If a large number of products fail, a revision of the manufacturing processes is required. If the number of failures in a screen test is negligible, the processes should be in control and any observed faults may be beyond the control of the design and production process. At the time the process matures and screening rejects decrease, the decision to screen is economical, because it may be appropriate to replace a 100% screen for statistical process control purposes (the preferred approach). High product reliability can be assured only through the use of robust product designs, capable processes that are kept in control, and qualified components and materials from qualified supplier processes.

The failure analysis requires some knowledge of the potential failure mechanisms. Techniques that may be used to identify failure site, failure mode, and failure mechanisms include visual examination, acoustic microscopy, fractography, chemical analysis, X-ray diffraction, and surface analysis techniques. These techniques are discussed in Chapter 6.

### 7.2.5 The Economy of Screening

The decision as to when and how to screen any product is largely influenced by the economics of screening. In making this decision, the following factors must be considered: the expected level of defects existing in the product; the cost of field failures (cost of not screening); the cost of screening; the potential cost of introducing new defects by screening; and, the reduction in useful life, if any.

In the early life-cycle of a product, failure mode effects techniques should be initiated. At each process step, the possible flaws that might be added to the product are identified and statistically based controls or monitors added to the process flow. Such techniques are generally more economical and more sensitive to detecting process changes.

Screening techniques should then be reviewed for inclusion based on engineering judgment and the history from similar products. A cost-effective screening procedure addresses all potential defects, employs screens that cause minimal damage to good products, and ensures that the screened products meet the service life requirement. The time required for the individual screen must be minimized to make it cost-effective. Screens can be applied either sequentially or simultaneously depending on the latent defects targeted, available hardware, and manufacturing constraints.

Products with standard designs and relatively mature processes need to be screened only if service failure returns indicate early failures, or as a check to ensure that the processes are under control. Screening is recommended for all new products, and those that do not employ mature manufacturing processes. Screening in such cases not only improves the reliability of the new products but also assists in process control.

### 7.2.6 Reduction or Elimination of Screening

Screening carries substantial penalties in capital, operating expense, and cycle time, and its benefits diminish as a product approaches based on its design, materials, and manufacturing robustness. Furthermore, with ever decreasing acceptable defect levels, traditional screening procedures, including burn-in, are being replaced by nonstress screens, such as visual inspection, with feedback for immediate corrective action before a defective product is produced. In high-reliability, high-volume production, neither screening nor statistical process control is an economical way to ensure the required low defect levels for two reasons: defects introduced into the product during the screening procedure may be significant relative to those expected in the product due to manufacturing processes, and the sample size required for ensuring low defect levels is large.

An effective means to decide the validity of a screen is to check the exposed defects and the feedback from failure analysis. Correlation between early field failures and the failures occurring during screening, if available, can be a valid justification for continuing to screen, reducing screens, or eliminating screens. For example, if there is a very large number of failures during screening, then there is a high probability that either the screen design is incorrect or the manufacturing process is out of control. The failures may not necessarily be those that would cause early field failures. Failure analysis can help in discovering if the cause of failure was a defect or not. Conversely, if a screen fails to precipitate early failures in the field, then the screen design is deficient in stimulating the appropriate defects. If a screen is found to degrade long term product reliability, the screen is probably causing damage to the good population and must be modified. Thus, the failure data provide an effective feedback measure, making the screen design an iterative process.

The critical decision of whether or not to screen is influenced by the desired quality level, a detected problem, the technology of the product (for example analog or digital devices), its package type (plastic-encapsulated or hermetically sealed), and its application. Certain screens, for example, static burn-in, are more applicable to analog devices than digital, due to susceptibility of analog circuits to parametric shifts that affect their performance. Similarly, vibration screening of unmounted plastic devices is not as effective as it is for hermetic packages because of the susceptibility of hermetic materials to cracking, and of loose bond-wire loops to fatigue.

The higher cost of materials and processing make the additional cost of screening hermetic devices a small percentage of the overall cost. Defects in plastic-encapsulated product are often attributable to poor material quality, out-of-control manufacturing processes, and contamination. Screening for hermetic product is a cost-effective and feasible approach in attaining higher reliability. For plastic-encapsulated product, better alternatives exist, such as statistical process control and closed loop feedback control, due to the automatic assembly and high volume of the product. The alternatives clearly fall in the domain of the product supplier. For a customer to ensure defect-free product, auditing the supplier's manufacturing process flow is necessary.

Apart from product defects due to defective material, defects can be different from one manufacturer to another, and from one process to another. Suppliers are expected to employ mature, controlled processes, and to produce reliable products. Although customers currently screen some plastic parts for high reliability applications, this is not desirable, and should be eliminated as soon as possible.

Table 7.3 shows three screening categories, based on expected manufactured defect density and on defect density introduced by the screening procedure. In category 1, the defect density that can be introduced by screening is negligible compared to the expected defect density in the product, and the fraction of the cost due to screening is small compared to the cost of the product. In this category nonstress as well as stress screens can be conducted provided the reduction in life due to screening is acceptable. In category 2, the cost of screening is also a small fraction of the product cost, and the expected defect density is low enough (10 to 100 ppm) that stress screening may cause

**Table 7.3 Definition of product categories for screening.**

| Product category 1 | Product category 2 | Product category 3 |
|---|---|---|
| • expected defect density in the product is much greater than the defect density likely to be introduced by the screen, and<br>• cost of screening is an acceptable fraction of the total cost of the product | • expected defect density in the product is greater or equal to the defect density likely to be introduced by the screen, and<br>• cost of screening is an acceptable fraction of the total cost of the product | • expected defect density in the product is comparable to the defect density likely to be introduced by screen |
| • nonstress and stress screening | • nonstress screening | • no screening, use effective process control |

significantly more defects due to physical probing or contamination during the screening process. Here, only nonstress and noncontact screens are recommended. In category 3, the expected defect density in the product is so low (200 ppb to 1 ppm) that screening cost becomes a large fraction of the product cost and may not be practically possible. In this case, no screening or a benign visual screen is recommended.

## 7.3 STATISTICAL PROCESS CONTROL

Statistical process control is a methodology to analyze a process and its outputs, so as to continually reduce variation in processes and products. The goal is to stop defects from occurring in order to cost-effectively provide product meeting customer requirements.

Statistical process control involves conducting measurements on critical process parameters of the products. Control charts are plotted, and upper and lower control limits are established. The control charts reveal when the process is drifting out of control, and steps are taken to bring the process within control limits before it drifts out to some unacceptable value. The control parameter can be either a variable or an attribute. Accordingly, there are two basic types of control charts: control charts for variables (x-bar charts, range charts) and control charts for attributes (*p* charts, *c* charts). The generic steps in statistical process control include evaluation of process behavior by means of control charts, determination of process variability, and corrective actions to ensure that the process is in control. This real-time feedback can be used to stop a process before a defect occurs. Defects that can be introduced by means other than systematic drifts in a process are not the target of statistical process control.

## 7.4 SUMMARY

Early failures cause disruption both in the factory equipment and in the field. The associated costs include incoming tests, burn-in, functional tests, system tests, system installation, and field tests, with costs increasing substantially after each new stage. Defects in the plastic packages can occur at any manufacturing step, from wafer scribe to die assembly. Since the probability of failures decreases rapidly with time, screening procedures typically can weed out the marginal devices responsible for early failures. With advances in the current packaging technology, plastic-encapsulated devices can be designed to last well beyond their service life before failing. Life testing and product qualifications ensure that the plastic packages do not fail when used properly within the design ratings.

Screening has to be effective in identifying potential failures that are intrinsic either to the packages or to the manufacturing processes. Screens that consistently yield no failures serve no useful purpose, while those that induce a high percentage of dropouts can be quite costly. Temperature cycling accompanied by a gross functional and continuity test, for instance, can be effective in detecting gross mechanical defects in plastic packages. Or a screening life test can provide information on chemical diffusive processes and material mainly on the use environment and the potential cost incurred. However, as process control methodology improves continuously, a point will eventually be reached when screens may become totally unnecessary.

## REFERENCES

Adams, J. Process Control in SMT Soldering. *Surface Mount Technology* (October 1988) 37-39.

ANSI/IPC-SM-786. Recommended Procedures for Handling of Moisture Sensitive Integrated Circuits Packages. *The Institute for Interconnecting and Packaging Electronic Circuits*, Lincolnwood, IL (1990).

Blish, R.C., and Vaney, P.R. Failure Rate Model for Thin Film Cracking in Plastic Integrated Circuits. *Proceedings 41st International Reliability Physics Symposium* (1991) 22-29.

Carlson, R.O., Yerman, A.J., Burgess, J.F., and Neugebauer, C.A. Voids, Crack, and Hot Spots in Die-Attach. *Proceedings 21th International Reliability Physics Symposium* (1983) 138-141.

Clatterbaugh, G.V., and Charles, H.K. The Effect of High Temperature Intermetallic Growth on Ball Shear Induced Cratering. *Proceedings 39th IEEE Electronic Components Conference* (1989) 428-437.

Ditali, A., and Hasnain, Z. Monitoring Alpha Particle Sources During Wafer Processing. *Semiconductor International* (June 1993) 136-140.

Gallo, A.A. Effect of Mold Compound Components on Moisture-Induced Degradation of Gold-Aluminum Bonds in Epoxy Encapsulated Devices. *Proceedings 40th International Reliability Physics Symposium* (1990) 244-251.

Hasnain, Z., and Ditali, A. Building-in Reliability: Soft Errors - A Case Study. *Proceedings 30th International Reliability Physics Symposium* (1992) 276-280.

Kessel, C.G.M. van, Gee, S. A., and Murphy, J.J. The Quality of Die-Attachment and Its

Relationship to Stresses and Vertical Die-cracking. *Proceedings 33rd Electronic Components Conference* (1983) 237-244.

Manko, H.H. *Solders and Soldering.* McGraw Hill, New York (1992).

Millard, D.L. Solder Joint Inspection. *Electronic Materials Handbook Packaging* Merrill L. Minges, ASM International, Ohio (1989) 735-739.

Okikawa, S., Sakimoto, M., Tanaka, M., Sato, T., Toya, T., and Hava, Y. Stress Analysis of Passivation Film Crack for Plastic Molded LSI Caused by Thermal Stress. *Proceedings International Symposium on Test and Failure Analysis* (1983) 275-280.

Pecht, M.G. *Soldering Processes and Equipment.* Wiley, New York (1993).

Vroonhoven, J.C.W. Effects of Adhesion and Delamination on Stress Singularities in Plastic-Packaged Integrated Circuits. *Transactions of the ASME* 115 (1993) 28-33.

Zelenka, R.L. A Reliability Model for Interlayer Dielectric Cracking During Temperature Cycling. *Proceedings, 41st International Reliability Physics Symposium* (1991) 30-33.

# 8

# QUALIFICATION AND ACCELERATED TESTING

Qualification is the validation of a product's capability to function in its intended application. Qualification includes auditing the product in terms of its nominal design variables, nominal manufacturing process parameters, and nominal material characteristics. Qualification is thus the evaluation of a product's capability, and not its ongoing quality.

Successful qualification of a given PEM does not assure that all PEMs made by the same manufacturer to the same specifications will also meet the qualification requirements. Quality must be assured by other methods, such as statistical process control, in-process monitoring, and, when necessary, screening. These methods are covered in Chapter 7.

Qualification should be conducted by the manufacturer, although the customer may do so for specialized applications. Data from all possible sources should be used in qualification. These sources include material and component suppliers' test data, qualification data from similar items, and accelerated test data from subassemblies, materials, and components.

Ideally, qualification should be conducted during the development stage, using analytical tools, simulation, and prototype testing. However, schedule constraints, small production volumes, and lack of knowledge about likely application failures, can make it impossible to avoid overlaps of qualification and production.

The most appropriate qualification procedure is the one that cost-effectively gives the maximum assurance that the item will meet requirements. While formal procedures are discussed in this chapter, the emphasis is on efficient and effective procedures to assure qualified products.

## 8.1 THE QUALIFICATION PROCESS

The qualification process includes the following steps:

- Determine the aim of the specific qualification process in terms of required nominal reliability.
- Determine the environmental and operational stresses at the upper and lower design limits. Typical qualification stresses include time and spatial dependent electrostatic discharge, current, and voltage.
- Identify the likely failure mechanisms and modes (during manufacture, system assembly, transportation, storage, and service), and determine the relevant acceleration models and factors.
- Conduct tests and collect the necessary failure data to assess reliability and durability of the product. Typically this involves accelerated testing. A sample size is chosen to achieve the qualification goals; Complete electrical measurements may be required before and after each qualification test to uncover any failures.
- Interpret data, and report results and conclusions with feedback for continuous improvement.

Objectives of testing can be to evaluate the effectiveness of new materials, processes, and design; to supply routine information on the quality of a product; to develop information on the integrity of a device and its structure; and to estimate its expected service life. Qualification tests estimate expected life and design integrity of a device. They are destructive by nature. Most tests are not conducted at the application conditions, but incorporate accelerated levels of stress to accelerate failure mechanisms, often at known sites in a device.

## 8.2 TAILORING TO THE APPLICATION

Application requirements define what the product is expected to do in the expected environment during its useful life. They are usually stated in terms of nominal and tolerance ranges for electrical outputs, mechanical strength, corrosion resistance, appearance, moisture protection, duty cycle, etc. Often, there are multiple requirements, which could compete, e.g., many products are expected to be both strong and light, or thermally conductive and electrically resistive.

All of the requirements must be acceptable by both the supplier and the customer in order for reasonable qualification tests to be defined. This is often difficult, and if all the relevant requirements cannot be known, then reasonable assumptions must be made. Sometimes, the customer's requirements are not well defined, or the customer is reluctant to give assent to some assumptions. In this case, the only recourse available to the manufacturer is to make the best assumptions possible and to inform the customer. It is never acceptable to simply ignore a requirement.

A well-designed product should survive the loads applied during operation, handling, and storage. Further, for good yield during manufacturing, the subassemblies of the product should be able to survive subsequent manufacturing and assembly loads. Qualification procedures should simulate all these environmental conditions that a package is expected to be subjected to and meet its reliability, durability, and operating requirements. Of course, these procedures should be reasonable. For example, there is no logic to subjecting a PEM, which is fully moisture saturated, to wave soldering conditions, if there are specifications that this not be done in the manufacturing process.

Environmental loads define the conditions under which the product must operate. They include such conditions as mean temperature, temperature limits, number and limits of typical temperature cycles, humidity, chemical contaminants, vibration, humidity, vibration level, mechanical shocks, radiation, contaminants, and corrosive environments. The level of these loads along with net change, rate of change, and the duration of exposure are important determinants of the magnitude of stresses induced in a product. A controlled environment, such as in a telephone exchange, where the device case temperature is relatively constant and the humidity is maintained at a desired level, is considered benign for a microelectronic device compared to the loads experienced by a device mounted in a missile system that can experience severe temperature, humidity, and high levels of pyrotechnically produced vibration and mechanical shock during a launch.

In application conditions where the environment is not controlled, the load profiles of temperature, humidity, vibration, contamination, and radiation level as a function of time must be predicted based on past experience. Primary stresses experienced by devices mounted in systems operating in some typical environments are listed in Table 8.1. Worldwide temperature and humidity conditions can be found in military standard MIL-STD-210C titled "Climatic Extremes for Military Equipment." Automotive environmental conditions can be found in J1211 "Recommended Environmental Practices for Electronic Equipment Design" and in JASO D001 "General Rules of Environmental Testing Methods for Automotive Electronic Equipment. SAE Recommended Practice J1879 "General Qualification and Production Acceptance Criteria for Integrated Circuits in Automotive Applications" provides further detailed information for IC qualification testing.

Certain manufacturing process-imposed load conditions must be accounted for in the design and qualification of a product for good manufacturing yield. Some manufacturing conditions that may be of relevance and that may be considered in the design and qualification of a device include electrostatic discharges or (static electricity), soldering temperature, rate of change of temperature during soldering, duration of exposure to solder heat, flux used for soldering, cleaning agents and solvents, mechanical agitation used during cleaning operation, liquid quenching after soldering, and terminal forming operations. Often rework conditions such as repair of neighboring components or devices that include exposure to hot air, cleaning agents, and solvents are important and require explicit testing for device reliability. Storage conditions are especially important for plastic-encapsulated memory devices.

As more and more customers begin to specify reliability in terms of expected field performance, product manufacturers must develop methods to understand the products and their behavior in operating environments and to assure that they will meet the reliability requirements.

## 8.3 ACCELERATED TESTING

For most microelectronic products with minimum service lives of several years and short manufacturing lead times, it is not practical or economical to run qualification tests at normal operating conditions. For long-life and high-reliability products, test periods can become prohibitive at operating conditions. For example, to demonstrate a failure rate of 0.1% per 1000 hr with zero observed failures at a 60% confidence level would require 915,000 device-hours of operation assuming a constant failure rate, say, by testing 915 devices for 1000

**Table 8.1 Life cycle loads in key applications (the cyclic temperature swings are not the difference between the maximum and minimum temperatures that can be experienced; they are significantly smaller and move between the extremes primarily due to seasonal and geographic variations)**

| Application | Use environment | | | | | | | Manufacturing process (°C) | Storage (°C) |
|---|---|---|---|---|---|---|---|---|---|
| | Minimum temperature (°C) | Maximum temperature (°C) | Rate of temperature change (°C/hour) | Time duration for temperature cycle (hr) | Cycles/ year | Humidity | Vibration | | |
| Consumer goods | 0 | +60 | 35 | 12 | 365 | low | low | +25 to +215 | -40 to +85 |
| Computers | +15 | +60 | 20 | 2 | 1460 | high | low | +25 to +260 | -40 to +85 |
| Telecommunications | -40 | +85 | 35 | 12 | 365 | high | medium | +25 to +260 | -40 to +85 |
| Commercial aircraft | -55 | +95 | 20 | 2 | 3000 | high | high | +25 to +260 | -55 to +125 |
| Military avionics | -55 | +95 | 60 | 2 | 500 | high | very high | +25 to +260 | -60 to +70 |
| Automotive under the hood | -40 | +105 to 150 | 100 | 1 | 300 to 2200 | high | high | +25 to 215 | -40 to +150 |
| Space | -40 | +85 | 35 | 1 | 8760 | low | high (launch) | +25 to +260 | -40 to +85 |
| Missile | -65 | +125 | 100 | 1 | 1 | low | pyrotechnic shock, acoustical noise | +25 to +260 | -60 to +70 |

hr or 92 devices for 10,000 hr. Neither of these methods is economic or time effective. Tests must therefore be performed at accelerated stress conditions to compress the time to failure. Because failures are distributed in time, statistical methods are employed to analyze failure data, with the desired level of statistical confidence obtained by controlling the sample size. Sometimes a Bayes approach [Pollock 1989] is applied to continuously update failure assessment based on past data and newly acquired available information.

Accelerated testing is conducted to cause the life aging process of products to occur at a rate faster than would be obtained under normal operating conditions. Steps involved in accelerated testing include the following:

- selecting stress(es) to be accelerated;
- determining the level of the stress(es) to be applied;
- designing the test procedure, such as multiple-level acceleration or step-stress acceleration;
- extrapolating the test data to the application conditions.

Table 8.2 lists various failure mechanisms observed in PEMs and the corresponding acceleration stresses. Care must be taken in specifying the accelerated conditions so that failure modes or mechanisms are neither introduced nor removed. Excessive acceleration of a stress may trigger a failure mechanism that may be dormant at service loads. This failure mechanism shifting may provide misleading service-life predictions. Each stress may also cause multiple failure mechanisms to be accelerated with differing sensitivities. For example, temperature accelerates electromigration, ionic contamination, and surface charge spreading, but at different rates. Conversely, a particular failure mechanism may be activated by multiple stresses. For example, corrosion is accelerated by both temperature and humidity. In light of these complexities, accelerated qualification testing should not be employed without a thorough understanding of how the test correlates with service conditions.

The acceleration factor is defined as the ratio of the time or cycles necessary to obtain a stated number of failures for two different sets of stress conditions keeping the failure modes and mechanisms the same:

$$\textit{Acceleration factor} = \frac{\textit{Time to failure at normal stress}}{\textit{Time to failure at accelerated stress}} \qquad (8.1)$$

**Table 8.2 Failure mechanisms and acceleration stresses**

| Failure mechanisms | Acceleration parameters |
|---|---|
| Fatigue crack initiation | • step load or displacement<br>• thermal shock |
| Fatigue crack propagation | • cyclic load displacement or temperature |
| Diffusion | • absolute temperature<br>• concentration, gradient |
| Interdiffusion | • absolute and cyclic temperature |
| Deadhesion and delamination | • absolute temperature<br>• relative humidity<br>• contaminants |
| Corrosion | • absolute temperature<br>• relative humidity<br>• contaminants |
| Electromigration | • current density<br>• absolute temperature and temperature gradients |
| Electron-hole pair generation | • radiation dose and dose rate |
| Popcorning | • relative humidity followed by thermal shock |

In an ideal case, the acceleration factor should be computed from an acceleration transform that gives a functional relationship between the accelerated stress and the application life, based on the failure models that address observed failure mechanism(s) tailored to the failure site. Fundamental constitutive and damage properties of the materials should be used in these models to be of value in new designs, new material combinations, and for new applications.

A qualification test to demonstrate acceptable reliability must be conducted for a long enough clock or calendar time to show that the application requirements have been met. For example, if the application requirement is for a time-to-first-failure (TTFF) of 10 years, and the qualification test is conducted under conditions which accelerate the relevant failure by a factor of 20, then a statistically valid sample size must be tested for at least a half a year before the first wearout failure occurs.

A good way to design tests can be found in the book by Montgomery [1991] on the use of fractional factorial arrays used in Design of Experiments. Since failures are often related to higher level (i.e. system) design and manufacturing factors, qualification of PEM's should be conducted on samples assembled using representative assembly materials and processes, as well as representative higher level design practices. Data should be collected from sufficient device-hours and clock or calendar time to cover anticipated failure rate and useful life requirements. If the tests are designed to be comprehensive, then accelerated test models can then be used to apply the data thus obtained to a range of product requirements.

## 8.4 QUALIFICATION TESTS

When determining which tests to use for qualification, some important factors should be considered:

- There can be significant differences among manufacturers of the same part number;
- There can be significant differences among part numbers from the same manufacturer;
- Distributors methods can impact reliability;
- Printed wiring board assembly processes can impact reliability;

This section overviews some of the key qualification tests for PEMs.

The reader is referred to Chapter 6 for a more detailed discussion on PEM failure mechanisms. For more generic information on failure mechanism modeling, the interested reader is also referred to the tutorial series on failure mechanisms in the IEEE Transactions on Reliability [Dasgupta and Pecht 1991].

Often, the reliability requirements will change from one product application to another. For example, if a product is to be used in a ground application in tropical areas, then humidity-related failures may be considered most likely, and the appropriate data for qualification must be derived from accelerated humidity testing, or combined temperature and humidity. If the same product is to be used in a space satellite with a low earth orbit, humidity is not an issue, unless low humidity storage is a potential problem.

### 8.4.1 Steady-State Temperature Tests

***High-temperature storage test.*** This test accelerates temperature-induced failures such as interdiffusion, Kirkendall voiding, and depolymerization. In electrically programmable read only memory (EPROM) devices, the test accelerates charge loss from floating gates in the device which is limited by gate oxide layer defects. Devices are stored in a controlled elevated temperature (typically around 150°C) for extended times (of more than 1000 hr) without electrical bias. Interim electrical parametric measurements and final measurements are conducted at the conclusion of the test. The electrical measurements include contact test, parametric shifts, and at-speed functional tests. Damage, such as package cracking, junction thermal resistance increase, or depolymerization, may also be considered a failure.

***High-temperature operation test.*** This test evaluates the capability to withstand maximum power operation at elevated temperatures. Electrical configurations include steady-state reverse bias, forward bias, or a combination of the two. Electrical frequency is set to maximum operating design level or to the limit of the test equipment. The test is usually conducted at a junction temperature, $T_j$, below the glass transition temperature of the plastic encapsulant. A simple model to describe junction temperature with respect to the case and ambient temperature is

$$T_j = T_a + (\theta_{jc} + \theta_{ca})P \tag{8.2}$$

where $\theta_{ca}$ is the case-to-ambient thermal impedance, which depends on the system equipment, whether it is forced air or still air cooling, $\theta_{JC}$ is the junction-to case thermal impedance, which depends on the thermal impedance of the plastic-encapsulant package, and $P$ is the power.

### 8.4.2 Thermal Cycling Tests

This section discusses temperature cycling, thermal shock, and combined power and temperature cycling tests. For PEMs, the high-temperature limit should be kept below the glass transition temperature of the molding compound.

***Temperature cycling.*** The temperature cycling test consists of the application of some temperature variation of a specified amplitude about a mean value. Temperature is usually varied at a fixed rate followed by a dwell period, subjecting interfaces of dissimilar materials in a device to mechanical fatigue.

In a plastic-encapsulated microelectronic device, test results are affected by encapsulant thickness, die size, die passivation integrity, wirebond, die cracks, and adhesion at the interfaces, including passivation to encapsulant, die pad to encapsulant, and leadfingers to encapsulant. For a device interconnected to a substrate (or circuit board), the fatigue endurance of the device-to-board interconnection structure can also be evaluated.

Temperature cycling is conducted in an environmental chamber equipped with a temperature-control device, a heating unit, and a cryogenic cooling unit with sufficient thermal capacity so that the sample can be heated and cooled within the specified time span in dry, flowing air. The dwell time at each extreme is the minimum needed to establish thermal equilibrium with the sample load, and sufficient stress relaxation (if this is a key parameter of the failure mechanism of concern). Poststress examination includes electrical parametric and functional tests and inspection for mechanical damage such as package cracking.

***Thermal shock.*** This test is conducted to validate unit integrity under extreme temperature gradients. Sharp temperature gradients can cause die cracking, delamination, die-passivation cracking, deformation of interconnections, and encapsulant cracking. The shock is applied by cyclic immersion of specimens in suitable fluids maintained at specified temperatures. The fluids are chemically inert, stable at high temperatures, nontoxic, noncombustible, of low viscosity, and compatible with package materials. Testing is concluded after the desired number of cycles has been completed, generally with the final immersion being in the "highest-stress," cold bath. Specimens are warmed to room temperature prior to measurements. End tests include electrical measurements and inspection for mechanical damage.

***Power and temperature cycling.*** The purpose of this test is to expose samples to worst-case temperature conditions during operation. The failures are generally those that are observed in a temperature cycling test. The test chamber is similar to that used for temperature cycling, except that mounting sockets with electrical feed-throughs are provided for the application of bias to test devices. Power may be cycled on and off, with or without synchronization with the temperature cycle. Electrical measurements to check for functionality and parametric limits, and visual inspection to detect mechanical damage should be conducted. Standard test conditions for this test are specified in JESD-22, method A105.

The severity of the test may be influenced by the temperature at which residual stresses are zero, usually governed by the differential between the glass transition temperature and the temperature limit. In a temperature cycling test, the temperature limit that is farther from the glass transition temperature should control the severity of the test. The glass transition temperature of encapsulants lies in the 150 to 180 °C range; therefore, the lower temperature limit of a temperature cycle test dictates the severity of the test. The distribution and magnitude of residual stresses can be altered by humidity-induced swelling of the encapsulant.

Power cycling can change the amount of moisture adjacent to the die, and thereby influences test severity. In PEMs, the influence is enhanced by the low thermal conductivity of the plastic, which increases the die temperature under the electrical bias. The die temperature is highest when the power is applied continuously, accelerating failures related to the die and reducing failures related to the plastic packaging. If the power ON/OFF cycle time is such that moisture escapes the plastic during the ON period, and the OFF period is shorter than it takes for moisture to ingress to the die-encapsulant interface, then it is expected that the total number of failures would be decreased, due to reduced humidity-related package failures during the ON state, and reduced bias-related failures during the OFF state [Shirley and Hong 1991]. The optimum failure acceleration due to the ON/OFF cycle time depends on the package.

Since a successful test is usually one in which no failures occur, there is always the question as to whether or not appropriate stresses, and levels thereof, have been applied to adequately test the product. It is also unknown whether the product barely passed, in which case there is a question regarding the reliability margin; or whether it would have passed a much more stressful test, in which case the product may be overdesigned. There is also the question of relating the results of tests to service conditions and performance. For this reason, at least some customers are placing more emphasis on achieved field reliability, and leaving it to the supplier to develop and implement an acceptable plan.

### 8.4.3 Tests That Include Humidity

This section discusses the following tests that include humidity: autoclave, combined temperature-humidity-bias, highly accelerated stress testing (HAST), and combined temperature-humidity-voltage cycling. A test to classify moisture sensitivity is also presented.

***Autoclave test.*** This test is also known as a "pressure cooker" test to evaluate moisture resistance, and is the simplest of the accelerated humidity tests. Typically, specimens are stored in saturated steam, at a pressure of 103±7 kPa (15±1 psig) and 121°C in a sealed autoclave containing deionized water. The devices are suspended at a minimum height of 1 cm above the initial water level in the chamber.

Severe conditions of pressure and temperature, not typical of actual operating environments, can be used to accelerate moisture penetration through the package. Galvanic corrosion is the major failure mechanism. Ionic contaminants in the encapsulant and phosphorus in the passivation are factors that accelerate corrosion in PEMs. However, test chamber contaminants can produce spurious failures, which are evidenced by external degradation of a package such as corroded device terminals/leads or the formation of conducting matter between the terminals.

***Temperature-humidity-bias stress tests.*** The temperature-humidity-bias test is one of the most commonly applied accelerated tests. Its purpose is to test for moisture ingress and corrosion of leads, bond pads, and metallization, with attendant high leakage current in the integrated circuits. Specimens are subject to a constant temperature and constant elevated relative humidity under electrical bias. Depending on the device type, bias may be applied either constantly or intermittently. Applying a continuous DC voltage to low power complementary metal-oxide semiconductor (CMOS) devices maximizes the chance of formation of electrolytic cells, because individual device power dissipation is minimized. Voltage cycling is applied to high-power devices, because continuous electrical power, dissipated as heat, can drive away the moisture needed for electrolytic corrosion. The cycled conditions must be optimized for individual device types to induce a maximum opportunity for failure.

This test is usually conducted at 85 ± 2 °C with relative humidity of 85±5% at a maximum rated operating voltage. Electrical measurements are usually conducted upon removal of the test devices. In some cases, samples may be dried before testing. The failure criteria are parametric shifts outside of specific limits, nonconformal functionality under specified nominal and worst-case conditions, or any other performance irregularities.

***Highly accelerated temperature and humidity stress test.*** The highly accelerated stress test (HAST) consists of elevated temperature and humidity, and can be coupled with electrically biasing with controlled current. The test can cause corrosion of metallization and bond pads delamination of interfaces, increased intermetallic growth, wirebond failures, and reduction in isolation resistance.

In a typical HAST test, unsaturated steam with constant relative humidity in the range of 50 to less than 100% at a constant temperature, usually greater than 100°C, is used. Details of the test method are described by Gunn et al. [1983] and Danielson et al. [1989].

***Cycled temperature, with constant humidity and bias.*** In this test, a device is subjected to thermal cycling under a high level of moisture with electrical bias. The failure mechanisms targeted by this test include electrolytic as well as galvanic corrosion, deadhesion, delamination, and crack propagation. The potential failure sites include interfaces between the leadfingers and the encapsulant, ball-bond, stitch-bond, bond-pads, and metallization corrosion.

This test is conducted in an environmental chamber capable of maintaining a controlled relative humidity level and a heat-cool cycle while electrically biasing the test units. Deionized water is used in the chamber as a moisture source. Typically, test units are subjected to temperature cycles of 30 to 65°C with heating and cooling times of 4 hr each and a dwell time of 8 hr at each temperature extreme. Relative humidity is generally maintained at a constant value in the range of 90 to 98%. A chart-type recorder with suitable chamber monitoring instrumentation is provided for continuous recording of chamber temperature and relative humidity.

***Test to classify moisture sensitivity level for surface-mount devices.*** Molding compounds, used for manufacturing surface-mount plastic electronic packages, which absorb moisture, can result in cracking of the plastic material and/or its delamination from the chip or the leadframe when a package is exposed to high temperature during reflow soldering. The failure mechanism, called popcorning, represents the worst-case situation that sometimes takes place because of unfavorable combinations of high moisture content, elevated thermal stress, and package design features, leading to low structural strength and propensity for crack initiation and propagation. The reader is referred to Chapter 4 for more information.

There are three possible moisture sensitivity levels for surface-mount plastic integrated circuits:

- moisture insensitive: there is no need to keep the integrated circuits in dry pack;
- moisture sensitive: the integrated circuits must be stored in dry pack until the dry pack is opened in a factory environment of 30°C (max) at 60%RH (max), where it can be left for unlimited time prior to PCB assembly and soldering;
- moisture sensitive: the integrated circuits must be stored in dry pack until the dry bag is opened in a factory environment of 30°C (max) at 60%RH (max), where it can be left for a maximum of one week prior to PCB assembly and soldering.

To classify surface mount packages into these different levels, PEMs are subjected to the following moisture preconditioning levels: (1) 85°C/85%RH/168 hr, (2) 85°C/60%RH/168 hr, and (3) 30°C/60%RH/168 hr. After moisture preconditioning, the packages are then exposed to PCB assembly simulation sequences test. The assembly simulation test consists of infrared (IR) reflows at a typical temperature of about 220 to 240°C and a number of chemical exposures (cleaning). At the conclusion of this test, the packages are visually inspected under a 40X microscope for external cracks. The C-mode scanning acoustic microscopy (C-SAM) can also be used to observe any internal surface delamination or cracks in the devices. Complete electrical measurements must then be carried out at the appropriate points.

Subsequently, the packages are subjected to the temperature-humidity-bias stress (THB) test. The purpose of this test is to accelerate moisture ingress, as well as the effect of corrosion on metallizations.

### 8.4.4 Solvent Resistance Test

This test evaluates the capability of samples to withstand detrimental effects such as swelling, cracking, deadhesion of the encapsulant, and corrosion of the leads caused by chemicals used in assembly processes. Chemicals included in this test are those used in the solder reflow and flux cleaning, such as fluxes and solvents. A procedure for a solvent resistance test proposed by Lin and Wong [1991] for PEM assemblies is presented in Table 8.3.

### 8.4.5 Salt Atmosphere

This test evaluates corrosion resistance of leadframes exposed to seacoast environments. Failure mechanisms that can be accelerated include electrochemical degradation, such as pitting, pinholes, blistering, and flaking of the package leads. This test is usually conducted following the lead-bending operation.

Specimens are placed in an environmental chamber in a flowing salt fog at 35°C for a specified duration. The salt fog is generated by atomizing a 0.5 to 3.0 % by weight sodium chloride solution with compressed air to create a mass flux of 7 to 35 $mg/m^2/min$. ($0.8 \times 10^{-3}$ to $4 \times 10^{-3}$ $oz/ft^2/min$.). The pH of the solution is maintained between 6.0 and 7.5. Care is taken to avoid the use of a chamber contaminated by prior tests. The specimens are placed in the chamber to achieve maximum exposure to the salt fog. Test duration is typically 96 hr.

**Table 8.3 Encapsulant chemical resistance test procedure [Lin et al. 1991]**

| Chemical | Exposure | Cleaning | Failure criteria |
|---|---|---|---|
| Alpha-100, EC-7, and 1,1,1-trichloroethane | 2-hr soak in each chemical in sequence | soak in isopropanol for 10 min; rinse in running distilled water for 15 min; dry in a clean oven at 120 °C for one hr | examine under microscope for swelling, crack, deadhesion, and corrosion |
| Polyalphaolefin | 96-hr soak | | |

### 8.4.6 Flammability and Oxygen Index Test

These tests evaluate flammability of mold compounds. Flammability is the property of a material whereby flaming combustion is prevented, terminated, or inhibited following application of a flaming or nonflaming source of ignition, with or without subsequent removal of the source. Underwriters Laboratories material flammability standard UL-STD-94 assigns a flame-retardant grade to a material based on its burning rate. The various grades are HB, V-0, V-1, V-2, and 5V, where HB indicates the highest and 5V the lowest burning rate.

A numerical value to flammability is assigned by the oxygen index. The oxygen index is the percent oxygen in an oxygen-nitrogen mixture that will just sustain combustion of a material. The oxygen index is obtained in compliance with test method ASTM D-2863. Decreasing flammability is indicated by a higher oxygen index. When flammability and oxygen index tests are not feasible, the EEC "glow wire" and "needle flame" tests may be used. Flammability of encapsulants can be reduced significantly by compounding with halogenated compounds, phosphate esters, and antimony trioxide.

### 8.4.7 Solderability

This test evaluates the susceptibility of leads on a device to solder wetting. The wettability of a metal surface depends on the integrity of the corrosion resistant coating, contamination-free surface, solder temperature, specific heat of the lead material, and lead design [Davy 1990].

The technologies used in solderability assessment tests are either dip-and-look or wetting balance. In a dip-and-look approach the surface to be soldered is dipped into a molten solder bath. The extent of surface coverage upon removal is visually judged in order to estimate solderability of the surface. The apparent enhancement of solderability due to solder freezing upon removal from the solder bath and undetected microscopic dewetting are of concern with dip-and-look tests.

A wetting balance test measures the force of attraction between the surface and the molten solder with time. It is insensitive to dewetting of solder, because dewetting begins only upon removal of the surface from the molten solder. The criteria for product acceptance have yet to be standardized.

### 8.4.8 Radiation Hardness

This test is used only in qualification of devices intended for applications such as space missions, where the PEM must withstand γ-rays, cosmic rays, X-rays, α-particle radiation, and β radiation. The radiation hardness test exposes a device to a specified total ionizing dose of radiation followed by parametric electrical tests.

Radiation-caused failure mechanisms are electron pair generation and lattice displacement. Traces of radioactive elements in package materials, such as uranium and thorium from inorganic fillers, are intrinsic sources of ionizing radiation that can cause errors in programmable devices. In logic and memory devices, radiation exposure can cause soft errors in dynamic memory devices, electrical parametric upsets, and latchup.

## 8.5 CONTINUOUS QUALIFICATION

Continuous qualification, or requalification, focuses on weaknesses that are likely to be introduced in a previously qualified product when certain changes are made. The changes that require requalification are those that affect product performance, reliability, length of service life, functionality, manufacturing, and assembly, in some unknown manner. Examples of changes in PEMs that could require requalification tests are presented in Table 8.4.

## 8.6 INDUSTRY PRACTICES

Manufacturers often formulate their own qualification procedures in keeping with market practice and customer specifications. As an example, Appendix A at the end of this chapter, is part of the U.S. Automotive Electronic Council document on stress test qualification for automotive-grade integrated circuits.

A wide variety of accelerated tests, involving various stress levels and combinations of stresses, are used by device manufactures. Table 8.5 provides a listing of typical stresses, levels, and durations [Nguyen et al. 1993, Motorola 1993, Intel 1989]. However, the test conditions are often dictated by customer

**Table 8.4 Requalification requirements for changes to a qualified plastic-packaged microelectronic device (X denotes tests required for changes to a qualified product)**

| | Area of change | | | | | | | | |
|---|---|---|---|---|---|---|---|---|---|
| Test to uncover potential failure mechanisms due to change | Die overcoat or preparation | Die-attach assembly | Molding compound | Leadframe material or dimensions | Lead plating or finish | Wirebond assembly | Device size | Encapsulation process | Package marking |
| Visual | | | X | X | X | | | X | X |
| Autoclave | | | X | | X | | X | X | |
| Temperature cycle | X | X | X | X | X | X | X | X | |
| Highly accelerated stress test | X | X | X | X | | X | | | |
| Solder heat (surface mount only) | X | | X | X | X | | X | X | |
| Solvent resistance | | | X | | | | | X | |
| Solderability | | | | | X | | | | |
| Salt atmosphere (for marine applications) | X | | X | | X | | | | |
| Lead integrity | | | | X | | | | X | |
| Wirebond strength | | | | | | X | X | X | |

requirements, which are not derived from physics of failure. Harsh stress levels are imposed during testing, based on the rationale that 'if the packages last longer under more stringent conditions, then the parts must be of better quality' [Nguyen et al. 1993]. However, this approach fails to consider the fact that subjecting the devices to harsh stresses may induce other failures, or may not be cost-effective.

To demonstrate the device capabilities in order to meet customer requirements, tests must be chosen that will accelerate time-to-failure in a predictable and understandable manner. The stress conditions must be selected in such a way that failures at accelerated test are those that will be expected to occur at normal device operation.

Whenever a change in material, package technology, or process technology is introduced, each test must be reevaluated to determinate its effectiveness in uncovering potential weaknesses.

The appropriate approach is to first identify the "root causes of failure" for the devices and develop qualification tests that focus on those particular causes. This is the essence of the physics-of-failure approach to qualification, which, by testing for the primary cause behind a failure, precludes the possibility of failures occurring. When qualification testing results in unexpected failures, then further investigation is necessary to determine the root causes behind them.

Table 8.6 lists various test methods that may be used to detect the failures. Some of these tests are applicable only to specific use environments and manufacturing conditions. For example, solder heat resistance evaluates a package's ability to resist "popcorning" during solder reflow operation for surface-mount devices. The radiation hardness test is suggested only for devices (i.e. memory) that are susceptible to and are expected to be exposed to either extrinsic or intrinsic radiation.

## 8.7 SUMMARY

Not all manufacturers and assemblers of PEM are the same: they use different encapsulants and additives, leadframes, and die passivation materials, and different assembly processes and materials. Manufacturers of PEM must implement qualification procedures tailored to evaluate and monitor the capability of their product to meet desired service life in the expected applications procedures to assure that products made with PEM are reliable.

PEM differ from hermetic devices in circuit board assembly processes and materials, electronic equipment designs, and environmental operating conditions. Qualification and process control procedures must be developed, taking into account the features of PEM that are different from hermetic devices.

**Table 8.5 Example test conditions used in industry [Nguyen 1993, Motorola 1993, Intel 1989]**

| Test | Motorola | Intel | Texas Instruments | Signetics | Micron | American Micro Devices | National Semiconductor |
|---|---|---|---|---|---|---|---|
| Temperature cycle | -65 to 150 °C, 1000 cycles | -55 to 125°C, 500 cycles | -65 to 150 °C, 500 cycles | -65 to 150 °C, 500 cycles | -65 to 150 °C, 1000 cycles | -65 to 150 °C, 1000 cycles | -65 to 150 °C, 1000 cycles |
| Autoclave | 121 °C, 15 psig, 96 hr | 121 °C, 96 hr | 121 °C, 15 psig, 240 hr | 127 °C, 20 psig, 336 hr | 121 °C, 15 psig, 96 hr | 121 °C, 15 psig, 168 hr | 121 °C, 15 psig, 500 hr |
| Temperature and humidity with bias | 85 °C, 85% RH, 1008 hr | 85 °C, 85% RH | 85 °C, 85% RH, 1000 hr | 85 °C, 85% RH, 2000 hr | 85 °C, 85% RH, 1000 hr | 85 °C, 85% RH, 2000 hr | 85 °C, 85% RH, 1000 hr |
| Operating life | Max. rated= 1,008 hrs. | | 125 °C, 1000 hr | 150 °C, 2000 hr | 150 °C, 1000 hr | 125 °C, 168 hr | 125 °C, 1000 hr |
| High temperature storage | | 200 °C, 48 hr | | 175 °C, 2000 hr | 150 °C, 1008 hr | 125 °C, 2000 hr | 150 °C, 1000 hr |
| Solder heat resistance | | | 260 °C, 10 sec | | | | 260 °C, 12 sec |
| Low temperature life | | | | -10 °C, 1000 hr | -10 °C, 1008 hr | | -40 °C, 1000 hr |

**Table 8.6 Possible test methods and conditions for tests used in qualification of PEM's**

| Test | Test conditions | Simulated environment | Applicable standards |
|---|---|---|---|
| Low-temperature operating life | -10°C / $V_{cc}$ max / max frequency / min 1000 device hr/ outputs loaded to draw rated current | field operation in sub-zero environment | JEDEC-STD-22 TM-A 106 |
| High-temperature operating life | +125 or 150°C / $V_{cc}$ max / max frequency / min 1000 device hr / outputs loaded to draw rated current | field operation in normal environment | MIL-STD-883 method 1005 |
| Temperature cycling | 500 cycles, -65°C to +150°C at a ramp rate of 25°C/min and with 20 min dwell at each temperature extreme | day-night, seasonal, and other changes in environment temperature | MIL-STD-883 method 1010 C |
| Temperature cycling, humidity, and bias | 60 cycles, $V_{cc}$ ON/OFF at 5-min interval, 95% RH, +30 to +65°C with heating and cooling time of 4 hr each and a dwell of 8 hr at each temperature extreme | slow changes in environment conditions while device is operating | JEDEC-STD-22 TM-A 104 |
| Power and temperature cycling | only on devices experiencing rise in junction temperature greater than 20°C; min 1000 cycles of -40 to +125°C | changes in environment temperature while device is operating | JEDEC-STD-22 TM-A 106 |
| Thermal shock | 500 cycles of -55 to +125°C | rapid change in field or handling environment | MIL-STD-883 method 1011B |
| High-temperature storage | 150°C, for 1000 hr min | storage | MIL-STD-883 method 1008 |
| Temperature humidity bias | $V_{cc}$ max / 85°C / 85% RH / 1000 hr min | operation in high-humidity environment | JEDEC-STD-22 TM-A 108 |

**Table 8.6 (cont.)**

| Test | Test conditions | Simulated environment | Applicable standards |
|---|---|---|---|
| Highly accelerated stress test | $V_{cc}$ max / 130°C / 85% RH / 240 hr min | operation in high-humidity environment | JEDEC-STD-22 TM-A 110 |
| Pressure cooker or autoclave | precondition for surface mount devices, 15 psig / 121°C / 100% RH / 240 hr | high-humidity environment; moisture ingress through cracks | MIL-STD-883 method 1005 JEDEC-STD-22 |
| Lead integrity | as appropriate for package type and lead configuration | manufacturing environment | MIL-STD-883 method 2004, JEDEC-STD-22 TM-B 105 |
| Bond strength | as appropriate for wirebond or lead configuration | manufacturing environment | MIL-STD-883 method 2011C/D |
| Die shear | consistent with the device specifications | characterization of die-encapsulant interface | MIL-STD-883 method 2019 |
| Mechanical shock | 5 shock pulses of g-level and duration as per device specification along a particular axis | avionics or spacecraft launch environment | MIL-STD-2002 |
| Vibration variable frequency | logarithmic variation from 20 to 2000 Hz and back to 20 Hz in a time duration of more than 4 min; 4 times along each applicable axis | avionics or spacecraft launch environment | MIL-STD-2007 |
| Flammability | with or without removal of ignition source | characterization of encapsulant flammability | UL-94-V0 or V1 |

**Table 8.6 (cont.)**

| Test | Test conditions | Simulated environment | Applicable standards |
|---|---|---|---|
| Solvent resistance | postsolder cleaning agents; example: 2 hr soak in trichloroethane, 10 min in isopropanol, DI 15 min, dry at 120°C for 1 hr, e.g., solder flux, flux cleaning solvents, and liquid coolant | cleaning; assembly environment | MIL-STD-883 method 2015 |
| Electrostatic discharge sensitivity test | human body model: 1 A (for a 1500 V) exponentially decreasing with time constant 300-400 nsec. OR charged device with model: 1500 $V_{rms}$, 15 A, 4-5 oscillations in 15 nsec | handling; damage by electrostatic discharge | MIL-STD-883 method 3015 |
| Oxygen index | sustained combustion of the material | characterization of encapsulant flammability | ASTM D-2863 |
| Radiation hardness test (RA) | Co-60 and alpha irradiation under use temperature and bias for total dose and dose rate; specified total ionizing dose of radiation | radiation environment space and high-radiation environment | - |
| Solder heat resistance | 260 ± 10°C for 10 sec | vapor phase or reflow solder heat | JEDEC-STD-22 TM-B 100 |
| Solderability | dip and look or wetting balance | characterize solderability as manufactured or after storage | MIL-STD-883 method 2003 |
| Latchup | appropriate charging voltage | Voltage excursion | MIL-STD-883 METHOD 1020 |

MIL-STD-883 refers to a military standard, "Test Methods and Procedures for Microelectronics."
JEDEC refers to EIA JEDEC Standard No. 22, "Test Methods and Procedures for Solid State Devices Used in Transportation / Automotive Applications."
ASTM refers to American Society for Testing Materials.

# APPENDIX

## 1 GENERAL REQUIREMENTS FOR AUTOMOTIVE QUALIFICATION

### 1.1 Objective

The objective of this qualification program is to ensure that the device to be qualified meets a minimum set of automotive-grade qualification requirements.

### 1.2 Test samples

**Lot requirements.** Test samples shall consist of a representative device from the qualification family. Where multiple lot samples are specified in Table 1, test samples must be composed of approximately equal numbers from three non-consecutive wafer lots from different molding operations in three different weeks of production. If manufactured on the same line, three different shifts of production (morning, afternoon, evening) is also acceptable.

**Production requirements.** All qualification parts shall be produced on tooling and processes at the manufacturing site that will be used to support part deliveries at projected production volumes.

**Reusability of test samples.** Devices that have been used for nondestructive qualification tests may be used to populate other qualification tests. Devices that have been used for destructive qualification tests may not be used any further except for engineering analysis.

**Sample size requirements.** Sample sizes used for qualification testing and/or generic data submission must be consistent with the specified minimum sample sizes and acceptance criteria in Table 1. Acceptance criteria for larger generic sample sizes can be calculated using the method in Section 1.4. All generic data applicable to the device to be qualified must be submitted.

**Pre and poststress test requirements.** All endpoint test temperatures (room, hot, and/or cold) are specified in the "Additional Requirements" column of Table 2 for each test. The specific value of temperature must address the worst case temperature extremes and designed product life for the application for at least one lot. For example, if the supplier designs a device intended solely for use in a benign environment (-40°C to 85°C), the endpoint test temperature extremes need only address those application limits. Qualification to applications in harsher environments (-40°C to 125°C) will require testing of at least one lot using these additional endpoint test temperature extremes.

**Table 1 Qualification test definitions**

| Stress | Abbreviation | No. | Note | Sample size per lot | Number of lots | Accept on no. failed |
|---|---|---|---|---|---|---|
| Pre-and post-stress electrical test | TST | 1 | HPNG | All qualification parts submitted testing | | 0 |
| High temperature operating life | HTOL | 2 | HPDG | 77 | 3 | 0 |
| High temperature bake | HTB | 3 | HPDG | 77 | 1 | 0 |
| Preconditioning | PC | 4 | PSNG | All surface-mount qualification parts to be subjected to THB, TC, AC, PTC | | 0 |
| Temperature humidity bias | THB | 5 | PDG | 77 | 3 | 0 |
| Autoclave | AC | 6 | PDG | 77 | 3 | 0 |
| Temperature cycling | TC | 7 | HPDG | 77 | 3 | 0 |
| Power temperature cycling | PTC | 8 | HPDG | 77 | 1 | 0 |
| Mechanical shock | MS | 9 | HDG | 39 | 3 | 0 |
| Vibration variable frequency | VVF | 10 | HD | Performed as a sequential test for mechanical integrity of hermetic packaged devices | | |
| Constant acceleration | CA | 11 | HD | | | |
| Gross/fine leak | GFL | 12 | HD | | | |
| External visual | EV | 13 | HPNG | All qualification parts submitted for testing | | 0 |

**Table 1 (cont.)**

| Stress | Abbreviation | No. | Note | Sample size per lot | Number of lots | Accept on number failed |
|---|---|---|---|---|---|---|
| Physical dimensions | PD | 14 | HPD | 30 | 1 | $P_{pk} \geq 1.66$ or $Cpk \geq 1.33$ |
| Lead integrity | LI | 15 | HPD | 45 leads from a min. of 5 devices | 1 | 0 |
| Lid torque | LT | 16 | HD | 5 | 1 | 0 |
| Bond pull strength | BPS | 17 | HPD | 30 bonds from a minimum of 5 devices | 1 | 0 and $P_{pk} \geq 1.66$ or $C_{pk} \geq 1.33$ |
| Bond shear | BS | 18 | HPD | 30 bonds from a min. of 5 devices | 1 | 0 and $P_{pk} \geq 1.66$ or $C_{pk} \geq 1.33$ |
| Die-shear strength | DSS | 19 | HPD | 5 | 1 | 0 |
| Electrostatic discharge | ESD | 20 | HPD | 3/pin combo model | 1 | 0 |
| Latchup | LU | 21 | HPD | 6 | 1 | 0 |
| Internal water vapor | IWV | 22 | HD | 3 | 1 | 0 |
| Solderability | SD | 23 | HPD | 15 | 3 | 0 |

**Table 1 (cont.)**

| Stress | Abbreviation | No. | Note | Sample size per lot | Number of lots | Accept on number failed |
|---|---|---|---|---|---|---|
| $E^2$PROM data:<br>Endurance test<br>Retention test | ET | 24 | HPD | 10 | 1 | 0 |
| Early life failure rate | ELFR | 25 | HPN | 800 | 3 | 0 |
| Generic leakage | GL | 26 | PN | 6 | 1 | 0 |
| Electrical distribution | ED | 27 | HPD | 30 | 3 | pre: $P_{pk} \geq 1.66$<br>post: $P_{pk} \geq 1.00$ |

### 1.3 Definition of test failure after stressing

Test failures are defined as those devices not meeting the individual device specification, criteria specific to the test, or the supplier's data sheet, in the order of significance. Any device that appears physically damaged is also considered a failed device. If the cause of failure is agreed (by the manufacturer and the user) to be due to mishandling or EOS, the failure shall be discounted, but reported as part of the data submission.

### 1.4 Criteria for passing qualification

Passing all appropriate qualification tests specified in Tables 1 and 2, either by performing the test (acceptance of zero failures using the specified minimum sample size) or demonstrating acceptable generic data (using an equivalent total percent defective at a 90% confidence limit for the total required lot and sample size), qualifies the device per this document. Any unique reliability tests or conditions requested by the user and not specified in this document shall be negotiated between the supplier and user requesting the test.

When submitting monitor or qualification test data from generic products to satisfy the qualification requirements of this document, the number of samples and the total number of defective devices occurring during those tests must satisfy the following mathematical test:

$$\frac{Chi^2}{ss_{gen}} \le \frac{4.61}{(ss \times \#\ lots)_{TI}} \tag{1}$$

Where,

$$ss_{gen} \ge (ss\ xx\ \#\ lots)_{TI}\ and\ \#\ lots_{gen} \ge \#lots_{TI} \tag{2}$$

Where $Chi^2$ is the chi-square goodness-of-fit test statistic at a 90% level of confidence (a =0.90). A partial table of values is included below in Table 3. The number of degrees of freedom is D.F. = 2(c+1), where c is the number of failures contained in $ss_{gen}$; ss is the total number of samples in the generic sampling base ($ss_{gen}$) or from the requirement in Table 1 ($(ss \text{ x } \#\text{ lots})_{T1}$) for the stress test in question.

For example, if a supplier had data from 700 samples of temperature cycling, the maximum number of failures that this sampling would allow for the device

**Table 2 Table of methods referenced**

| Stress | Abbreviation | No. | Reference | Additional requirements |
|---|---|---|---|---|
| Pre- and poststress electrical test | TST | 1 | user of supplier specification | test is performed as specified in the applicable stress reference and the additional requirements |
| High-temperature operating life | HTOL | 2 | JA108 | 150°C $T_a$ for 408 hr or 125°C $T_a$ for 1008 hr (junction temperature not to exceed 175°C) at $V_{cc}$(max) and static or dynamic bias (dynamic bias (per engineering spec); equivalent time-temperature combinations are acceptable; tri-temp TST and ED before and after HTOL |
| High-temperature bake | HTB | 3 | JA103 | 150°C/1000 hr for plastic and 250°C/10 hr or 200°C/72 hrs. for ceramic packaged parts; TST before and after at room and hot temperatures |
| Preconditioning | PC | 4 | JA112<br>JA113 | performed on surface mount devices only; PC performed before THB, AC, and TC stresses; perform JA112 to determine at what preconditioning level to perform in the actual preconditioning stress JA113; the minimum acceptable level for qualification is level 3; any plastic delamination from the die surface is unacceptable; any replacement of parts must be reported; TST before and after at room temperatures |
| Temperature humidity bias | THB | 5 | JA101<br>JA110<br>M1004 | PC before THB for surface mount devices, 85°C/85%RH/1000 hr or 130°C/85%RH/72 hr (HAST); TST before and after THB at room and hot temperatures |
| Autoclave | AC | 6 | JA102 | PC before AC for surface mount devices, 121°C/15 psig/96 hr, TST before and after AC at room temperature |

**Table 2 (cont.)**

| Stress | Abbreviation | No. | Reference | Additional requirements |
|---|---|---|---|---|
| Temperature cycling | TC | 7 | JA104 | PC before TC for surface mount devices, condition C (-85 to 150°C) for 500 cycles or (-50 to 150°) for 1000 cycles, TST before and after TC at hot temperature; three gram-force bond pull strength (BPS) after decap on five parts on corner bonds (2 bonds per corner) and one midbond per side |
| Power temperature cycling | PTC | 8 | JA105 | test is performed only on devices with maximum rated power $\geq$ 1 W -40°C to 125°C, 1000 cycles, TST before and after PTC at room and hot temperatures |
| Mechanical shock | MS | 9 | M2002 | Y1 plane only, 5 pulses, 0.5-msec. duration, 1500g peak acceleration; TST after CA |
| Vibration variable frequency | VVF | 10 | M2007 | 20 Hz to 2 KHz to 20Hz (logarithmic variation) in > 4 min, 4X in each orientation, 50g peak acceleration, TST after CA |
| Constant acceleration | CA | 11 | M2001 | Y1 plane only, 30-Kg force for < 40 pin packages, 20-Kg force for 40 pins and greater; TST at room temperature |
| Gross/fine leak | GFL | 12 | M1014 | any single-specified fine test followed by any single-specified gross test |
| External visual | EV | 13 | M2009 | |
| Physical dimensions | PD | 14 | M2016<br>JB100 | see applicable JEDEC standard outline and individual device spec for significant dimensions and tolerances |

**Table 2 (cont.)**

| Stress | Abbreviation | No. | Reference | Additional requirements |
|---|---|---|---|---|
| Lead integrity | LI | 15 | M2004<br>JB105 | |
| Lid torque | LT | 16 | M2024 | |
| Bond pull strength | BPS | 17 | M2011 | Condition C or D |
| Bond shear | BS | 18 | USAEC-300-001 | see attached procedure for details on the acceptance criteria and how to perform the test |
| Die shear strength | DSS | 19 | M2019 | |
| Electrostatic discharge | ESD | 20 | USAEC-300-002 | see attached procedure for details on how to perform test |
| Latchup | LU | 21 | USAEC-300-003 | see attached procedure for details on how to perform test |
| Internal water vapor | IWV | 22 | M1018 | |
| Solderability | SD | 23 | M2003<br>JB102 | if burn-in screening is performed on the device, samples or SD must first undergo burn-in (8 hr steam aging prior to testing) |
| $E^2$PROM data:<br>Endurance test<br>Retention test | ET | 24 | USAEC-<br>300-004<br>300-005 | for devices that contain $E^2$PROM devices only; sequential test 005 followed by 006; TST before and after at room and hot temperatures |

**Table 2 (cont.)**

| Stress | Abbreviation | No. | Reference | Additional requirements |
|---|---|---|---|---|
| Early life failure rate | ELFR | 25 | JA108 | $T_a$=125°C for 48 hr performed after standard postproduction flow unless supplier can demonstrate low initial failure rate (as agreed to by the user); generic data is applicable; TST before and after at room and hot temperatures |
| Generic leakage | GL | 26 | USAEC-300-008 | TST before and after at room temperature; must be performed for evaluation purposes on all qualifications until it becomes official on March 1995 |
| Electrical distributions | ED | 27 | user or supplier specification | supplier and user to mutually agree upon electrical parameters to be measured |

Note H, required for hermetic packaged devices only; P, required for plastic-packaged devices only; N, nondestructive test, devices can be used to populate other tests or they can be used for production; D, destructive test, device are not to be reused for qualification or production; S, required for surface mount devices only.

Methods M, MIL-STD-883, the most current revision and notice; J, JEDEC JESD22, the most current method; No, number of the attached procedure. All electrical testing before and after the qualification stresses are performed to the limits of the individual device specification in temperature and limit value.

in question to be considered qualified for this test requirement is:

$$Chi^2 \le \frac{(700)(4.61)}{(3)(77)} = 13.97 \qquad (2)$$

*and* $c \le 3$ *failures from the table above*

If the qualification test needs to be performed, use the accept on zero criteria in the table above with the appropriate sample size. If there is applicable generic data as defined, acceptance will be based on the chi-square test statistic applied to the total percent defective of the data that meet or exceed the equivalent specified in this document for accept on zero at the specified sample size.

Devices that have failed the acceptance criteria of specific tests require the supplier to satisfactorily determine root cause and corrective action to assure the user that the failure mechanism is understood, contained, and prevented.

## 1.5 Alternative testing requirements

Any deviation from the test requirements listed in Table 1 and the test conditions listed in Table 2 must be approved by the user through supporting data presented by the supplier demonstrating equivalency. Three deviations will be clearly reported when the results of the qualification are submitted to the user for approval.

## 1.6 Definition of disqualification

When the number of failures for any given test in Table 1 exceed the acceptance criteria using the procedure in Section 2.4, the device shall be disqualified until the root cause of the failure(s) is (are) determined and the corrective and preventive actions are confirmed to be effective. New samples or data may be requested to verify the above.

**Table 3**

| c | Chi$^2$ | c | Chi$^2$ | c | Chi$^2$ | c | Chi$^2$ |
|---|---|---|---|---|---|---|---|
| 0 | 4.61 | 5 | 18.5 | 10 | 30.8 | 15 | 42.6 |
| 1 | 7.78 | 6 | 21.1 | 11 | 33.2 | 16 | 44.9 |
| 2 | 10.6 | 7 | 23.5 | 12 | 35.6 | 17 | 47.2 |
| 3 | 13.4 | 8 | 26.0 | 13 | 37.9 | 18 | 49.5 |
| 4 | 16.0 | 9 | 28.4 | 14 | 40.3 | 19 | 51.8 |

## 2 QUALIFICATION AND REQUALIFICATION

### 2.1 Qualification of a new device

The stress test qualification requirements for a new device qualification are listed in Table 1, and the corresponding test conditions are listed in Table 2. For each qualification, the supplier must present data for ALL of these tests, whether it is stress test results on the device to be qualified or acceptable generic data. A review should be made of other parts in the same generic family to ensure that were no common failure mechanisms in that family. Justification for the use of generic data, whenever it is used, must be demonstrated by the supplier and approved by the user.

### 2.2 Requalification of a changed device

Requalification of a device will be required when the supplier makes a change to the product and/or process that impacts the form, fit, function, quality, and/or reliability of the device.

*Process change notification.* A change to a product can be implemented only after approval by all users affected by the change. The supplier should submit a projection to the users of all forecasted process changes. Information required to be submitted to the user will include the following as a minimum:

1) benefit to the user (value, time and quality);
2) for each user part numbers involved in the change, the following information is required:
   (a) supplier part number
   (b) the estimated date of the last production lot of unchanged parts
   (c) the estimated final order date and final ship date of unchanged parts
   (d) the first projected shipment date and date code of changed parts
3) a detailed description of the change in terms of the materials, processes, characteristics, rating, die layout, circuit design, die size and wafer size, as applicable;
4) technical data and rationale to support proposed changes;
5) an electrical performance characterization comparison (between the new and original product) of all significant electrical parameters over temperature, voltage, and frequency extremes that could be affected by the change; Changes in median and dispersion performance shall be noted even though conformance to specification limits is still guaranteed;

6) the supplier shall submit an updated Certificate of Design, Construction and Stress Test Qualification along with information required by this section plus any changes impacting the original Appendix 2 information;
7) the results of completed supplier requalification tests of the changed device(s);

Items 1, 2, 3, and 4 are background information needed up front to evaluate the impact of the change on supply and reliability and to come to agreement on a qualification plan acceptable to the supplier and suer. Items 5, 6, and 7 must be submitted prior to any final approval to implement any change on the user's product.

## 3 QUALIFICATION TESTS

### 3.1 General tests

Test details are given in Tables 1 and 2. Not all tests apply to all devices. For example, certain tests apply only to ceramic packaged devices, others apply only to devices with EPROM memory, and so on. the applicable tests for the particular device type are indicated in the "Note" column of Table 1 and the "Additional Requirements" column of Table 2. The "Additional Requirements" column of Table 2 also serves to highlight test requirements that supersede those described in the referenced test.

### 3.2 Device specific tests

The following tests must be performed on the specific device to be qualified for all hermetic and plastic devices; generic data are not allowed for these tests:

- electrostatic discharge (ESD) - all product;
- latchup (LU) - all product;
- electrical distribution - the supplier must demonstrate, over the user's application temperature, voltage, and frequency range, that the part is capable of meeting parametric limits as agreed to between the user and supplier in the individual device specification; these data must be taken from at least three lots, or one matrixed (0 skewed) process lot, and must be of enough samples to be considered by the user and supplier statistically valid; post HTOL electrical distributions may not be required if the supplier can demonstrate stability over time on similar product structures and correlated to parameters specified for parts being qualified;
- other environmental stress testing (HTOL, THB, TC, etc.) - additional testing may be required by the user's experience with the supplier;

### 3.3 Wearout reliability tests

As a minimum, testing for the failure mechanisms listed below must be available to the user whenever a new technology or material relevant to the appropriate wearout failure mechanism is to be qualified. The data, test method, calculations and internal criteria need not be demonstrated or performed on the qualification of every new device, but should be available to the upon request.

- electromigration;
- time-dependent dielectric breakdown - for all MOS technologies;
- hot carrier injection - for all MOS technologies.

## REFERENCES

ASTM D-2863 Low Cycle Fatigue. *Special Technical Publication 942*, American Society for Testing Materials, Philadelphia (1988).

Danielson, D.D., Marcyk, G., Babb, E., and Kudva, S. HAST Applications: Acceleration Factors and Results for VLSI Components. *Proceedings of the 26th International Reliability Physics Symposium*, IEEE (1989) 114-121.

Dasgupta, A., and Pecht, M. Material Failure Mechanisms and Damage Models, *IEEE Transactions on Reliability* 40, No. 5 (1991) 531-536.

Davy, J.G. Accelerated Aging for Solderability Testing: A Review of Military Standards. *Proceedings, 6th National Conference and Workshop Environmental Stress Screening of Electronic Hardware*, Baltimore, MD (1990) 49-58.

Gunn, J.E., Camenga, R.E., and Malik, S.K. Rapid Assessment of the Humidity Dependence of Integrated Circuits Failure Modes by Use of HAST. *Proceedings of the 21st International Reliability Physics Symposium* IEEE (1983) 66-72.

Ilyas, Q.S.M. 10,000 hours of Accelerated Life Test and Reliability of 0.9 μm CMOS Devices in Plastic Packages. publishing pending (1994).

Ilyas, Q.S.M. et al. Evaluation of Moisture Sensitivity of Surface Mount Packages. *EEP-7, ASME* (1993) 145-156.

Intel. *Components Quality/Reliability Handbook* (1989).

Lin, A.W., and Wong, C.P. Encapsulant for Non-Hermetic Multichip Packaging Applications. *Proceedings, IEEE-CHMT* (1991) 820-826.

Montgomery, D. Design and Analysis of Experiments. Wiley, New York (1991).

Motorola. *Personal communications* (1993).

Nguyen, L.T., Lo, R.H.Y., Chen, A.S., Takiar, H., and Belani, J.G. Molding Compound Trends in a Denser Packaging World, II. Qualification Tests and Reliability Concerns. *SEMICON/Singapore 93*, Singapore World Trade Center, Singapore (1993).

Pollock, L.R. *A Wide Parametric, Bayesian Methodology for System-Level, Step Stress, Accelerated Life Testing*, Thesis, Florida Institute of Technology (1989).

Shirley, G.C., and Hong, C.E.C. Optimal Acceleration of Cyclic THB Tests for Plastic-Packaged Devices. *29th Annual Proceedings, International Reliability Physics Symposium* (1991) 12-21.

# 9

# DEFECT ANALYSIS TECHNIQUES

Microelectronics reliability is often affected by discontinuities within the component or an assembly. These discontinuities may grow with time or with thermal cycling, and when they reach a critical size, electrical performance may be affected — a connection may break or an electrical characteristic may change. To assure both electrical and mechanical reliability, failure analysis is a crucial factor in package design. Failures in plastic packages appear with a variety of morphologies (large/small, subtle/coarse), and can be located in the external package, internal package, die surface, die subsurface, or interface. To conduct effective failure analysis on electronic devices and packages, a disciplined step-by-step schedule must be followed to ensure that no relevant information is lost. The multitude of technologies demands a corresponding multitude of failure analysis techniques.

## 9.1 GENERAL DEFECT ANALYSIS PROCEDURES AND TECHNIQUES

For the majority of failed packages, there are several basic analysis steps:

- visual examination;
- decapsulation (package opening);
- internal examination;
- selective layer removal;
- location of the failure site;
- identification of the cause of the failure and its mechanism;
- simulation testing and final examination.

### 9.1.1 Visual Examination

Visual examination can be performed by the naked eye or with the help of an optical microscope or, sometimes, an X-ray microscope. X-ray radiography may change the electrical properties of devices, so the application of this technique should be reconsidered whenever the electric properties of package are to be tested.

### 9.1.2 Decapsulation (Package Opening)

Whether to open the package permanently or not is the first important decision to be made by the failure analyst. Some failure analysis techniques, such as X-ray microscopy and scanning acoustic microscopy, can be utilized without plastic removal; these techniques are widely applied as nondestructive failure analysis methods. Nondestructive techniques minimize decapsulation; unfortunately, they have certain resolution limitations. Defects of 1 μm or less are rarely detected by either X-rays or acoustic microscopy, but may be identifiable other techniques — for instance, electron microscopy.

Most failure analysis techniques require decapsulation to locate and examine defects inside plastic packages. Several methods are generally employed for plastic removal; these include chemical methods, such as sulfuric and nitric acid etching, thermomechanical methods [Byrne 1980], and plasma etching [Pfarr and Hart 1980]. The classic wet chemical technique is usually applied in several steps: heating the package to 220 to 250°C; periodically dripping sulfuric or nitric acid into a cavity milled in the mold compound until the die is exposed; rinsing the decapsulated device with cold fuming acid, then acetone, and finally, alcohol, and air drying. Heating accelerates the reaction of the epoxy decomposition and speeds decapsulation; however, it also removes contamination from the surface of the die, preventing chemical analysis. Jet-etching, another chemical decapsulation method, employs a jet of decapsulating acid. The acid jet hits the top of the package until the die is exposed. This is a quicker approach than drip methods, and is usually implemented for large-scale decapsulation, where the time required to open each device becomes critical.

The thermomechanical method, involving various combinations of grinding, heating, and breaking with force, has no chemicals to react with contaminants at the die surface, and therefore is typically used to identify failures expected to be caused by metallization corrosion. The primary disadvantage of this method is the loss of electrical continuity due to the removal of the wires.

When the plasma etching method is used for decapsulation, low-temperature plasmas of electrically excited oxygen emitted on the plastic encapsulants ash the materials off. The plasma treatment has proven valuable because of its selectivity, gentleness, cleanliness, and safety [Pfarr and Hart 1980].

However, the overall decapsulation time is too long for routine use and limits its application to the more critical failure analysis studies. Great care must also be taken to avoid artifact failure, which may be mistaken as a real failure mode. It is advantageous to mechanically remove as much material as possible above the area to be investigated prior to plasma processing.

### 9.1.3 Internal Examination

Internal examinations can be done with many failure analysis techniques, selected according to the possible location of the failure in the package, the expected size of the failure to be examined, and other factors.

### 9.1.4 Selective Layer Removal

Some defects, such as evidence of electrical overstress or pin holes in insulating films that can lead to short circuits, may be sub-surface and hence not visible. This necessitates individually removing the various component layers on the semiconductor. The two layer removal methods are wet etching and dry etching [Richards and Footner 1984].

Oxides and metallizations are usually removed using wet chemical etching. Nitride passivation layers are removed by etching with fluorine-containing plasmas or ion (Ar) milling.

### 9.1.5 Locating the Failure Site and Identifying the Failure Mechanism

Similar to internal examination, there many possible techniques to locate a failure site (see Section 9.6). Identifying the cause of the failure and its mechanism, the most important step in failure analysis, is done by fully considering the location of the failure, its morphology and history, and the conditions of the package's manufacture and application.

### 9.1.6 Simulation Testing and Final Examination

In some cases, failures such as large cracks, voids, or wire debonding are immediately evident. At other times, however, direct evidence of the mechanism is destroyed either by the failure itself or by the subsequent analysis. These cases require simulation testing to reproduce the observed failures. The package or device can be stressed to failure using environmental facilities or electrical loading logically related to in-use failure. Such environmental conditions can be selected as elevated temperature, temperature cycling, elevated relative humidity, and so on.

New failure analysis techniques continue to emerge. The following primary and specialized techniques apply to failure analysis in microelectronic packages:

- optical microscopy (OM);
- microprobing (MP);
- liquid crystal display (LCD);
- X-ray microscopy (XM);
- infrared microscopy (IRM);
- scanning acoustic microscopy (SAM);
- scanning electron microscopy (SEM);
- environment scanning electron microscopy (ESEM);
- transmission electron microscopy (TEM);
- scanning tunneling microscopy (STM);
- atomic force microscopy (AFM);
- scanning near-field optical microscopy (SNFLM);

- low-energy electron microscopy (LEEM);
- emission microscopy (EM).

Several advanced failure analysis techniques, including the nondestructive techniques of scanning acoustic microscopy and X-ray microscopy and the high-resolution techniques of electron microscopies, are illustrated in Sections 9.2, 9.3, and 9.4, respectively. Other techniques and their applications are briefly described in Section 9.5. Finally in Section 9.6, the proper selection of failure analysis techniques is discussed.

## 9.2 SCANNING ACOUSTIC MICROSCOPY

Assembly-related packaging defects in plastic-encapsulated integrated circuits are a major source of reliability problems. Although certain applications have relatively stringent requirements, the typical defects illustrated in Figure 9.1 include delamination of the molding compound (MC) from the leadframe, die, or paddle; MC cracks; die-attach disbonds; die tilt; and voids in the MC. Each of these can be nondestructively detected and viewed via acoustic microimaging (AMI) technologies.

AMI may be used to inspect many devices in order to screen out defective units prior to use or to ensure the output quality of a molding process in line. AMI techniques can respond to a full range of needs, from production volume screening to detailed laboratory analysis. It is important to fully characterize the defect and understand the failure mode. Used properly, AMI can provide valuable data not available by any other technique. Moreover, since it is nondestructive, components are not sacrificed. For reliability studies, the growth rate of a flaw can be monitored as the component cycles through environmental stresses or through its normal operating life.

AMI techniques can be classified into two methods: the scanning laser acoustic microscope (SLAM) and the C-mode scanning acoustic microscope (C-SAM). Both are routinely used to evaluate plastic-encapsulated devices and have been codified into standards [Semmens and Kessler 1988, ANSI/IPC-SM-786 1990].

### 9.2.1 Scanning Laser Acoustic Microscopy

SLAM is a through-transmission technique operating over a frequency range of 10 to 500 MHz [Kessler 1989]; however, due to the absorption of ultrasound by polymers, the highest acoustic frequencies are not needed for inspecting encapsulated integrated circuits. With SLAM, a plane wave of continuous

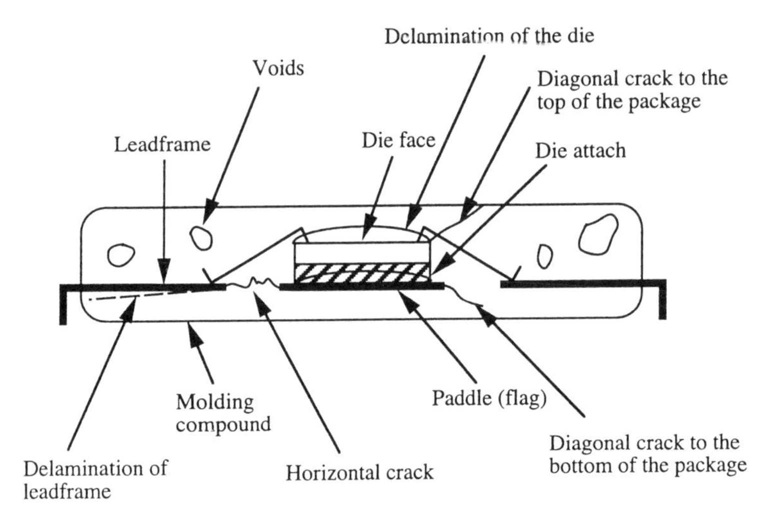


**Figure 9.1** *Potential defects in plastic-encapsulated microcircuit packages [ANSI/IPC-SM-786 1990]*

ultrasound is introduced to one side of the sample and travels through it to the opposite side. Variations in the ultrasound wave pattern due to differential attenuation in the sample are detected by a scanning laser beam. Since high-frequency ultrasound does not travel in a vacuum or in air, flaws characterized by air gaps block transmission to the detector in the affected area and appear dark in the acoustic image. The true real-time imaging capabilities of SLAM (typically 30 pictures/sec) make it a useful technique for production-line, high-volume screening.

The basic operating principle of SLAM is illustrated in Figure 9.2. Ultrasound illuminates the area of a sample to be inspected. The ultrasound passing through the sample causes tiny perturbations of a mirrored surface on a plastic coverslip mirror. A focused laser beam scans the surface of the plastic block and reads off the acoustic levels. SLAM images the entire sample thickness simultaneously. As a by-product of laser scanning, the SLAM can also produce optical images. SLAM produces images and data in several different modes. The most common "shadowgraph" mode images the structure throughout the thickness of the sample; this allows the distinct advantage of simultaneous viewing of defects anywhere in the sample, like X-ray radiography. In situations requiring focus on one specific plane, holographic reconstruction of the SLAM

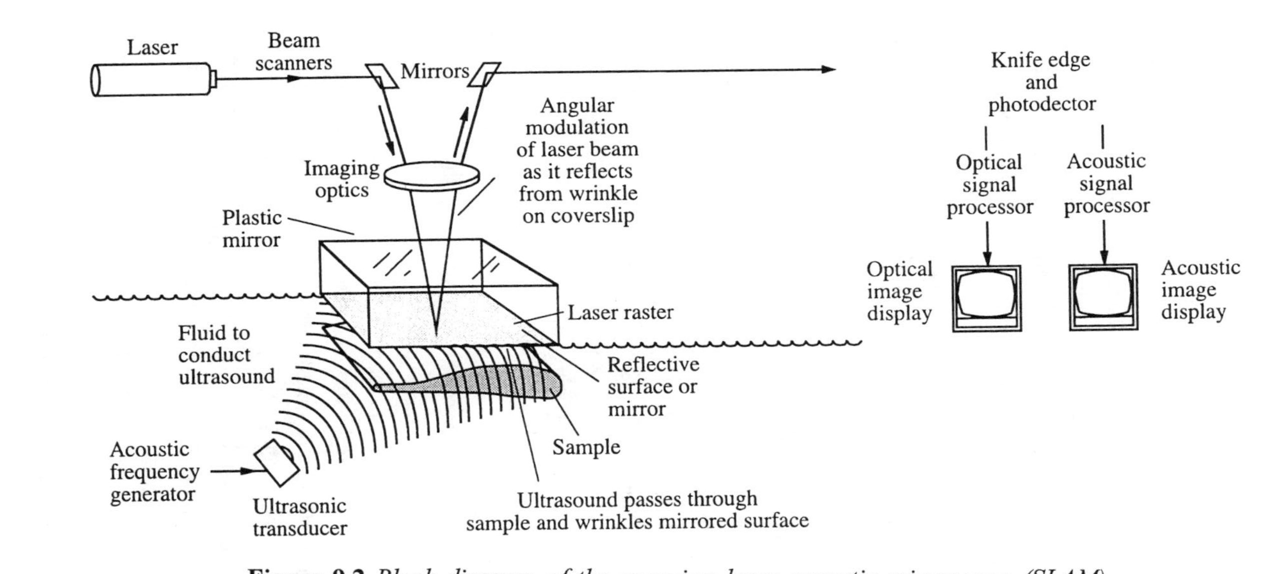


**Figure 9.2** *Block diagram of the scanning laser acoustic microscope (SLAM)*

data can be employed [Kessler 1989]. In the other, "interferogram," mode, "fringes" appear on the CRT that are related to the velocity of sound variations in the sample.

In addition to acoustic image capability, an optical image is produced by the direct laser-scanned illumination of the sample surface. The optical image serves as an operator reference for landmark information, artifacts, and positioning of the sample. With an auxiliary lens, the SLAM optical mode can be utilized as a high-resolution optical scanning laser microscope.

With SLAM, the brightness of the image corresponds to the acoustic transmission level. By removing the sample and restoring the image brightness level with a calibrated electronic attenuator, precise insertion loss data about the material composition can be obtained. The attenuation and velocity data are affected by the modulus of elasticity and the viscoelastic material behavior. In practice, void population, filler aggregation, and the degree of epoxy cure can also be determined.

C-SAM is useful for defining the exact nature of flaws. Unlike SLAM, which images all depths simultaneously, C-SAM can be used depth-selectively. The basic operating principles of C-SAM are illustrated in Figure 9.3 [Kessler 1989]. A focused spot of ultrasound is generated by an acoustic lens assembly at frequencies typically ranging from 10 to 100 MHz. The angle of the rays from the lens is generally kept small so that the incident ultrasound does not exceed the critical angle of refraction between the fluid coupling and the solid sample. C-SAM introduces a very short acoustic pulse into the sample; return echoes are produced at the sample surface and at specific interfaces within the part. The echo return time is a function of the distance from the interface to the transducer and the speed of sound in the sample. An oscilloscope display of the echo pattern, known as an A-scan, shows the echo levels and their time-distance relationships with the sample surface. An electronic gate can select information from a specific depth level while excluding other levels. The gated echo signal brightness modulates a CRT that is synchronized with the transducer position. To make an image, a mechanical scanner moves the transducer over the sample and produces the data in tens of seconds or less depending on the field of view.

The gray-scale image on the CRT can be converted into false color for contrast enhancement of the amplitude information. The images can also be color-coded with echo polarity or phase information [Cichanski 1989]. Positive echoes, which arise from reflections off a higher acoustic impedance interface, are displayed in a gray scale with one color scheme, while negative echoes, from reflections off lower acoustic impedance interfaces, are displayed in a different

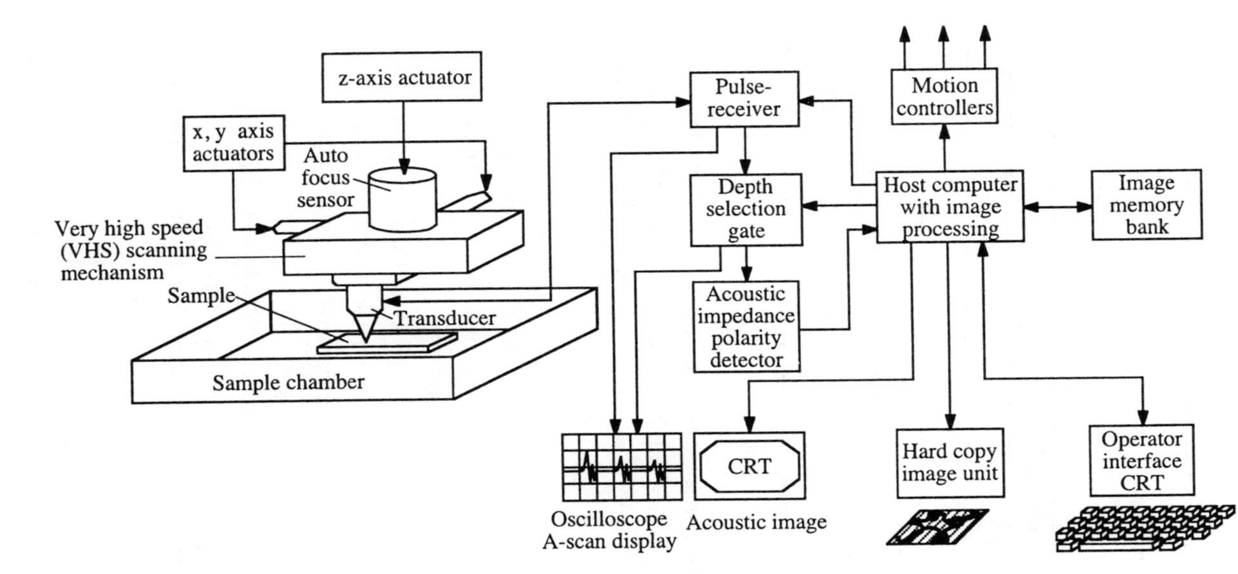


**Figure 9.3** *Block diagram of a C-mode scanning acoustic microscope (C-SAM)*

**Table 9.1 Typical acoustic impedance values**

| Material | Acoustic impedance ($10^6$ rayl) |
|---|---|
| Air (vacuum) | 0 |
| Water | 1.5 |
| Plastic | 2 to 3.5 |
| Glass | 15 |
| Aluminum | 17 |
| Silicon | 20 |
| Copper | 42 |
| Alumina (depends on porosity) | 21 to 45 |
| Tungsten | 104 |

color scheme. The acoustic impedance, $Z$ is a material characteristic related to acoustic wave propagation; In particular, $Z = pv_{sound}$ where $p$ is the mass density and $v_{sound}$ is the velocity of sound for the material through which the wave is propagating. The velocities of sound for typical materials can be found in Selfridge [1985]. When an acoustic wave is incident upon a boundary between two materials, part of the wave is reflected and part is transmitted. Typical acoustic impedance values for common materials are shown in Table 9.1.

Depending on the relative magnitudes of $Z_2$ and $Z_1$ the reflected wave can carry positive or negative polarities. In C-SAM operation, the A-scan mode oscilloscope trace displays the echo pattern of interfaces encountered by the acoustic pulse. Each echo has a characteristic amplitude and polarity, depending on the nature of the corresponding interface.

Focused acoustic wave transducers have limited depths of field over which accurate data can be obtained. Outside of this area, the apparent magnitude of an echo can be lower due to defocusing. Moreover, the shape of the echo might be distorted, since the angular rays may not all return to the transducer at the correct time. Therefore, polarity data might not be valid.

Acoustic impedance analysis of a sample allows quantitative determination of the nature of internal interfaces. For example, the echo amplitude from a plastic-silicon boundary (good bond) is close to that from a plastic-airgap boundary (disbond), but the echoes are 180° out of phase. This technique, acoustic impedance polarity detection (AIPD), allows the operator to determine whether a material boundary is being opposed by a "harder" or "softer" material on the other side (strictly speaking, acoustic impedance is neither hard nor soft, but this analogy is appropriate in some cases.) The technique of displaying a unified image, containing both amplitude and polarity information simultaneously [Cichanski 1989], has an important role in integrated-circuit defect detection.

### 9.2.2 C-mode Scan

With C-SAM, several distinct imaging techniques can be chosen to analyze a sample; the most common is the C-mode scan. In C-mode, a focused transducer is scanned over the planar area of interest and the lens is focused to some depth, as illustrated in Figure 9.4. The echoes arising from that depth are electronically gated for display. The electronic gate may be adjusted to be either narrow or wide, and the depth information content of the image will correspond to the thin or thick "slice." For example, in a plastic-encapsulated integrated circuit, a narrow gate can produce an isolated image of the die surface alone, whereas a wider gate can image the die surface, the perimeter of the paddle, and the leadframe simultaneously. In C-SAM, the gate can be used to nondestructively microsection the sample. By changing the configuration of the C-SAM instrumentation, other imaging modes that emphasize particular structural features and defects can be accessed. These are briefly described below.

### 9.2.3 Bulk Scan

By positioning the electronic gate or window between the internal interfaces of an integrated circuit, material texture, nonhomogeneity, and voids can be detected. For example, if the electronic gate is placed behind the top surface echo of an integrated circuit but in front of the leadframe echo, it is easy to identify agglomerations of filler material and voids in the encapsulant. Of course, the AIPD is needed to differentiate filler from void spaces, since both produce large-amplitude echoes.

***Transmission scan.*** Transmission images are made by recording the ultrasonic energy transmission through the entire component, instead of only that reflected from interfaces. In this mode, which can be accomplished by SLAM or C-SAM (Figure 9.5), defects anywhere in the part will block the ultrasound and cause dark features to appear in the image. With C-SAM, the transmission scan (THROUGH-scan) mode requires ultrasound transducers placed on both sides of the sample, one for producing the ultrasound and the other for receiving it.

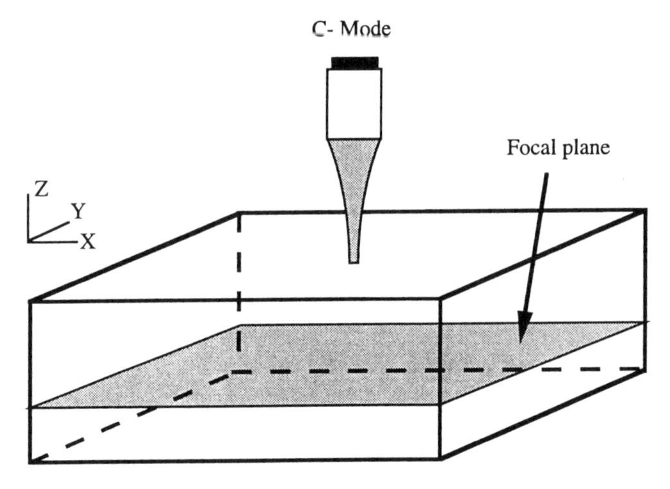


**Figure 9.4** *Schematic of a C-mode acoustic scan*

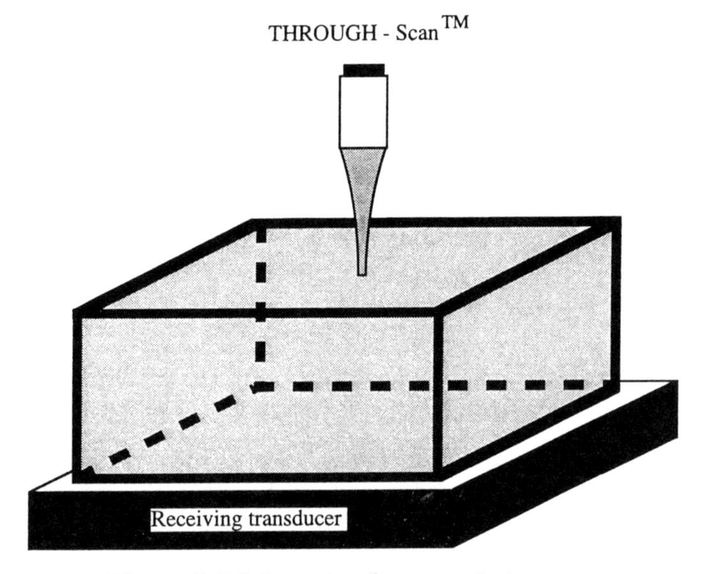


**Figure 9.5** *Schematic of a transmission scan*

THROUGH-scan can be used to determine with only one scan whether a defect exists in a sample. THROUGH-scan can also be of great help in confirming reflection-mode data; in particular, some highly absorbing samples can distort the ultrasonic pulse shape so severely that AIPD is difficult to use. Furthermore, with THROUGH-scan, the presence of delamination does not affect echo polarity.

The major difference between SLAM and C-SAM with regard to THROUGH-scan is the speed of imaging. C-SAM requires tens of seconds to mechanically scan the transducer over the area of interest, whereas SLAM produces 30 images per sec. Therefore, SLAM is generally used for higher-speed inspection, such as is needed for on-line screening applications.

***Time-of-flight scan.*** While in the reflection-mode techniques echo amplitudes are recorded as corresponding gray scales, in the time-of- flight scan (TOF scan), the arrival time of the echo is converted to a gray scale (see Figure 9.6). In this mode, the echo amplitude has no bearing on the gray scale, except that it must be large enough to be detected. This type of image is useful for a general overview of feature depth, and although the image may appear visually similar to a conventional C-mode image, the information content is quite different. When projected into three dimensions, the TOF image gives a perspective of the contour of the internal interface. TOF scans are particularly useful for profiling cracks in plastic-packaged integrated circuits.

***Quantitative B-scan analysis mode.*** In transmission and TOF scans, the planar location of each pixel on the CRT corresponds to a planar position on the sample. However, in the B-scan mode, the planar CRT pixels correspond to one dimension of the plane and the depth position in the sample. B-scan is analogous to the technology employed by the medical ultrasound scanners used in hospitals. In conventional B-scan, a vertical cross-sectional image is produced of the sample along any line across it, just as if the sample were cut open with a saw. The image is made by scanning the transducer across one dimension of the plane, recording echoes that return from all depths, and displaying these on the vertical axis of the CRT. Unfortunately, in conventional B-scan the echoes are not in focus at all depths due to the fixed transducer-lens focal position. Figure 9.7 illustrates a B-scan cross section of a sample in which the dark shading represents the limited focal zone of the transducer. In quantitative B-scan analysis mode (Q-BAM), however, the transducer is also indexed in the depth direction throughout the entire thickness of the component in order to ensure continuous uniform focus. The Q-BAM cross section is illustrated in Figure 9.8. In a Q-BAM image, the echo time-of-flight information is converted to metric depth data so that the operator can see a unified display of the cross section in the one dimension of the plane and depth plane, along with its planar location.

### 9.2.4 Case Study: Delaminations in a 40-Pin PDIP

Delaminations in 40-pin PDIPC-mode reflection images were made of a 40-pin PDIP from the top surface; the transducer was focused at the die face. The sample was then turned over, and the transducer was focused to the surface of

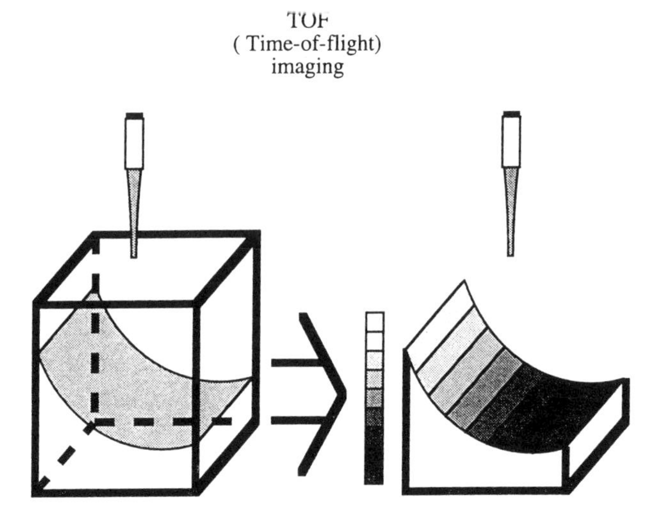


**Figure 9.6** *Time-of-flight scan where the arrival time of the echo is converted to gray scale*

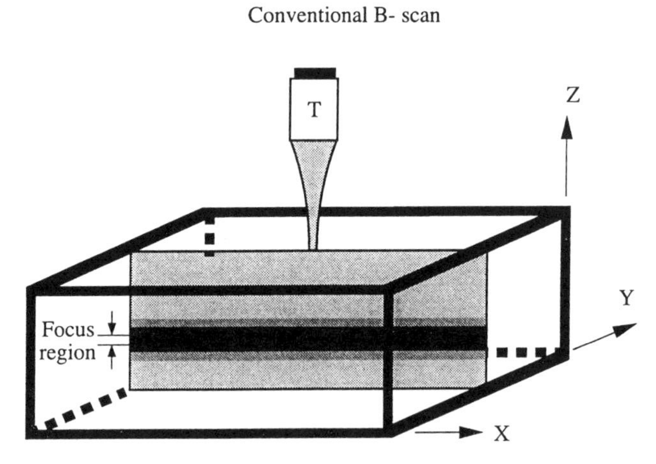


**Figure 9.7** *B-scan cross section of a typical sample; the dark shading represents the focal zone of the transducer wherein the acoustic data are most accurate*

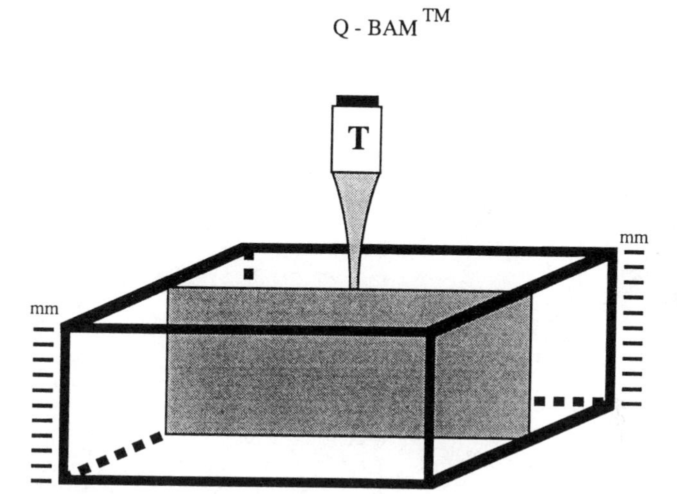


**Figure 9.8** *Quantitative B-scan analysis with accurate depth information over the entire depth range*

the die paddle. (Unfortunately, this text cannot reproduce the standard C-mode color display, so the figures present the data in various other ways.)

Figure 9.9a shows an amplitude image of the sample that presents the amplitudes of all echoes as brightness levels on the CRT, regardless of polarity. In this useful image all structures are revealed at the level of focus selected. This figure shows unusual brightness changes around the die and leadframe, presumably due to delamination. Using AIPD, the echo polarity data can be distinguished.

Figure 9.9b gives a delamination image that consists only of negative echoes. White features shown in this image are all disbonded from the molding compound.

Figure 9.9c is a black-and-white unified display that electronically codes bonds as bright and disbonds as black. Figure 9.9d is a black-and-white unified display of the same sample as viewed from the back side. This PDIP has extensive delaminations on both sides of the leadframe, as well as small delaminations on the die face and paddle area. Leadframe delaminations of this type are frequently a result of poor control of the molding process [Manzione 1990].

### 9.2.5 Case Study: Rapid Screening for Defects With Through-Transmission Imaging or THROUGH-Scan

To accurately determine whether a plastic-encapsulated integrated circuit has assembly defects, three or more reflection-mode or C-mode scans are usually

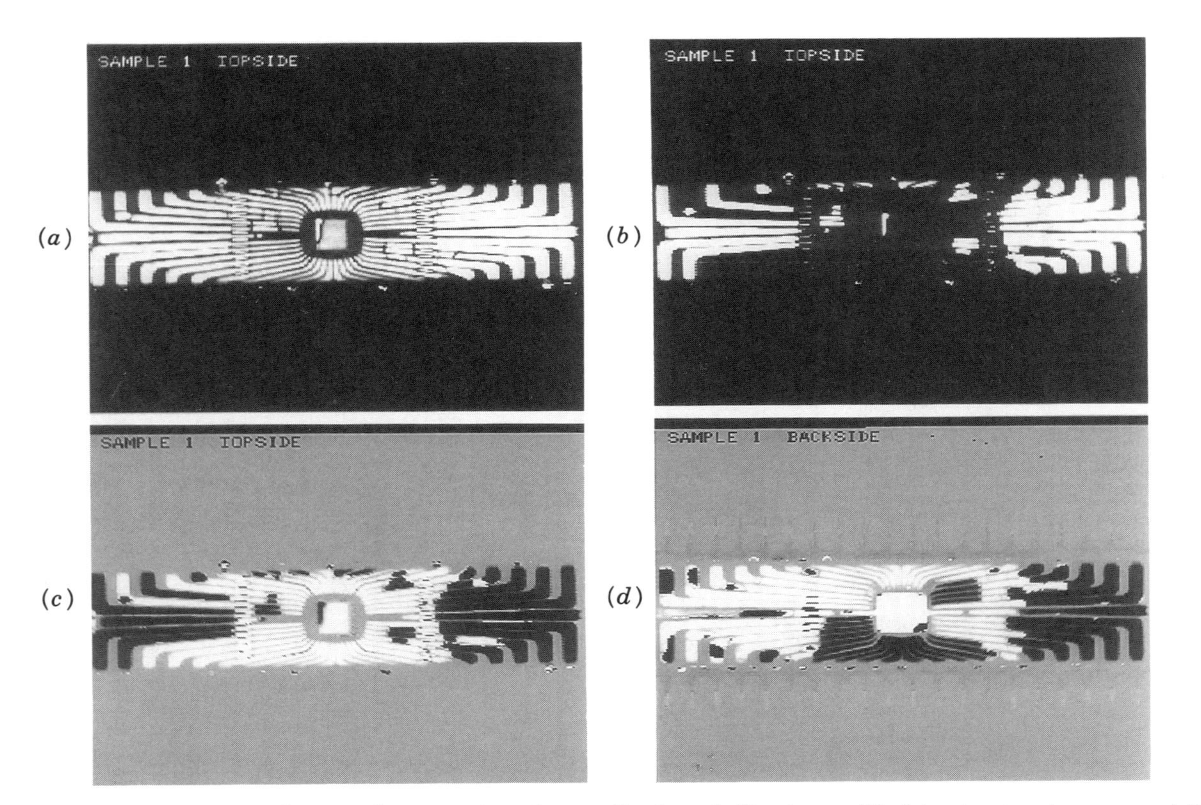


**Figure 9.9** *(a) Amplitude image of a sample presenting the amplitudes of all echoes; (b) delamination image consisting only of negative (disbond) echoes; (c) a black and white unified display, in which disbonds are dark; (d) a black and white unified display of the same sample as viewed from the back side*

performed on each component. Series of scans typically includes the following

- focusing from the top side of the integrated circuit to the die surface; if the leadframe is also in focus, it is included in this scan for all top-side delaminations, voids, and cracks;
- while still viewing the IC from the top side, refocusing the transducer to the back side of the die (if possible) to search for die-attach delamination and voids;
- turning the part over and focusing from the back side to the die paddle and leadframe to search for additional anomalies; since this procedure is very time-consuming, each component can be screened instead with the through-transmission modes of either SLAM or C-SAM.

In this case, just one scan reveals the presence of defects throughout the package. Although this does not necessarily determine the depth of the defect, it is quite accurate in the planar dimension.

Figure 9.10a is a through-transmission scan of a 72-pin plastic quad flatpackage that reveals a large acoustically opaque (black) central zone. Acoustic opacity results from a gap separation between layers of material, in this case caused by a popcorn crack. The reflection amplitude image of the same device is shown in Figure 9.10b. Although the polarity data are not shown here, the MC is delaminated from the die face. Turning over the part and scanning from the bottom surface, Figure 9.10c indicates delamination over the entire surface of the paddle as well. In addition, a dark circular region around the die paddle is due to a crack that extends from the die paddle out toward the package surface. The echoes from the crack interface occur earlier than those from the zone of focus, so the image of the leadframe appears to be missing within the circular zone.

Figure 9.11a is a through-transmission scan of another plastic quad flatpack that appears to have a problem in the die-paddle area. A close look at the image, which has been "saturated" or overexposed in the leadframe area to bring out detail in the center, suggests that the dark delamination area is spatially restricted to the die itself. This indicates the problem may be associated with poor MC die adhesion or poor die attach. To gain more certainty, top-side and back-side C-mode reflection scans were done; the amplitude images are shown in Figures 9.11b and 11c. Since the AIPD reveals no negative (disbond) signals, MC adhesion is good. Therefore, the die attach itself is the problem, since it is the only other interface in the IC that could block the ultrasound.

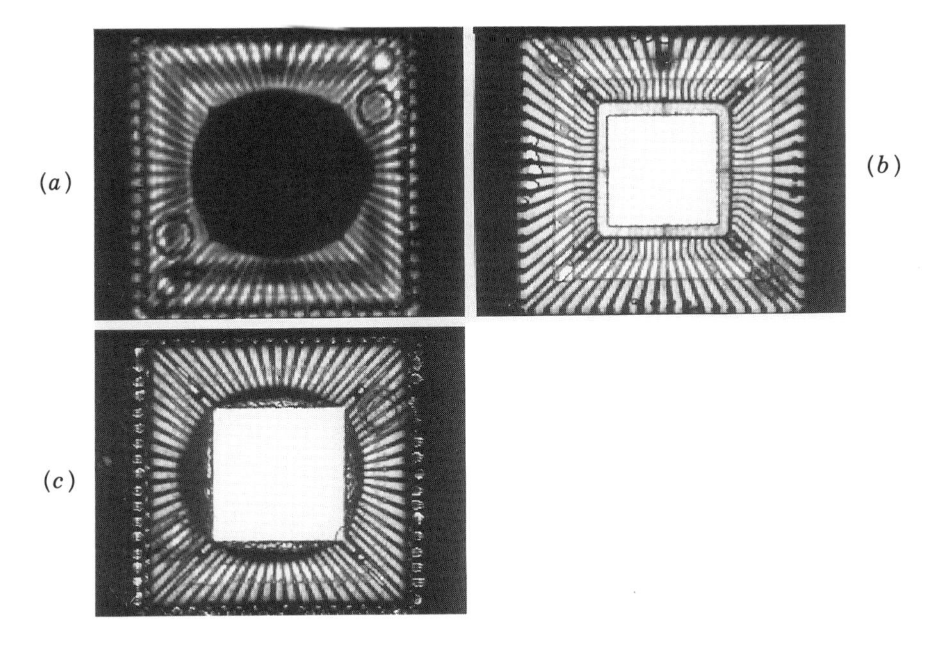


**Figure 9.10** *(a) Through-transmission scan of a 72-pin plastic quad flat package revealing a large acoustically opaque zone; (b) the reflection amplitude image of the device; (c) delamination over the entire surface of the paddle*

In addition to the die-attach problem, there are a number of tiny black spots in all three images. These correspond to MC voids that typically arise during the molding process if the packing pressure profile is not adjusted properly [Manzione 1990]. Voids such as these threaten reliability most when they are located on and around the silicon chip or the wirebonds. Because of the broad distribution of voids and the poor die attach, in Figure 9.11 this device could exhibit poor reliability.

The series of images in Figures 9.10 and 9.11 illustrate how THROUGH-scan can be used to detect the presence of defects and determine their spatial extent. The reflection mode is then used complementarily to further analyze the defect types and locations. The THROUGH-scan serves as a guide to distinguishing good and bad parts and provides an image of the lateral dimensions of defects.

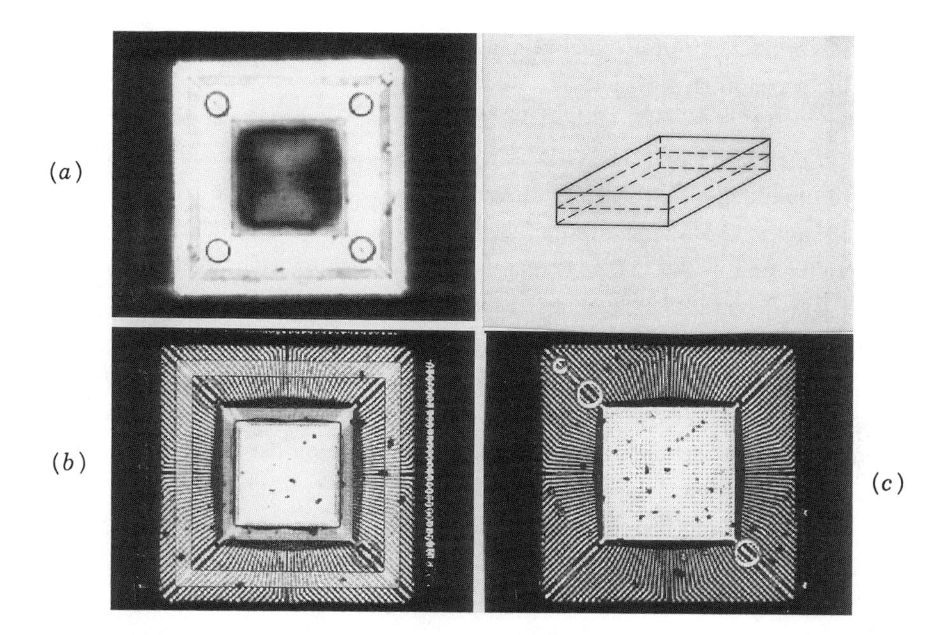


**Figure 9.11** *(a) A through-transmission scan of a plastic quad flatpack that has a problem in the die/paddle area; (b) reflection image from the topside; (c) reflection image from the back side*

### 9.2.6 Case Study: Nondestructive Cross-sectioning of Plastic-Encapsulated Devices

C-SAM images are generally made of an internal horizontal plane of the sample, located at a given depth. To obtain more complete depth information, a systematic series of planar scans can be made at different depths, or a quantitative B-scan (one dimension of the horizontal plane and the depth) can be made at any position along another dimension of the plane.

Figure 9.12a shows a Q-BAM image of a U.S. penny, imaged through the coin and focused at the rear surface. The cross section is, of course, acoustic not metallurgical. The top half of the CRT screen is a C-mode scan of the penny; the bottom half of the screen shows one dimension of the planar plane and the depth scan made across the coin's diameter. The bottom line of the C-mode image is also the precise location of the Q-BAM image; this gives the analyst a cross-reference for the data. The bottom surface of the coin appears to have dimensional changes associated with the embossed pattern on the surface, as well as some overall distortion of the surface flatness. Thus, the Q-BAM can be used to search the depth of a component to accurately locate features with respect to the top and bottom surfaces.

Figure 9.12b shows a 68-lead PLCC cross sectioned midway through a dark spot in the die region. The Q-BAM reveals a molding compound void 0.6 mm above the die surface, which is far enough above the die to be considered safe. Note that the Q-BAM shows the height differences between internal structures. On either side of the Q-BAM image, a scale is calibrated by both echo arrival time and distance of travel, as determined by the velocity of sound in the material.

Figure 9.12c shows another device, with a popcorn crack originating from the edges of the die paddle and traveling toward the surface of the package. The crack intersects the surface on the left side, but not on the right. Figure 9.12d shows a significant degree of paddle shift. This is typically caused by a rapid increase in packing pressure after the mold-filling step; the differential packing rates between the top and bottom zones of the leadframe lead to this type of anomaly (as well as to wire sweep problems). In this figure, the die can be seen above the leadframe, and a faint outline of the wires can be seen extending from the die surface to the leadframe fingers.

Figure 9.13 shows a time-of-flight (TOF) image of another popcorn-cracked integrated circuit. In this case, the image is produced by recording the echo arrival times. In addition, however, the data are projected isometrically, to produce a true three-dimensional perspective of the crack profile within the package. In combination with the Q-BAM images, both qualitative and quantitative perspectives of plastic package anomalies can be produced.

### 9.2.7 Case Study: Production Rate Screening Applications

The field of view can be enlarged to scan trays of parts and improve the efficiency of on-line screening. The larger field of view requires more time for scanning (with C-SAM), but the process of loading and unloading individual parts is circumvented.

Figure 9.14 shows 32 memory chips in the field-of-view at the same time. This is a reflection C-mode image, with a unified black-and-white display in which delaminations are coded black. Because of slight height differences between the devices, it is necessary to use an echo-tracking technique called front interface echo (FIE) tracking. In this method, the instrument senses the height of the top surface of each component and then adjusts the digital gates appropriately during the scanning process. Without this feature, some echoes

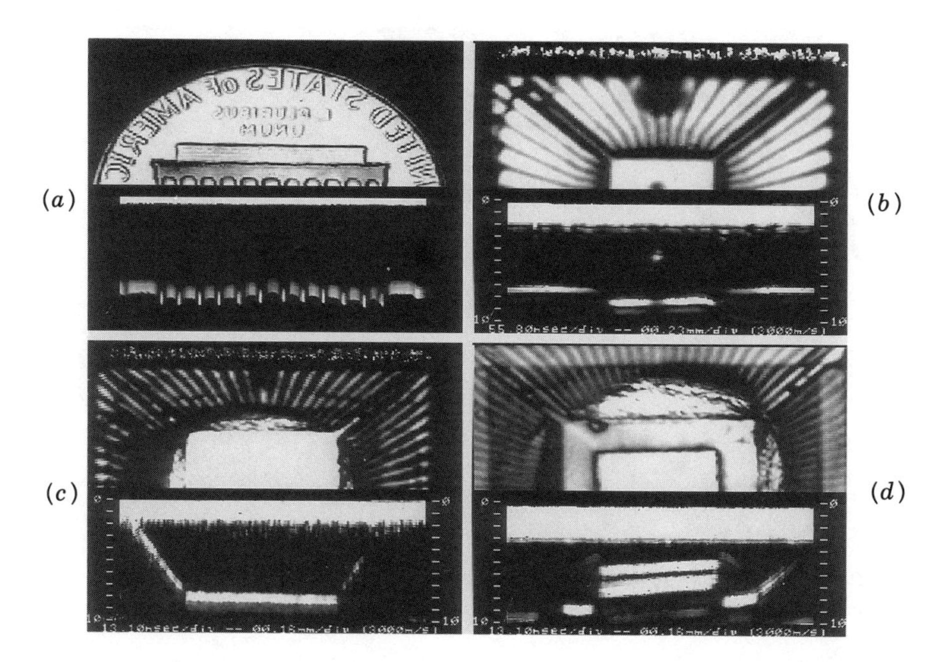


**Figure 9.12** *A Q-BAM image of a U.S. penny imaged through the coin and focused at the rear surface; (b) a 68 PLCC cross-sectioned midway through a dark spot; (c) device with a popcorn crack originating from the edges of the die paddle; (d) a significant degree of paddle shift*

could miss the electronic gate because of signal arrival time differences. FIE tracking is also useful for following the contours of curved or tilted devices.

### 9.2.8 Case Study: Molding Compound Characterization

Characterizing molding compound materials has generally been done from a chemical perspective; physical characterization has usually been limited to density, modulus/stiffness, thermal expansion, and moisture absorption. AMI offers the additional possibility of quantitatively measuring the homogeneity, degree of polymerization, porosity, and overall distribution of filler. The parameters that are measured acoustically are the velocity of sound and the attenuation coefficient. The absorption of ultrasound increases greatly with porosity, for example. It has also been observed that more flexible molding

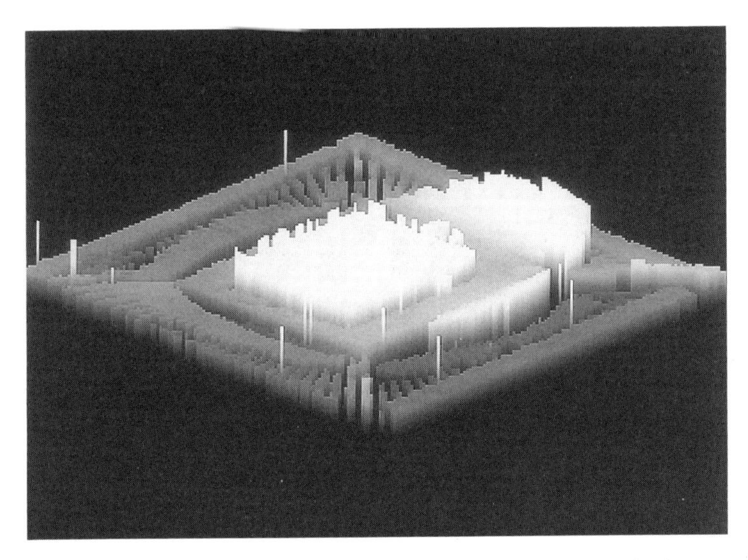

**Figure 9.13** *A time-of-flight (TOF) image of a popcorn cracked integrated circuit*

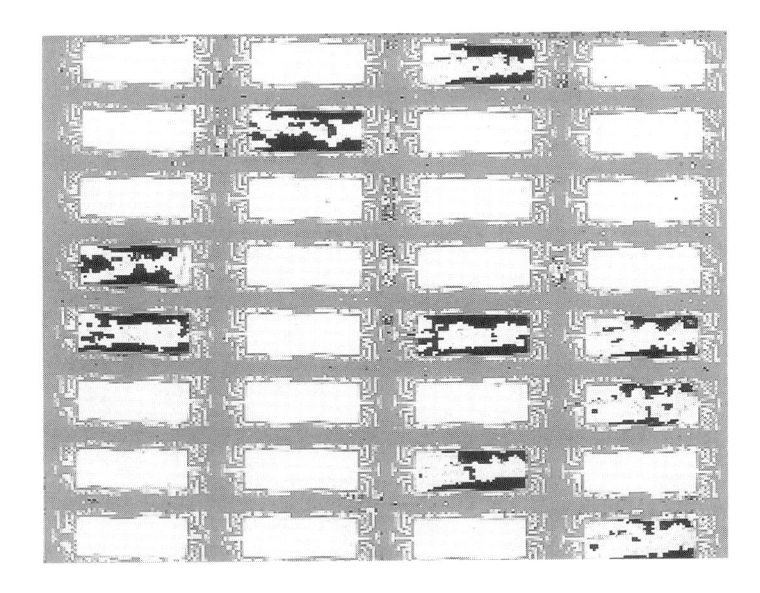

**Figure 9.14** *32 memory chips in the field-of-view at the same time*

materials are far more lossy (absorbing) than stiff materials. This is why lower acoustic frequencies are required to inspect devices containing large dies, which are subjected to far greater mechanical and thermal stresses. Data from typical molding compounds are shown in Figure 9.15.

***Evaluation Check List.***

- void content of molding compound
  - bulk scan top
  - bulk scan bottom
- paddle tilt
- die tilt
- die-surface delamination
- die-attach delamination
- leadframe delamination
- paddle delamination
- heat-spreader delamination
- heat-sink delamination
- package cracks
- Q-BAM acoustic cross section
- 3-D TOF crack profiling (for three-dimensional void and crack profiles)

## 9.3 X-RAY MICROSCOPY

The principal advantage of X-ray microscopy for failure analysis lies in the nature of the image contrast that X-rays produce as a result of differential absorption of the primary X-rays by the specimen. The degree of absorption depends both on the species of atoms in the specimen and on their atomic number. It is thus possible not only to reveal the presence of different microstructural features, but also to obtain information about their composition. Various X-ray microscopes, can reveal microstructural features: the X-ray contact microscope, projection microscope, reflection microscope, and diffraction microscope. Compositions are analyzed using various X-ray spectroscopes. Another advantage of the X-ray microscope is the comparatively deep penetration of X-rays into thick specimens, which enables detection of the internal structure

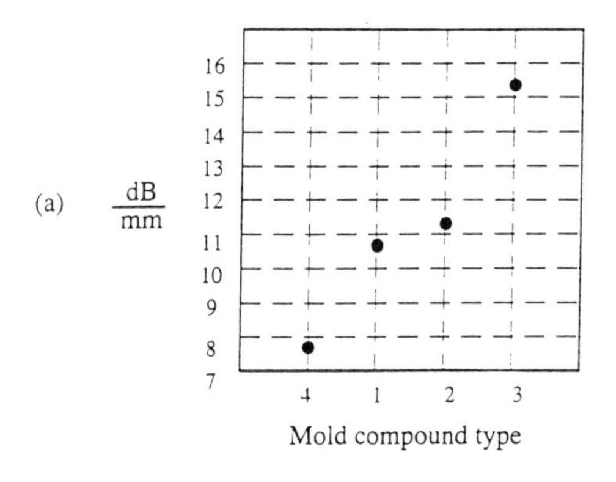


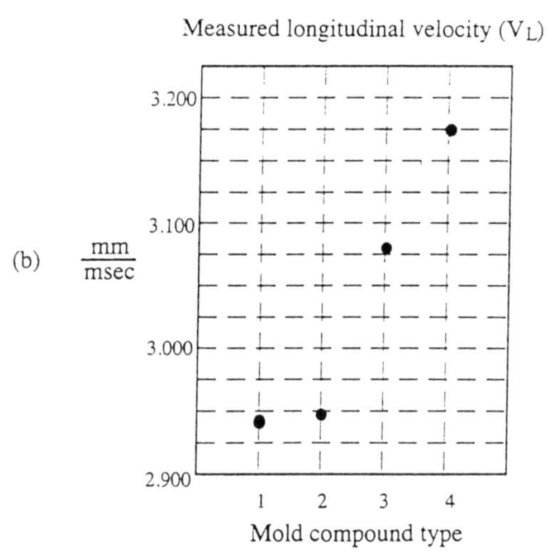


**Figure 9.15** *(a) Attenuation absorption rate measurements for molding compounds of the same basic formulation with variations in particle size and additives; (b) velocity of sound measurements for molding compounds of the same basic formulation with variations in particle size and additives*

of electronic packages or components. X-ray microscopy (also called microradiography) is a nondestructive technique that, like SAM does not require opening the package. A further advantage is that the package or component can be examined in its natural state; neither a conductive coating nor high vacuum is needed, as they are with conventional electron microscopes. No medium like water or oil is required either, while it is for SAM. However, X-ray radiation may change the electrical properties of microelectronic packages, so it should not be used when the electrical properties of the package are under investigation. An X-ray image is formed as a projection of contrast through the whole thickness of a specimen. Difficulties thus arise in analyzing an image of overlapped multiple components. Because of the short wavelength of X-rays, better ultimate resolution can be obtained using X-rays than can be obtained using light, although it is not as good as that obtained using electrons. In applying X-ray microradiography, the difficulty is finding a way to focus the X-rays (electrons, on the other hand, can be focused by a magnetic or condensing lens). X-rays carry no electric charge, so neither magnetic nor electrostatic lenses affect them. The ultimate resolution is usually around 1 μm, although a resolution in the 10-nm region can be achieved with effort. Attention must be paid to radiation safety when an X-ray microscope is used.

### 9.3.1 X-ray Generation and Absorption

X-rays are usually generated in an X-ray tube. Figure 9.16 shows a schematic cross section of an X-ray tube [Southworth 1975]. Under a high-voltage potential, electrons are produced by a heated tungsten filament and are accelerated and focused onto the anode, a metal (e.g., copper) target. The accelerated primary electrons are then decelerated as they collide with the atoms of the target material. X-rays are generated by the energy released by the collisions. Less than 1% of the energy of the primary electrons is actually converted to X-rays; most of the remainder takes the form of heat which is dissipated via a water-cooling system. Generated X-rays are emitted in all directions and allowed to escape from the tube through windows, as shown in Figure 9.16. X-ray tubes are usually evacuated down to a pressure of about $10^{-4}$ Torr in order to minimize the collision of electrons with gas molecules. Permanently sealed tubes are often used to avoid the need for a vacuum-pumping system. The quantity of X-rays emitted by the tube is controlled by varying the current heating the tungsten filament, while the wavelength of the X-rays is determined by the magnitude of the accelerating voltage. Two types of X-ray spectra, the continuous spectrum and the characteristic spectrum, are classified according to features of their wavelengths. The energy of a primary electron is given by $E = eV$ where $e$ is the charge carried by the electron and $V$ is the accelerating potential. If the electron is completely stopped by a single collision, the energy of the generated X-ray quantum can be given by $E = Kv_{quan}$ where $k$ is the Plank constant and $v_{quan}$ is the velocity of the quantum. Since $v = v_{light}/ \lambda$, where $v_{light}$ is the velocity of light and $\lambda$ is the wavelength of the X-rays, $\lambda = kc/qV$. Substituting the values of these constants, $\lambda$ = 1240/V, when $\lambda$ is measured in nm and $V$ is volts. Generally, only a small fraction of the primary electrons is completely stopped by a single collision. The majority of the electrons are decelerated by repeated collisions with other atoms. The X-rays generated by these collisions have lower energies and, thus, longer wavelengths. Consequently, a continuous spectrum with a broad range of wavelengths is generated. The minimum wavelength of the spectrum is determined by the accelerating voltage of the primary electrons.

The determination of a characteristic spectrum is a very different matter. A characteristic spectrum is determined by the target material. When an inner-shell electron in an atom of target material is knocked out by a collision with a primary electron, the atom is unstable. Another electron in the same atom will jump down to fill the vacancy by losing the energy, $\Delta E$, and an X-ray quantum is generated with the wavelength, $\lambda = kv_{light}/\Delta E$. Since $\Delta E$ is a specific quantity associated with the particular energy change occurring in this atom, the wavelength generated is characteristic of this atomic species. Several characteristic wavelengths coexist for a particular species, determined by the $\Delta E$ released by the electron jumping down between different shells in the atom. A characteristic spectrum of a species generally includes $K_{\alpha}$, $K_{\alpha 1}$, $K_{\alpha 2}$, $K_{\alpha\beta}$ radiation, with their specific wavelengths. Table 9.2 lists the wavelengths emitted by some typical target materials.

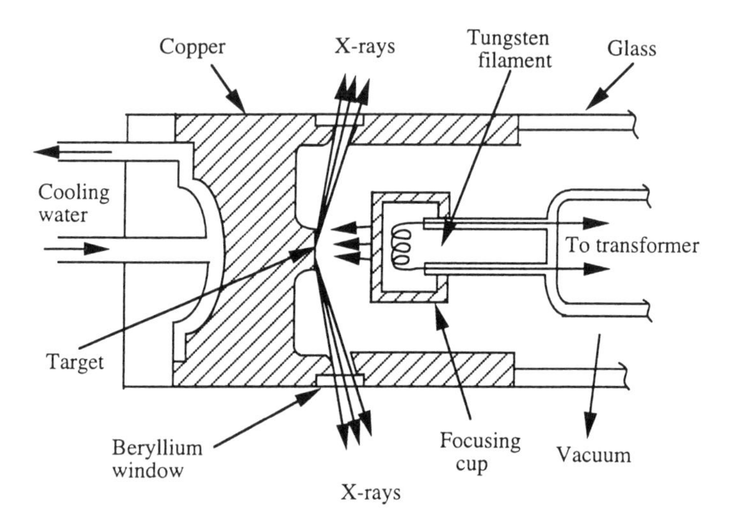


**Figure 9.16** *Schematic cross section of an X-ray tube [Southworth 1975]*

**Table 9.2 Characteristic X-ray wavelengths emitted by target materials**

| Element | $K_{\alpha2}$ (nm) | $K_{\alpha1}$ (nm) | $K_{\alpha B}$ (nm) |
|---|---|---|---|
| Cr | 0.229315 | 0.228962 | 0.208480 |
| Fe | 0.193991 | 0.193597 | 0.175653 |
| Co | 0.179278 | 0.178892 | 0.162075 |
| Ni | 0.166169 | 0.165784 | 0.150010 |
| Cu | 0.154433 | 0.154051 | 0.139217 |
| Mo | 0.071354 | 0.070926 | 0.063225 |
| Ag | 0.056378 | 0.055936 | 0.049701 |
| W | 0.021381 | 0.020899 | 0.018436 |

When an X-ray beam passes through a particular material, its intensity is reduced because the material absorbs some X-rays. The linear absorption coefficient is determined by the atomic number of the material and the X-ray wavelengths. Figure 9.17 shows the linear absorption coefficient versus the X-ray wavelength in the continuous region for a number of materials [Niemann et al. 1982]. Assuming that a 3.0-nm X-ray wavelength is considered, the $\mu$ value changes from approximately 25 $\mu m^{-1}$ for gold and 2.5 $\mu m^{-1}$ for polyimide to 0.15 $\mu m^{-1}$ for water. In other words, a given attenuation, $I = I_0 e^{-1}$, is generated by a thickness of 0.04 μm of gold, 0.4 μm of polyimide, or 7 μm of water. There is a difference of about one order of magnitude between the absorption coefficients for polyimide and gold; this difference provides a good contrast for imaging electronic packages. These data also provide a thickness ratio between different materials that the X-rays can penetrate.

### 9.3.2 X-ray Contact Microscope

Contact microscopes are widely used in the electronic packaging field in both manufacturing and research. In contact microscopy, the specimen is placed in contact with an X-ray image receptor, such as a film cassette or film pack, at a certain distance from an X-ray source (Figure 9.18). The X-rays transmitted through the specimen form an image on the receptor at, effectively, unit magnification. The resulting film is then examined under an optical microscope. The entire internal structure of the specimen can be viewed on the X-ray film. Selected areas may subsequently be enlarged photographically.

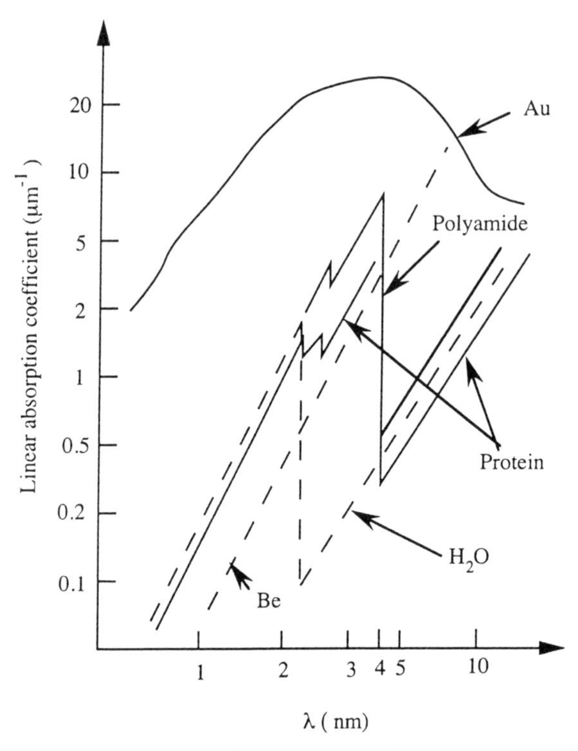


**Figure 9.17** *The absorption coefficient versus the wavelength of X-ray in the continuous region for a number of materials [Niemann et al. 1982]*

To operate a contact X-ray properly, several parameters and aspects should be considered:

- X-ray high-voltage setting and current setting;
- material and thickness of specimen;
- position of specimen;
- exposure time;
- type of film.

The intensity of the primary X-rays is a function of both high-voltage settings and current setting. The intensity of the X-rays penetrating the specimen is dependent on the intensity of the primary X-rays, the absorption coefficient of the specimen, the material, and the thickness of the specimen. The brightness and contrast of the radiographic image shown on the film is, in turn, controlled by the intensity of the penetrating X-rays, the X-ray exposure time, and the sensitivity

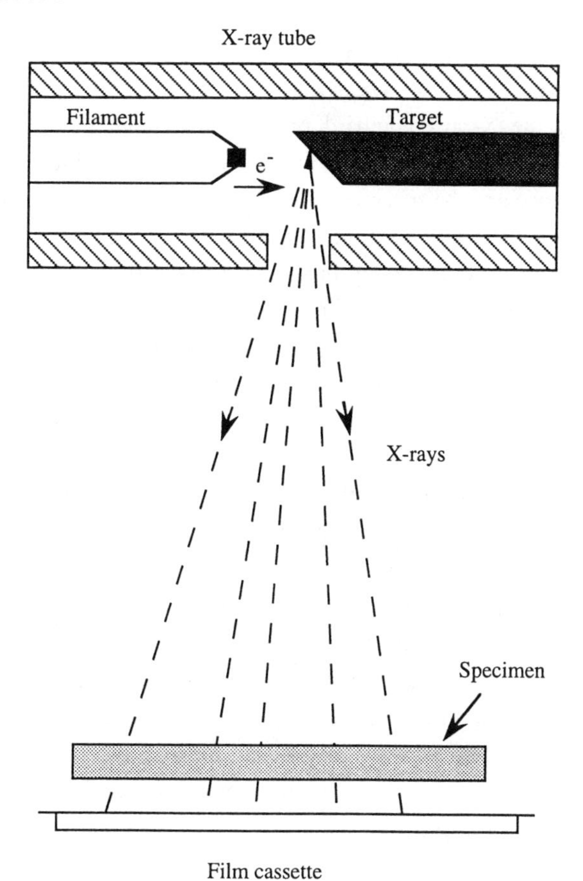


**Figure 9.18** *Schematic of X-ray contact microscopy*

of the film. Optimum parameter selections to obtain a good X-ray radiograph are usually based on empirical practice.

Aside from these parameter selections, lead shielding around the specimen and/or underneath the film helps to increase the image contrast. According to the Abbe diffraction theory, geometrical blurring in the image, gets worse as the specimen-to-film distance is increased. To obtain high resolution and keep the blurring minimal the film should be placed as close to the specimen as possible. Beyond this, the resolution of the X-ray photograph is dependent on that of the optical microscope used for later film-image examination.

### 9.3.3 X-ray Projection Microscope

Unlike the X-ray contact microscope, the X-ray projection microscope provides a primary magnification. The principle of this microscope is demonstrated in

Figure 9.19. The electron beam is focused by a set of magnetic lenses to a tiny spot on the target material. The specimen is placed within a millimeter of the target so that the X-rays bombard only a small area of the specimen. The magnification of the film image produced is calculated as the sum of the target-specimen distance and the specimen-film distance, divided by the target-specimen distance. Since the target-specimen distance is very small with respect to the specimen-film distance, the X-ray image has a magnification effectively equal to the ratio of the specimen-film distance to the target-specimen distance.

Like the image from an X-ray contact microscope, the primary X-ray photograph is viewed subsequently at higher magnifications under an optical microscope. Because the primary image already has a certain magnification, the resolution limit of the technique is determined by those of both the X-ray microscope and the optical microscope. Using the X-ray projection technique, resolutions of 0.1 to 1 μm can be achieved, with a primary magnification of up to about 1000x.

Image contrast with the X-ray projection microscope is achieved in the same way as with contact microscope. The applications and operations of the projection technique are also similar to those of the contact microscope. In addition to higher magnification, the projection technique also provides a greater depth of field, which allows images of stereoscopic pairs, even at high magnification. However, the X-ray projection microscope is much more complex and expensive than the contact microscope.

### 9.3.4 Case Study: Encapsulation in Plastic-encapsulated Packages

An X-ray contact microscope (MICRO RT Model B-510) was employed for this case study. Plastic-encapsulated packages were placed on a manipulator system that can be moved in three directions for adjusting specimen position in a specimen chamber. The adjustment was monitored via a TV imaging system. An X-ray-sensitive film was inserted between the packages and the manipulator system.

A good radiographic image was obtained by adjusting the specimen position and properly selecting the X-ray tube voltage and current, the film type, and the exposure time. Higher tube voltages, and sometimes higher current, are employed for examinations of thicker specimens. Lower tube voltages are usually selected if the films are more sensitive to X-rays. Xeroradiography, Kodak X-omat TL, and Polaroid Type 52, 53 and 55 are the common selectable films. A proper exposure time is determined by the X-ray absorption of the materials, the thickness of the specimen, the tube voltage and current, the

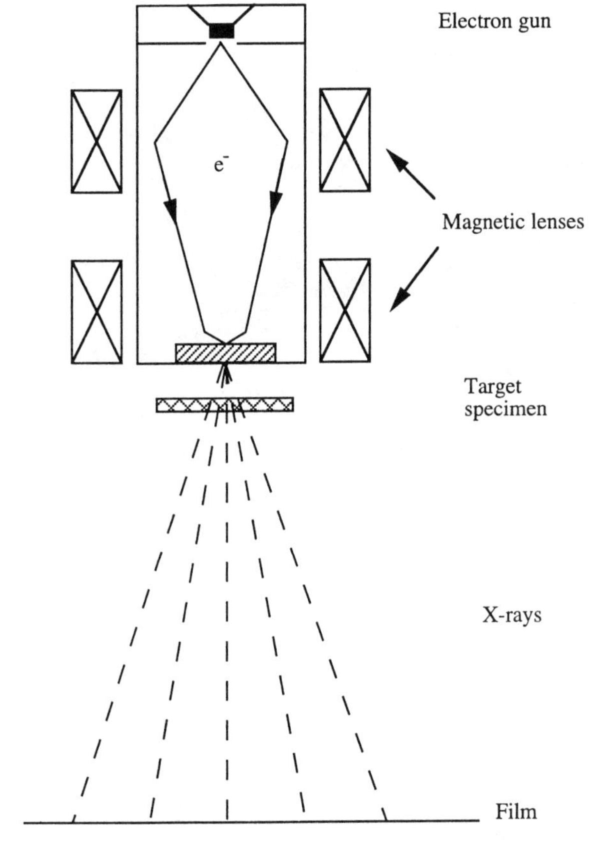


**Figure 9.19** *Principle of X-ray projection microscope*

specimen position and the film type. A comparative exposure guide can be found in the operators manual of the MICRO-RT Series Radiography System (Model B-510).

Figure 9.20 shows the top and side of two 18-pin DIP plastic packages. The packages were thermocycled 100 times between 25 and 200°C. The study focused on examining the internal structure of the packages after the thermal cycling test. Two packages were positioned, one flat and one sideways, on the manipulator. X-ray radiographies were captured by subjecting the packages to 70 kV tube voltage and 0.4-mA tube current, with Polaroid Type 55 film, for 2 min. To minimize the geometrical blurring, the films were placed as close as possible to the plastic packages, allowing one-to-one magnification of the specimens on the films. An optical microscope was then employed to enlarge the negative images captured on the films.

Figure 9.21 shows the enlarged X-ray image. The radiograph is, in fact, a

**Figure 9.20** *Top view and side view of the 18-pin plastic packages*

projective image of the specimen. It is sometimes difficult to interpret the image, especially when many components are overlapped in the packages. Although the top-view image in Figure 9.21 shows wires clearly, the central area image is a projection of molding compound, die, die attach, paddle, and molding compound again, as shown in Figure 9.22a. It is difficult to visualize the interfaces between any two components. The side-view image in Figure 9.21 is a rather simple case. The image shows the plastic/die interface and a dark strip between the molding compound and the die, indicating the plastic/die interfacial delamination.

Defects such as paddle/plastic delamination in QFP and SOJ packages are caused by the popcorn effect. Plastic molding compounds are inherently hygroscopic in nature and will absorb moisture to a level dependent on the ambient humidity and temperature. Moisture diffuses through the plastic, eventually accumulating at the die-plastic and paddle-plastic interfaces. With heating, such as vapor pressure solder reflow, the high temperatures and high heat transfer rate cause absorbed moisture to volatilize rapidly in the package. If the steam pressure is sufficiently high, the package will crack, or "popcorn" (see Section 9.2.14).

The side-view image in Figure 9.21 does not show the paddle/plastic interface clearly because the interface image was overlapped with the image of leads. Paddle/plastic interfacial delamination can be observed by turning the package

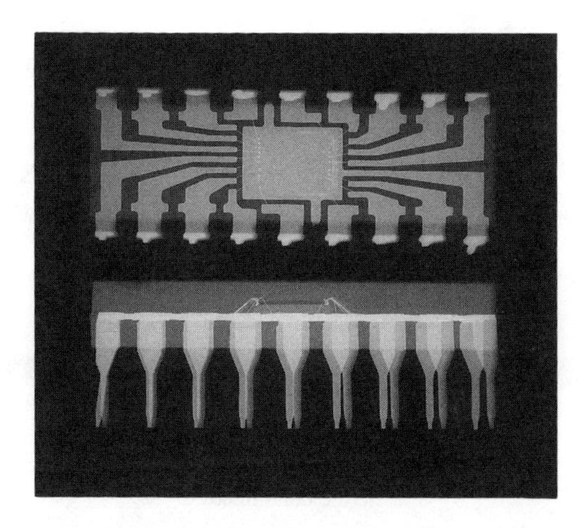

**Figure 9.21** *X-ray radiograph showing the internal structure of the 18-pin plastic packages*

on the other side. Both plastic/die and paddle/plastic interfacial delamination was verified by scanning acoustic microscopic examination.

## 9.4 ELECTRON MICROSCOPY

The first electron microscope was fabricated some sixty years ago. The basic idea arose from the understanding that the concentrating action of a magnetic field in focusing an electron beam was analogous to that of a lens used with visual light. Today there are hundreds of electronic microscope models, including numerous variations on the basic idea. In comparison with the optical microscope, scanning acoustic microscope, and X-ray microscopes, electron microscopes have far superior resolving power; a certain type of electron microscope can even reach atomic scale resolution. An additional feature is that electron microscopes are not restricted solely to providing microstructural information. It is also possible to obtain electron diffraction patterns that offer crystallographic information. Direct chemical analysis provides compositional information on specimens.

### 9.4.1 Electron-specimen Interaction

Electron microscopy has been developed with numerous microanalysis techniques based on the electron-specimen interaction. The conventional techniques are scanning electron microscopy (SEM), transmission electron microscopy (TEM), scanning and transmission electron microscopy (STEM), and electron

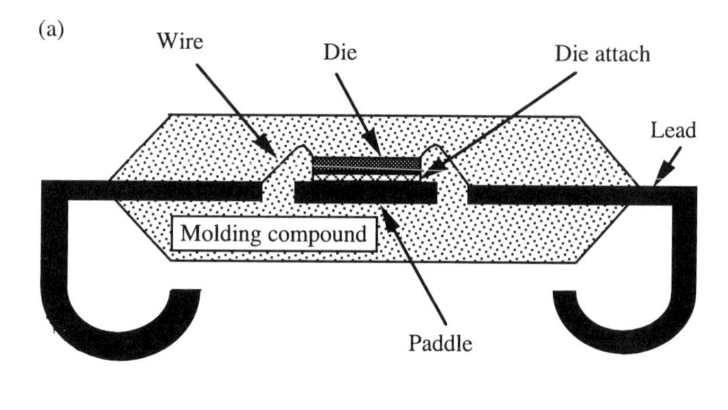


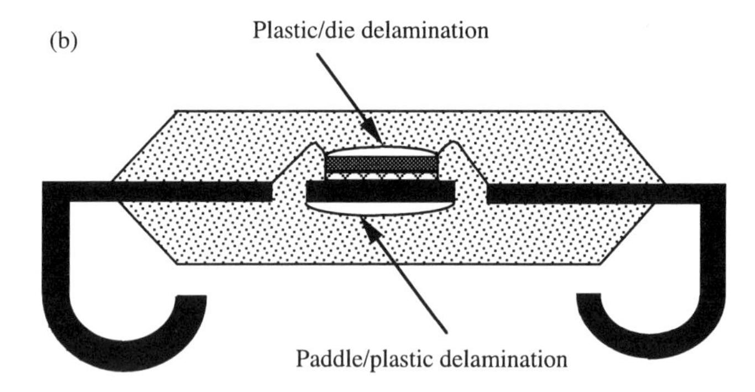


**Figure 9.22** *Popcorn effect in plastic packages: (a) moisture diffuses through the plastic; (b) moisture volatilizes in the package*

microscopies combined with other analysis techniques, such as X-ray and acoustic imaging. When an electron microscope is in operation, the electrons generated by high voltages, often part of the so called primary electron beam, are aimed at the specimen. As shown in Figure 9.23, the primary electron beam interacts with the specimen in a numbers of ways.

A portion of the primary electrons will be transmitted through the specimen if it is thin enough. The transmitted electrons can be both unscattered (coming out of the specimen along the direction of primary electron beam) and scattered (with angles apart from the direction of primary electron beam). Since both unscattered and scattered electrons are transmitted through the specimen, they carry the microstructural information of the specimen. Transmission electron microscopes collect the electrons with a comprehensive detection system and project the microstructural images onto a fluorescent screen. A TEM image is a projection of the microstructure of the specimen.

When the primary electron beam is targeted at the specimen, a portion of the

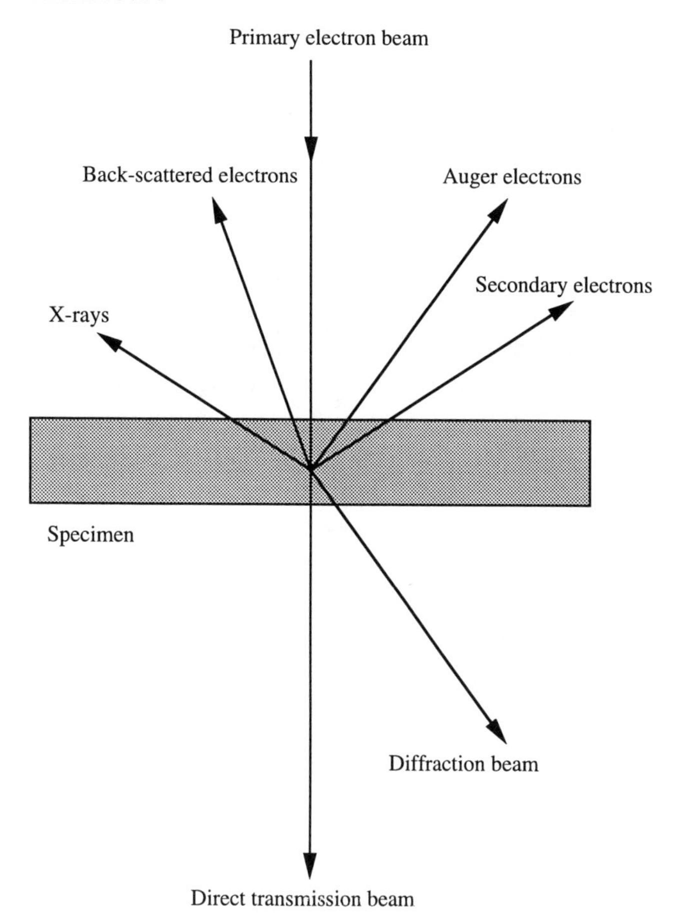


**Figure 9.23** *Signals created by the interaction of high-energy electrons with the specimen*

electrons are back-scattered from the upper surface of the specimen. The electrons in the specimen can also be excited and emitted from the upper surface; these are called secondary electrons. Both backscattered and secondary electrons carry the morphological information from the specimen surface. A backscatter electron detector and a secondary electron detector assembled in a scanning electron microscope will collect these electrons and transmit the signals to a CRT that is scanned synchronously. A SEM image provides morphological information on the specimen surface. A STEM can be understood as a combination of a SEM and a TEM.

In addition to the detectors for TEM and SEM images, further signal detection systems, can also be established with an electron microscope. X-rays and Auger electrons excited from the specimen carry information on the energy levels of the atomic electron orbital, and can be collected to identify chemical elements unambiguously. The energy analysis of X-rays can be carried out using either wavelength dispersion (WDX) or, more often, energy dispersion (EDX). In general, all elements with atomic numbers greater than Z=10 (neon) can be detected using EDX, although the technique can be extended down to boron (Z=5).

Auger electrons are collected by Auger electron spectroscopy (AES). Auger electrons have energies determined by the energy levels in the parent atom and are characteristic of that atom. Since Auger electrons have energies typically in the range 0 to 2000 eV, the distance that they can travel in a solid before losing energy is limited to approximately 1 to 2 nm, as shown in Figure 9.20. This characteristic gives the AES technique its high surface sensitivity.

### 9.4.2 Scanning Electron Microscopy

A modern analytical scanning electron microscope consists of electron optics, comprehensive signal detection facilities, and a high-vacuum environment. A schematic diagram of the scanning electron microscope is shown in Figure 9.24.

The electron optics include an electron gun, condenser lenses, scanning coils, and an objective lens. The electron source is usually a pointed tungsten or lanthanum hexaboride filament that emits a stream of electrons. The filament is kept at a high potential and the electron beam is accelerated through a small hole in the grounded anode before being focused on the specimen by means of a system of condenser lenses. This accelerating potential determines the wavelength of the electrons, $\lambda$. For an accelerating voltage of 10 kV, for instance, the electron wavelength is on the order of $10^{-2}$ nm. If the imaging system of the electron microscope could be made as effective as that of the optical microscope, then the limit of resolution would in theory be of the same order as the wavelength. Unfortunately, the condenser lens aberrations associated with the use of magnetic fields for focusing are so great that the practical resolution limit is on the order of $10^{0}$ nm. Normally, the accelerating voltage for a scanning electron microscope is in the range of 5 to 30 kV, though in some special cases 1 kV to 5 kV is used. The resolution limit is also affected by the conductivity of the specimens. The high resolution is achieved with specimens of high conductivity.

Two detection systems are generally used for imaging with a scanning electron microscope: the secondary electron detector and the backscattering electron detector. With lower energies (< 50 eV), secondary electrons travel in solids in a few nanometers. As a result, secondary electrons generated at more than a few

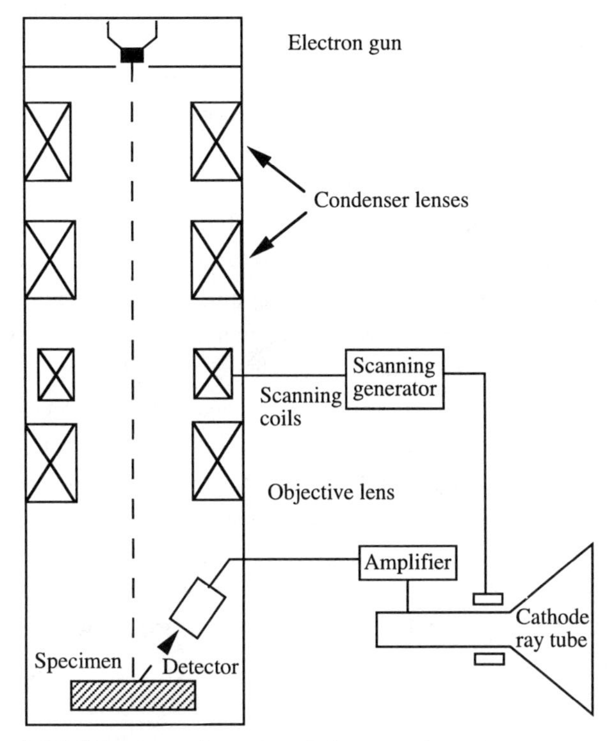


**Figure 9.24** *Schematic diagram of the scanning electron microscope*

nanometers fail to escape from the specimen. This feature makes secondary electron imaging low contrast but high resolution. Traveling distances of backscattering electrons (with energies > 50 eV) reach 10 to 100 μm, depending upon the specimen material. Strong signals are generated because of the large interaction volume in the specimen. Backscattering images have sharper contrast but lower resolution than secondary electron images.

A high-vacuum system is essential for conventional SEMs and TEMs, because passage of the electron beam requires a high vacuum in the microscope column and the areas immediately surrounding both the specimen and the photographic emulsion on which the image is recorded. The vacuum level must be sufficient to avoid collisions between electrons and gas molecules that could affect the routine interaction of primary electrons with the specimen by changing electron paths. Moreover, the residual materials from the collision reactions deposit on the specimen, objective aperture, and other microscope elements, causing contamination that obscures the microstructural observation and modifies the imaging properties of the instrument. In conventional electron microscopes, column pressures are on the order of $10^{-6}$ torr, and electron gun pressures can reach $10^{-9}$ torr (a pressure of 1 torr – 1 mm Hg). Three levels of vacuum system — a mechanical pump ($10^{-3}$ torr), a diffusional pump ($10^{-6}$ torr), and an ion pump ($10^{-9}$ torr) — are typically utilized to provide high-vacuum conditions.

Selection and adjustment of operational variables are critical in order to obtain SEM images with high quality. Accelerating voltage, beam current, and final aperture size are the primary variables in SEM operation. High accelerating voltage is essential to SEM resolution, as illustrated before. However, the effect of interaction volume and specimen charging must be considered. Higher voltage is usually applied to obtain backscattering images, while lower voltage is used to get secondary-scattering images. Specimen charging becomes more severe for a poorly conductive specimen when high accelerating voltage is applied. Beam current is often adjusted by selecting the spot size of the electron beam; a strong signal requires a large beam and thus a reduced resolution. Aperture size affects resolution, too. A small aperture gives good resolution and is often used for images with high magnification; a large aperture is needed to allow the passage of a large beam, and is suitable for X-ray spectroscopy. Aperture size also affects the depth of field. A large aperture is needed if a low magnification limit is approached.

Specimens for conventional SEM analysis must be covered with a thin, conductive coating to avoid electrical charging. A number of coating materials and techniques are available. The most common materials used are carbon, gold, gold-palladium, platinum, and aluminum.

The SEM technique is widely applied as a powerful tool in both semiconductor device and package inspection. Small defects, such as pinholes and hillocks on semiconductors, voids caused by electrostatic discharge, short circuits induced by electrical overstress, hairline fractures of passivation layers, open circuits due to electromigration in metallization, dendrite growth and bonding failures, can be visually inspected easily. SEM can also be used as an electrical testing instrument. When bias is applied to the devices, the SEM image contrast is enhanced by the magnitude of the bias voltage; this technique is called voltage contrasting. The negatively biased areas in the device appear bright, while positively biased areas appear dark. Defective sites can be detected by comparing the contrast of the biased areas.

### 9.4.3 Environmental Scanning Electron Microscopy

ESEMs are a special type of SEM that work under controlled environmental conditions and require no conductive coating on the specimen. The pressure in the sample chamber of an ESEM can be adjusted from 1 to 20 torr or from 1 to

50 torr in terms of different models, that is, only 1 to 2 orders of magnitude lower than the atmosphere. In comparison, the pressure in the sample chamber of a conventional SEM has to be $10^{-5}$ torr or less, which is 6 to 7 orders of magnitude lower than the atmosphere. The pressure condition allows the researcher to examine unprepared, uncoated specimens that are free of surface charging and high-vacuum damage. This makes it possible to examine specimens in their natural states. The environment in an ESEM can be selected from among water vapor, air, $N_2$, Ar, $O_2$, etc. Dynamic characterization of wetting, drying, absorption, melting, corrosion, and crystallization can be performed using ESEM.

ESEMs are able to work with certain pressures and without surface charging because the secondary electron detector is designed on the principle of gas ionization. This is illustrated in Figure 9.25. As primary electrons are emitted from the gun system, the secondary electrons on the specimen surface are accelerated toward the detector, which is biased by a moderate electric field. The collisions between the electrons and gas molecules liberate more free electrons, and thereby provide more signals. Positive ions created in the gas effectively neutralize the excess electron charge built up on the specimen. Proper operating pressure controls the specimen surface charging. Depending on the environmental requirements, the pressure source can be water vapor, air, argon, nitrogen, or other gases. Four-stage vacuum levels (electron gun chamber, optical column, detector chamber, and specimen chamber) are maintained from $10^{-7}$ to $10^{1}$ torr by a computer-controlled vacuum system.

The specimen chamber pressure is one of the important ESEM parameters, in addition to the parameter settings for conventional SEMs. Adjusting the specimen chamber pressure helps provide stronger signals, and consequent sharper contrast for ESEM imaging, because the adjustment controls the number of gas molecules between the specimen and the secondary electron detector. The environmental relative humidity in the specimen chamber can also be varied by adjusting the chamber pressure. Relative humidity can be varied in the range of 20 to 90% at around room temperature; however, this is not an easy procedure, because other parameters, such as electron beam size, voltage, image contrast, and brightness, must be adjusted appropriately to maintain a good image.

The low vacuum level required for the specimen chamber allows a large working space in the chamber. A temperature stage, mechanical test system, in either tensile, compression, four-point bending, or shear mode, and a micromanipulator/microinjector system can be installed in the specimen chamber to enable various dynamic investigations, such as the simulation of different failure mechanisms on electronic packages. Another advantage of ESEMs is that, working under controlled environmental conditions, they can still function as well

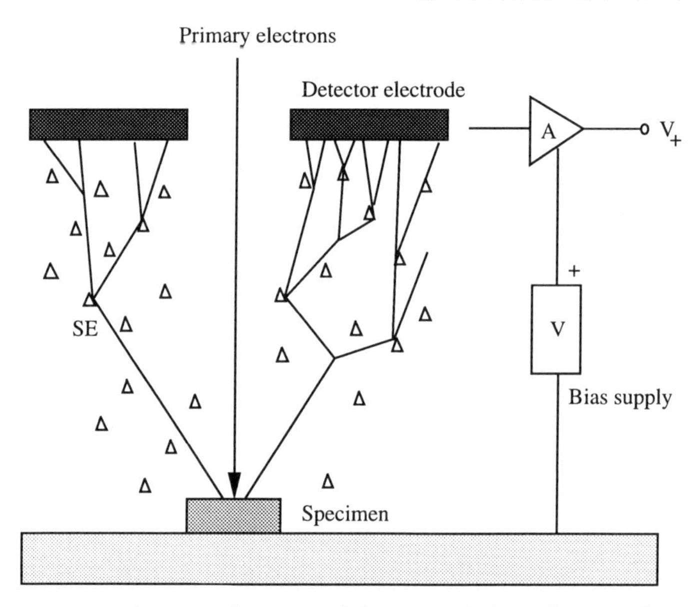


**Figure 9.25** *Schematic diagram of the transmission electron detector*

as ordinary SEMs without such trade-offs as a loss of resolution.

### 9.4.4 Transmission Electron Microscopy

TEM is composed of comprehensive electron optics, a projection system, and a high-vacuum environment. A schematic diagram of TEM is given in Figure 9.26. The electron optics consist of an electron gun, condenser lenses, objective lens, and a projector lens system. High vacuum is also a critical issue for successive imaging and the prevention of contamination. Higher voltage is applied to the electron gun than in SEM, so the primary electrons can carry sufficient energies to penetrate the specimen. The ultimate voltage for a TEM can generally be from 100 to 1000 kV, depending on the requirement of resolving power for a particular TEM. TEMs with higher ultimate voltage have higher resolving power and handle thicker specimens. The ultimate resolution of TEMs with ultrahigh voltage can reach subangstrom levels (1 angstrom = $1 \times 10^{-10}$ m).

Alternative operating modes can be utilized in applying TEM to failure analysis of electronic materials and packages. Bright-field and dark-field modes can be selected to obtain different diffraction contrast images; high-resolution mode can be used to get phase-contrast images, especially useful for failure analysis in semiconductor/oxide/metallization interfacial failure modes. These three modes are discussed below.

As discussed in Section 9.4, electrons transmitted through the specimen are

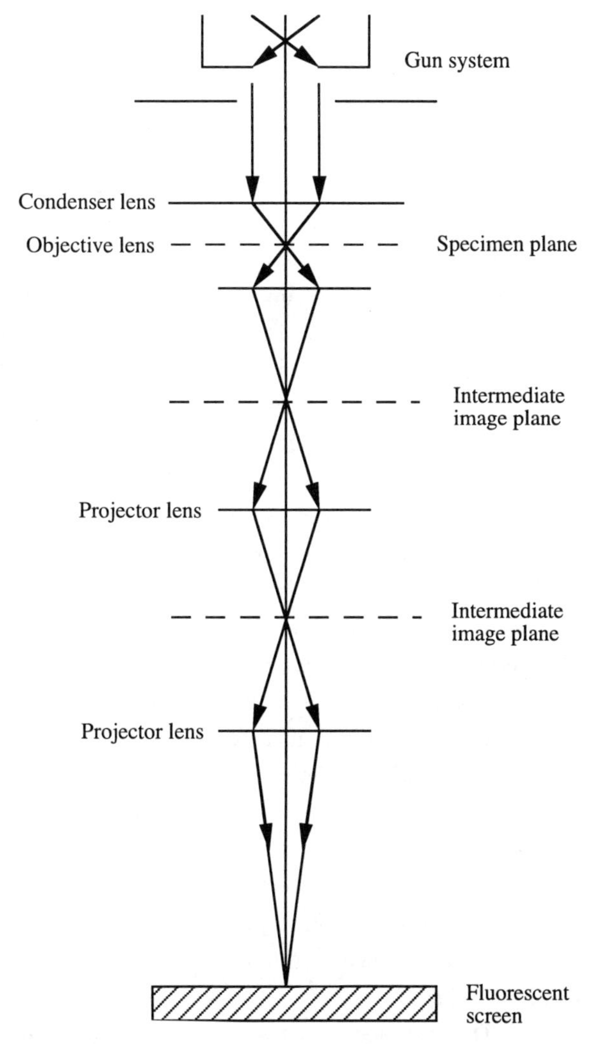


**Figure 9.26** *A schematic diagram of transmission electron microscopy*

divided into unscattered electrons and scattered electrons. The former moves parallel to the primary beam and is called the direct beam; the latter is scattered from the primary beam direction and is called the diffracted beam if the electrons are scattered without energy loss. The angle that the diffracted electrons make with the direct beam, defined according to Bragg's law, is $2\theta$. In practice there are, of course, a number of these diffracted beams for any crystalline specimens. Normally, all the diffracted beams are stopped by the objective aperture, and only the direct beam contributes to what is known as a bright-field image, as shown in Figure 9.27a. The bright-field imaging mode is most often selected in TEM operation.

As an alternative to bright-field imaging, the direct beam may be obstructed and one of the diffracted beams allowed to form the image. This dark-field image is illustrated in Figure 9.27b. Dark-field imaging can be obtained by displacing the objective aperture, as shown in the figure, leading to poorer resolution since the electron rays are now off-axis. As an alternative, the electron gun may be tilted so the primary beam strikes the specimen at an angle. Since the diffracted electrons are now used for imaging, the contrast will be the reverse of that seen under bright-field imaging. The dark-field technique is particularly useful when imaging defects or particles, since the electrons contributing the image are diffracted from the defects or particles alone.

Dark-field and bright-field contrast together are known as diffraction contrast. While the diffraction contrast can reveal many microstructural features, it cannot reveal the periodic crystal structure itself. A high-resolution technique creates the possibility of revealing phase contrast, which can provide periodic crystal structure information. Phase contrast is obtained from the phase difference between a direct beam and diffracted beams. Atomic planes are imaged by opening up the objective aperture to allow one or more diffracted beams, in addition to the direct beam, through. Obviously, only the planes with spacing greater than the resolution limit of the microscope can be imaged. One popular application of this high-resolution technique in electronic packaging is as a test of the interface mismatch between the metallization thin film and the silicon substrate.

TEM specimen preparation is a destructive and time-consuming procedure. The chip is micro-sectioned into disks typically 0.5 mm thick and 3 mm in diameter. The disk is then polished to a thickness of several micrometers. An ion-etching technique mills the center of the disk to a thickness a tenth of a nanometer or less. The last step is the same conductive coating required in conventional SEM specimen preparation. Temperature substage and strain substage are also employed in TEM investigations. In the failure analysis of plastic packages, SEM is more popular than TEM. In terms of operation techniques, environmental control is involved in ESEM operation, along with the conventional SEM operation routines. The case studies illustrated in this section are related to ESEM failure analysis.

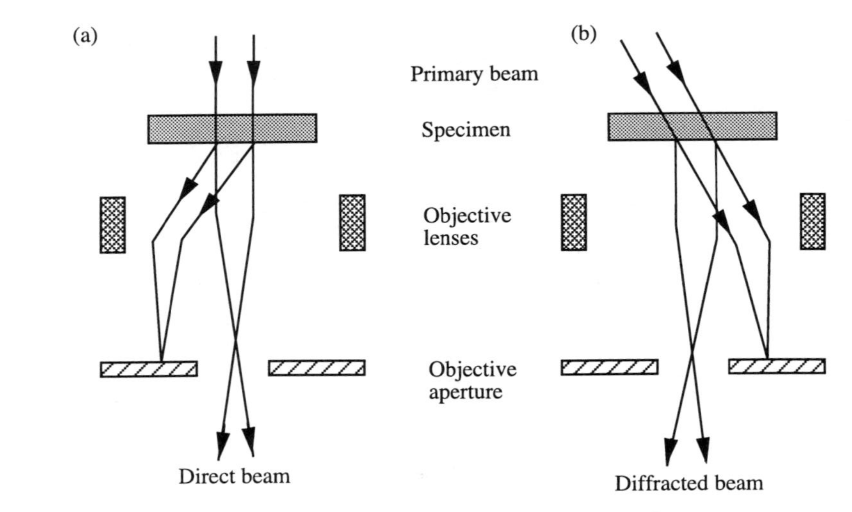


**Figure 9.27** *Illustration of diffraction contrast: (a) bright-field and (b) dark-field illumination*

## 9.5 OTHER TECHNIQUES

Various other failure analysis techniques are available in the microelectronic field. Among these techniques, optical microscopy, microprobing, liquid crystal display, and infrared microscopy are primarily employed in microelectronics industries and research laboratories because they are conventional, cheap, and easy to use. As a result of rapidly decreasing device sizes in VLSI and ULSI technology, new analysis tools are required to detect failure modes scaled from the submicrometer to the atomic level. New techniques have been developed, including scanning tunneling microscopy, atomic force microscopy, scanning near-field optical microscopy, low-energy electron microscopy, and emission microscopy.

### 9.5.1 Optical Microscopy

Optical microscopy (OM) is a major tool for failure analysis in different levels of packages because optical microscopes are cheap, convenient, and easy to use. OM consists mainly of a light-illuminating system; an objective system that determines objective magnification, numerical aperture, resolution, depth, and curvature of field; and an eyepiece system with selectable magnifications. Specimens can be examined in reflected light, or transmitted light if the specimen is made of transparent material. Besides bright field, advanced optical microscopy techniques include dark field, phase contrast, interference contrast, polarized light, and others. These techniques are illustrated in detail in Southworth [1975]. In comparison with electronic and other microscopes, optical microscopes work without requiring a high vacuum or a conducting material. The disadvantage is that resolution is limited by the wavelength of visible light.

### 9.5.2 Microprobing (MP)

Microprobing (MP) is widely applied to test electrical properties of packages and, more commonly, monolithic integrated circuits after package opening. MP is straightforward and relatively inexpensive. An MP facility consists primarily of electronic equipment, a probe station with optical microscope, manipulators, and probeware. The electronic equipment includes an oscilloscope or curve tracer, a pulse generator (for dynamic measurements), and a digital voltmeter and spectrum analyzer (for noise measurements). During the measurement, the electronic measuring equipment is connected through microprobes to a circuit node and presents the measurement results. The curve tracer provides more information than a voltmeter about the nature of the failure mechanism beyond the failure mode, and plays an important role in the decision-making segment of analysis. Examples of curve interpretation can be found in Patterson [1978] and Schwartz [1980]. The probe station is usually designed for packaged devices with interchangeable sockets so that various integrated circuit packages can be accommodated. The optical microscope helps locate the pins or circuit nodes to be tested and the failure sites. Due to the increasing complexity of semiconductor devices, the probe diameter must be smaller, so the probe sharpening technique becomes more critical. Control of probe shape and diameter during sharpening is discussed in Hiatt [1980].

### 9.5.3 Liquid-Crystal Display

Liquid-crystal display (LCD) is a conventional failure analysis technique that produces a static optical display of circuit-node logic levels for identifying the location of hot spots or shorts by color change. In some cases, using LCD eliminates the need for microprobing. The basic LCD involves procedure chip-surface coating with a thin carbon film; adding a drop of liquid-crystal solution, cooling, applying power to the integrated circuits; and identifying short locations with a microscope. Since the liquid crystal and carbon film can be removed with a solvent such as pentane, benzene, or chloroform, LCD is a nondestructive technique for failure analysis of microelectronic packages and devices. LCD is particularly advantageous when the application of a mechanical microprobe becomes difficult. Specimen preparation, display-cell characterization, device control, and limitations are discussed in Burns [1978] and Goel and Gray [1980].

### 9.5.4 Infrared Microscopy

Infrared microscopy (IRM) is often applied after package opening. In thermal emission detection and analysis, using IR radiation with a wavelength of ~10 μm, the temperature distribution on the device surface can be displayed using a scanning infrared microscope and a gray scale to differentiate the temperatures and pinpoint hot spots. Measurement accuracy depends to a considerable extent on magnification, average temperature, and emissivity. Since silicon is essentially transparent at IR wavelengths of 0.8 to 1.3 μm, a microscope using special IR transmitting optics, combined with an IR camera and monitor, can readily be employed to investigate failure modes in die subsurfaces. Failures at the interface between a gold ball bond and an aluminum metallization pad, sub-surface corrosion of the aluminum metallization, silicon precipitates, and damage due to electrostatic discharge in a device are readily diagnosable. Although the

IRM technique is considered very useful for the failure analysis of plastic-encapsulated components, it is not fully satisfactory because of its resolution limit. IRM is restricted to the failure analysis of submicrometer structures. Moreover, many packaging materials are not IR transparent, so that the transmission IR technique is limited to a few materials, such as silicon and gallium arsenide.

### 9.5.5 Scanning Tunneling Microscopy

Scanning tunneling microscopy (STM) is a newly developed surface examination technique based on the principle of electron tunneling [Hansma et al. 1987, Silverman 1989]. Electrons can cross a vacuum gap between two solids; the number of electrons that jump in a given time, defined as the tunneling current, depends on the separation distance and the bias. The bias is very small, ranging between 2 mV and 2 V. Because one of the solids is the specimen surface and the other a probe electrode, the tunneling current will vary with the roughness of the surface as the probe is scanned across it. The STM is designed to maintain a constant current using a feedback loop. Piezoelectric elements move the tip of the probe up and down during scanning. A topographical map of the surface is deduced from changes in the piezo-driver voltage. The resolving power of STM reaches 0.01 nm vertically and 0.6 nm horizontally. STM has some limitations: only conducting materials can be examined, and STM is very sensitive to vibration from thermal drift because the distance between probe and specimen surface has to be stabilized on a scale <0.01 nm.

### 9.5.6 Atomic Force Microscopy

Atomic force microscopy (AFM) uses the same technology as STM, but can be employed in applications for both conductive and nonconductive specimens, presenting a major advantage over STM. AFM uses a vibrating tip very close to a specimen surface, tracked by laser heterodyne interferometry [Binnig 1986, Martin et al. 1987]. The tip motion is disturbed by forces associated with either topography and/or environments determined by surface atoms, so the surface can be profiled and characterized. Other advantages of AFM are that more types of materials can be examined and a high vacuum is not required.

### 9.5.7 Scanning Near-field Optical Microscopy

As with conventional optical microscopy, scanning near-field optical microscopy (SNFOM) utilizes visible light as an optical source. The resolution of a conventional optical microscope, however, is limited by the wavelength of visible light, usually measured in micrometers. SNFOM, in fact, has broken this limit and become a new branch of optical microscopy. In a SNFOM, a very fine optical antenna (an optical fiber) is used to probe the near field of the specimen, which contains structural information limited in resolution only by the structural details of the specimen itself and by the distance between the tip of the antenna and the specimen. The interaction between the two controls the light emission of the antenna, and thus provides signals for topographic imaging. Scanning near-field optical microscopes have been developed with a resolution between 20 and 50 nm laterally and about 1 nm vertically, which is two to three orders of magnitude higher than that of conventional optical microscopes. Convenient, easy to use, and capable of high resolution, SNFOM promises continuing advances in failure analysis at the semiconductor wafer level, especially for failures occurring during processing.

### 9.5.8 Low-energy Electron Microscopy

Low-energy electron microscopy (LEEM) utilizes low-energy electrons as an optical source [Bauer et al. 1987]. At low energies, electrons interact strongly with the first few atomic layers of the specimen and are backscattered and collected by detectors to form an image of the surface (much like the behavior of backscattered electrons in a scanning electronic microscope with a back-scattering electron detector). The LEEM technique, however, does not require electron scanning line by line, so it has the advantage of rapid imaging, like the collection of reflected light in an optical microscope. The changes of a surface can be followed in real time at video rates; moreover, the surface steps of atomic height can be seen in good contrast. LEEM is superior to the scanning tunneling microscope in its relatively large field of view and in its image acquisition time, both of which are critical in practical failure analysis. LEEM images have high magnification (10,000x) and high resolution (15 nm). The disadvantage of LEEM is that an ultrahigh vacuum is required for insuring a very clean specimen surfaces.

### 9.5.9 Emission Microscopy

Emission microscopy (EM) is another exciting development in failure analysis technology. EM offers nondestructive opportunity to image defects in dielectric films. It combines the imaging of photon emission with various modes of electrical testing [Khurana and Chiang 1986, 1987]. As current flows through thin dielectric films, faint light is emitted. If the film is damaged, the local current flow is increased, and more photons are thus emitted. The light in the local area is modified, showing bright spots that indicate the presence of defects. The defects can be process- or structure-induced oxide breakdown, electrostatic discharge damage, electromigration voidage, and so forth. Hot electron problems and oxide defects in VLSI products have been successfully detected using EM, and emission behavior has been observed on several dielectric films (silicon oxide, nitride, and oxynitride). The resolution of EM reaches 0.5 μm. The disadvantage of EM is that a limited range of materials can be examined. Spiked junctions, metal short circuits, very heavily damaged oxide, or defects lying under extremely wide metal layers cannot be detected using an emission microscope.

## 9.6 SELECTION OF FAILURE ANALYSIS TECHNIQUES

The failure analyst must decide which of these various primary and advanced failure analysis techniques should be used to detect failure modes in a particular case for a particular package. Misusing analysis techniques will affect the analysis and may result in an incorrect conclusion. Proper technique selection is usually based on a consideration of both the status of the package to be analyzed and the performance of the failure analysis instrument. Significant factors for the status of the package or component include

- history (process, application, environment);
- materials;
- structure size;
- geometry;
- possible failure modes; and
- possible location of the failure.

Factors reflecting the performance of the analysis tool include

- resolution;
- penetration;
- method (destructive or nondestructive);
- specimen preparation requirement;

- cost; and
- time.

To analyze a failed package the processing and application history of the package should first be obtained. This will help to indicate the proper failure analysis technique and quickly locate the failure. Electrical bias, temperature, relative humidity, vibration conditions, radiation, and so on are all considered environmental historical conditions of the package. If this information is not available, experience becomes critical. Knowing the construction materials of the package is also important. For instance, metallic materials may cause electromigration-induced failure; polymeric materials may lead to moisture absorption-related failure; and a package of multimaterials may have a mismatch between coefficients of thermal expansion causing interfacial failure.

Consideration of the size of the components in a package also helps in selecting the proper failure analysis tool. For example, contamination failure on the surface of a plastic package can be identified easily by a conventional optical microscope, while that on the gate surface of an FET with a submicrometer gate length has to be detected by an analysis tool with high resolution, like a scanning electron microscope. Considering the geometry of a package, possible failure modes, and possible locations of the failure is essential for decisions such as, whether it is necessary to open the package; whether a non-destructive failure analysis technique can be utilized, and so forth.

Reviewing the package status should enable a failure analyst to understand the relevant failure analysis requirements of a particular failed package and select the proper analysis technique that matches the requirements.

Each technique has its own resolution and penetration limits. Figure 9.28 shows an overall comparison of lateral resolution versus depth of penetration for the various failure analysis techniques. X-ray microscopy offers the best depth of penetration, followed by scanning acoustic microscopy; a 1-μm lateral resolution seems to be the boundary. Among these techniques, only electronic microscopies (TEM, SEM, and E-SEM) are able to pass the boundary; transmission electronic microscopy provides the best resolution. Oxide layers on semiconductor substrates usually are less than 1 μm thick (ranges from a few nanometers to hundreds), so oxide failures are best detected by those techniques.

Specimen preparation and destructive requirements also must be taken into account. The use of electronic microscopes, except for the environmental scanning electronic microscope, usually requires destructive specimen preparation, for example, sectioning, layer removal, permanent coating, and package

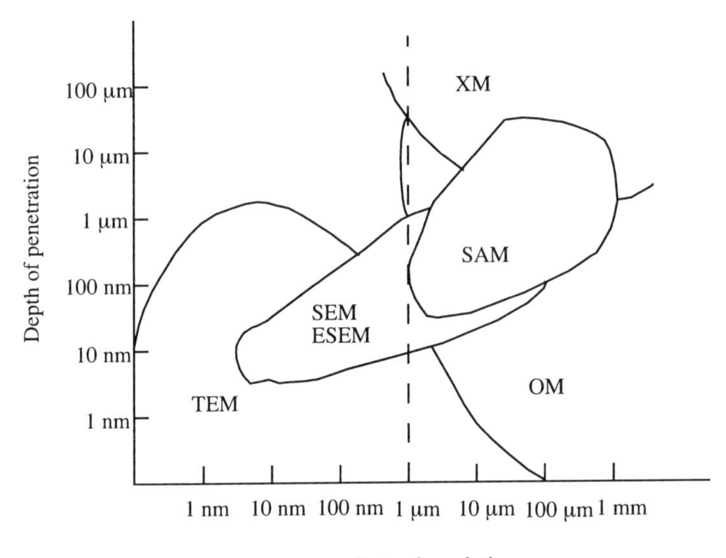


XM - X-ray microscopy
SAM - Scanning acoustic microscopy
OM - Optical microscopy
TEM - Transmission electron microscopy
SEM - Scanning electron microscopy
ESEM - Environmental SEM

**Figure 9.28** *Schematic comparison of lateral resolution versus depth of penetration of various failure analysis techniques*

decapsulation. Transmission electronic microscopy requires specimen sectioning, while conventional scanning electronic microscopy requires at least conductive coating. Some failure modes in microelectronic packages, such as microcracking, interfacial delamination, and debonding, can be introduced by specimen preparation, confusing the analyst.

Fortunately, a number of nondestructive analysis techniques, such as scanning acoustic microscopy and X-ray microscopy, can be used to detect internal failure modes. These techniques, however, cannot provide images with high resolution. A compromise needs to be achieved between resolution and nondestructive methods. Among the previous failure analysis techniques, it is evident that the environmental scanning electronic microscope and atomic force microscope perform with relatively high resolution and without requiring layer removal or permanent coating, but E-SEM and AFM detect only failures exposed on the specimen surface; they cannot image internal failures in plastic packages without decapsulation.

Generally, the failure analysis with the higher-resolution techniques will cost more, and will take longer if specimen preparation is required. Transmission electronic microscopy (0.2-nm resolution) usually costs 30 to 40 times more than optical microscopy (1-μm resolution). There are, however, some exceptions, especially for newly developed failure analysis techniques. Commercialized atomic force microscopy competes in resolution with scanning electronic microscope (2 to 5 nm), and costs only about 20% of the latter. The cost also depends on other factors. Scanning acoustic microscopy gets similar results, or even lower if a low-frequency transducer is employed. Its resolution is comparable to that of optical microscopy.

For reference, the common failures in plastic packages and the proper failure analysis techniques are listed in Table 9.3. External and package failure modes can be inspected without decapsulation. Internal failure modes are often located at the die level or other places inside the package, and usually require decapsulation for failure inspection, unless nondestructive techniques like scanning acoustic microscopy or X-ray microscopy are utilized. In most cases, more than one technique is listed in Table 9.3 for analyzing a particular failure mode. These provide alternatives when resolution or location of the failure must be taken into account. Scanning acoustic microscopy, for instance, is the proper selection for detecting interfacial delamination because of its nondestructive penetration ability. However, the resolution of SAM is not high enough to inspect interfacial delamination in devices with a submicrometer scale, such as VLSI and ULSI devices. ESEM is recommended for those cases, but care must be taken because the cross-sectioning of the specimen may be required. Often, two or more analysis techniques are employed in order to confirm a particular failure mode.

## 9.7 SUMMARY

Microprobing and liquid-crystal display, combined with conventional optical microscopy, are still the primary failure analysis techniques in the microelectronic industry. The use of these techniques will continue, because they are cheap, convenient, and easy to use.

Nondestructive failure analysis techniques play key roles in the analysis of failures not visible on the specimen surface. Optical microscopy and infrared microscopy can work in this way, but only for transparent materials and a few limited materials. Scanning acoustic microscopy and X-ray microscopy are the best techniques for this task. The examples presented in this chapter demonstrate

**Table 9.3 Selection of failure analysis techniques for plastic package failures**

| Failure mode | Failure analysis technique |
|---|---|
| External and package failure | |
| Package mechanical damage | OM, SLAM, SAM |
| Corrosion, contamination | OM, ESEM, SEM, EDX, SLAM |
| Plastic cracking | SAM, IRM, ESEM, SEM, OM, SLAM |
| Plastic/die delamination | SAM, IRM, SLAM |
| Internal failure | MP, LCD, IRM, OM, ESEM |
| Electrical overstress | MP, LCD, IRM, OM, ESEM |
| Electrostatic discharging | OM, ESEM, SEM |
| Electromigration | XM, SAM |
| Die-attach voiding | SAM, SLAM |
| Die-attach delamination | OM, SEM, ESEM, EDS, SLAM, SAM |
| Corrosion, contamination | SAM, ESEM |
| Interfacial delamination | SAM (high frequency), SLAM |
| Die cracking | ESEM, SEM, SAM |
| Metallization microcracking | EM, ESEM, SEM |
| Oxide breakdown | TEM, ESEM, SEM |
| Si or Ga-As defect | |

the wide utility of acoustic microimaging for detecting and characterizing flaws in plastic packages. The specific information these techniques provide on the presence and character of a defect is essential for failure analysis, device screening, and process control feedback.

The use of electronic microscopies, with their high resolution, has increased with the booming popularity of VLSI and ULSI. Conventional SEM and TEM require high-vacuum conditions and specimen coating preparation. Environmental scanning electronic microscopy allows investigation of microscopic morphologies under controlled environmental conditions without conductive coating, and has become an attractive technique for microstructural characterization of microelectronic packages. This advantage of E-SEM particularly facilitates investigation of devices with dielectric materials, such as polymers and ceramics. Thermal and sorptive environmental effects and environmental cycling effects on different package levels can be dynamically investigated using E-SEM.

New, proven failure analysis techniques — specifically, atomic force microscopy and emission microscopy — are being developed that may continue to be, useful into the future. A unique failure analysis technique is still to be developed that can perform not only with high resolution, but also without the destruction of packages.

## REFERENCES

ANSI/IPC-SM-786, Recommended Procedures for the Handling of Moisture Sensitive Plastic Integrated Circuits Packages. *Institute for Interconnecting and Packaging Electronic Circuits (IPC)* (December 1990).

Bauer, E. Telieps, W., and Turner, G. Low-Energy Electron Microscopy. *National Symposium of the American Vacuum Society* (November 1987).

Binnig, G., Quate, C.F., and Gerber, C. Atomic Force Microscope. Review Letters 56 (1986) 930.

Burns, D.J. Microcircuit Analysis Techniques Using Field Effect Liquid Crystals. *IEEE 16th Annual Proceedings of the Reliability Physics Symposium* (1978) 101-107.

Byrne, W.J. Three Decapsulation Methods for Epoxy Novalac Type Packages. *IEEE 18th Annual Proceedings of the Reliability Physics Symposium* (1980) 107-109.

Cichanski, F.J. Method and System for Dual Phase Scanning Acoustic Microscopy, U. S. Patent 4,866,986 (September 1989).

Goel, A., and Gray, A. Liquid Crystal Technique as a Failure Analysis Tool. *IEEE 18th Annual Proceedings, Reliability Physics* (1980) 115-120.

Hansma, P.K. and Tersoff, J. Scanning Tunnelling Microscopy. *Journal of Applied Physics* 2 (1987) R1-R23.

Hiatt, J. Microprobing. *IEEE 18th Annual Proceedings of the Reliability Physics Symposium* (1980) 116-120.

Kessler, L.W. Acoustic Microscopy. *Metals Handbook, Ninth Edition, Nondestructive Evaluation and Quality Control,* ASM International, Materials Park, OH. (1989) 465-482.

Kessler, L.W., and Yuhas, D.E. Acoustic Microscopy. *Proc. IEEE,* 67(4) (April 1979) 526-536.

Khurana, N., and Chiang, C.L. Analysis of Product Hot Electrons by Gated Emission Microscopy. *IEEE IRPS* (1986) 189-194.

Khurana, N., and Chiang, C. L. Dynamic ImaginFERENCESrent Conduction in Dielectric Films by Emission Microscopy. *IEEE IRPS* (1987) 72-76.

Lin, Z.C., Lee, H., Wade, G., Oravecz, M.G., and Kessler, L.W. Holograph Image Reconstruction in Scanning Laser Acoustic Microscopy, *Transactions IEEE UFFC-34* (No.3) (May 1987) 293-300.

Manzione, L.T. Plastic Packaging of Microelectronic Devices. Van Nostrand Reinhold, New York (1990) 273-279.

Martin, Y., Williams, C.C. and Wickramasinghe, H. K. Atomic Force Microscope: Force Mapping and Profiling on a Sub 100Å scale. *Journal of Applied Physics* 61 (1987) 4723-4729.

Niemann, B., Schmahl, G., and Rudolph, D. X-ray Microscopy: Recent Developments and Practical Applications. *Proceedings SPIE* 368 (1982) 2-8.

Patterson, J. M. Developing an Approach to Semiconductor Failure Analysis and Curve Tracer Interpretation. *IEEE 16th Annual Proceedings of the Reliability Physics Symposium* (1978) 93-100.

Pfarr, M., and Hart, A. The Use of Plasma Chemistry in Failure Analysis. *IEEE 18th Annual Proceedings of the Reliability Physics Symposium* (1980) 110-114.

Richards, B.P., and Footner, P.K. Failure Analysis in Semiconductor Devices: Rationale, Methodology and Practice. *Microelectronics Journal* 15 (1984) 5-25.

Schwartz, R. A. Electrical Testing in Failure Analysis: The Role of the Curve Tracer. *IEEE 18th Annual Proceedings, Reliability Physics* (1980) 101-109.

Selfridge, A.R. Approximate Material Properties in Isotropic Materials. *IEEE Transactions on Sonics and Ultrasonics* SU-32, No. 3 (May 1985) 380-394.

Semmens, J.E., and Kessler, L.W. Nondestructive Evaluation of Thermally Plastic Integrated Circuit Packages Using Acoustic Microscopy. *Proceedings of the Intl. Symposium on Testing and Failure Analysis, ASm International* (1988) 211-215.

Southworth, H. N. Introduction to Modern Microscopy. Wykeham, London (1975).

# 10

# TRENDS AND CHALLENGES

Microelectronic packaging technology has to address both front-end and back-end evolutionary trends in the system. Front-end technology trends include integrated devices and their fabrication techniques, while back-end trends include interdevice interconnection technology and circuit board assembly. This chapter addresses those trends and challenges in circuit technology, plastic-encapsulated microcircuit (PEM) packaging technology, and card assembly technologies, which have implications on PEMs. The chapter concludes with a section on the trends in the requirements and standards specific to microcircuit packaging. Table 10.1 provides an overview of the trends from the point of view of the 1993 Semiconductor Industries Association (SIA) and Nitto Denko roadmaps.

**Table 10.1 Technology roadmap using data taken from SIA [1993] and Nitto Denko [1993]**

| | | | 1992 | 1994 | 1996 | 1998 | 2000 |
|---|---|---|---|---|---|---|---|
| Devices | Width of pattern (μm) | | 0.5 | 0.4 | 0.3 | 0.25 | 0.2 |
| | Integrated density | DRAM | 16M | 64M | | 256M | |
| | | SRAM | 4M | 16M | | 64M | |
| | Size ($mm^2$) | Logic | 250 | 400 | | 600 | 700 |
| | | DRAM | 132 | 200 | | 320 | 400 |
| | Frequency (MHz) | off chip (entire package) | 60 | 100 | | 175 | 200 |
| | | on chip (die only) | 120 | 200 | | 350 | 400 |
| | Power (W) | micro computer | 3 | 4 | | 4 | 4 |
| | | mini computer | 10 | 15 | | 30 | 35 |
| | | main frame | 15 | 30 | | 40 | 100 |
| Bonding | Pitch (μm) | wirebond | 100 | 70 | | 50 | |
| | | TAB | 100 | 70 | | 50 | |
| | | flip chip | 200 | 200 | | 150 | 100 |

**Table 10.1 (cont.)**

| | | | 1992 | 1994 | 1996 | 1998 | 2000 |
|---|---|---|---|---|---|---|---|
| Package | Small-thin | thickness (mm) | 1.4 | | 1 | | |
| | | pin count | < 80 | | < 208 | | |
| | | pin pitch (mm) | 0.3 to 0.5 | | 0.3 to 0.5 | | |
| | | power (W) | < 1 | | < 2 | | |
| | High pin count | pin count | 200 to 300 | | 300 to 400 | bare chip/ MCM-L | |
| | | pin pitch (mm) | 0.5 | | 0.4 | | |
| | | power (W) | 2 to 5 | | 5 to 10 | | |
| | Area array | pin counts | 200 to 300 | | 500 to 600 | 750 | 2000 |
| | | pad pitch (mm) | 0.2 to 1.5 | | 0.2 to 1.0 | 0.15 to 0.8 | 0.1 to 0.5 |
| | | power (W) | 2 to 5 | | 5 to 10 | 10 to 15 | 40 |
| Application | Workstation | | 1 chip, pin grid array | | | | |
| | Portable | | MCM (4 to 8 chip) | | | (8 to 16 chip) | |
| | | | 1 chip / surface mount | | | | |
| | | | chip on board | | | | |
| | | | MCM (area array and thin type) | | | | |
| | | | chip on glass and chip on flex | | | | |

## 10.1 TRENDS IN CIRCUIT TECHNOLOGY

Integrated-circuit technology trends are expected to impact plastic packaging needs. This section focuses on how integrated circuit density, feature size, die size, clock rate, power dissipation, and I/O count may impact PEM trends.

### 10.1.1 Minimum Feature Size, Integration Density, and Die Size

Since the discovery of integrated circuits, continuous evolution of novel and precision wafer-fabrication processes has led to smaller feature sizes. This capability has encouraged higher designed-in functional integration and packing density. For example, MOS integration levels have increased an average of 35 to 50% per year, a trend that has held up over the past twenty years. Typical production wafer sizes have increased from about 18 mm (0.75 in.) to about 200 mm (8 in.) in diameter, producing a fairly constant ~ 300 dies per wafer. The minimum feature size has decreased for a typical production device from about 5 μm in 1980 to about 1.1 μm in 1990, for a 12% decrease every year. This trend is expected to continue, with the development resolution reaching 0.2 μm before the late 1990s.

As the number of transistors per die have escalated, average die sizes have increased at about 13% per year. This trend is forecasted to continue at this rate and will be limited only by the availability of lithographic exposure field size.

The principal concerns arising from reduced feature size for plastic-encapsulated modules are higher power dissipation, faster clock rates and signal rise times accompanying reduced feature sizes in a constant electric field, and the deformation of the circuit interconnect pattern due to shrinkage of the molding compound after curing. With present plastic encapsulation materials, this differential thermal shrinkage can cause die metallization deformations roughly equal to or exceeding the space width between the interconnect lines, causing intralevel electrical shorts and metal line lift-offs, with shorts and/or continuity losses. The problem is more acute near the high-stress corners and edges of the chip and on conductor lines that run perpendicular to the shrinkage direction [Isagawa et al. 1980].

The challenges of plastic packaging for larger devices also originate from higher differential thermal shrinkage forces, because the dimensional differences in shrinkage between the die and the molding compound will be greater. Larger dies require larger paddle support (die-mounting substrate), making paddle shift problems more likely. Since larger leadframes are needed in large-die designs, the dimensional differences between the die and the leadframe can warp the assembly, causing internal buckling. A compliant die-attach material could absorb a good part of the dimensional disparities.

The differential thermal shrinkage stresses associated with oversized dies can also cause passivation layer cracking. This is often associated with ball-bond lift-off and ball-bond shearing, because the close proximity of these structures to the passivation layer may damage the edges of the chip. The same compliant cover coat material, such as silicone rubber or polyimide, that mitigates interconnect deformation can significantly decrease the stresses transmitted to the passivation layer. However, the adhesion between the buffer layer and molding compound needs to be assured to avoid a delamination at the interface. Passivation layers in multi-level metallization architectures (all VLSI logic devices) are particularly vulnerable to thermal shrinkage-induced stress cracking because of their more nonplanar topography. Interestingly, lowering the coefficients of thermal expansion of the molding compound to decrease metallization deformation and passivation cracking leads to other stress problems with copper leadframes. An acceptable trade-off is thus required.

### 10.1.2 Clock Rate and Power Dissipation

The recent availability of low-latency, high-speed signal processing systems for real-time computation and communication is a direct result of the development of high-clock-rate, very high-speed integrated circuits. These high operational speeds and fast gate-access times have been achieved primarily through low-parasitic design of small device features. Such increases in device speed place great demands on the dielectric properties of the plastic packaging materials for the preservation of signal speed and minimization of signal dissipation. The package can degrade device performance if the clock rate is faster than the signal propagation rate through the package. In cases where the overall resistance of the conductor is low, signal propagation is dependent on the surrounding dielectric medium, as well as the geometry of the circuit layout. Plastics have superior dielectric properties, compared to ceramics, and lower parasitic capacitance. Thus, plastic-packaged integrated circuits may find greater acceptance in traditional high-performance systems. However, the effects of moisture and temperature on the dielectric properties of packaging materials will have to be considered.

The figure of merit of a high-speed, high-density integrated circuit is expressed in gate-hertz/$cm^2$. For higher figures of merit, more gates or memory elements are being squeezed into smaller areas, with an increasing number of features per unit area. The constant voltage scaling law dictates that power consumption by the chip will go up by the square of the feature size scaling parameter or reduction ratio. Even with a 3.3 V operating voltage, silicon chips dissipating 40 to 50 W may not be uncommon by the end of the century.

Because of higher power dissipation and increased speed, gallium arsenide devices also pose a challenge in package thermal management. Consequently, the management of higher power dissipation is a potential barrier affecting the future of low-cost plastic packaging. Some current state-of-the art devices require active thermal management with plastic packaging. Heat spreaders and heat radiators embedded in the encapsulant, currently used for thermal management of high-power dissipating devices, may evolve into higher heat-dissipating configurations.

One such example is the surface-mount, plastic-packaged, 600-MHz, 557-pin, 350 K gallium arsenide gate-array by Vitesse Semiconductor that dissipates 44 W [Watson 1992]. The oversized copper-tungsten heat spreader is in direct contact with the chip and extends outside the molded plastic body. A key component of the package is a molded Kevlar-filled polyethersulfone chip carrier seated on a strip-line printed-wiring substrate containing nine conductor layers. Another example is the EDQUAD family of plastic packages developed by ASAT, Inc. (Section 10.2.2).

In general, if the trends predicted by McCall [1989] are true, thermal management of heat dissipating devices in plastic packages will require materials with thermal conductivities on the order of four to six times those in present-day molding compounds using fused silica fillers (15 cal/cm sec °C). This could be achieved with available higher thermal-conductivity filler materials. However, many of these materials (e.g., silicon carbide or boron nitride) increase the viscosity of the molding compound out of proportion to their loading, probably due to their particle shapes and wettability by the molten components of the compound. Some of these fillers can also interact with the curing agent or catalyst, and adversely affect the molding cycle. Moreover, compared with fused silica, they are significantly less cost-effective. Metallic or diamond cold plates and/or active microchannel fluid cooling may also play a role in thermal dissipation for higher-power chips.

### 10.1.3 Input-Output Pad Counts

To avoid or minimize chip-to-chip delays, integrated circuits manufacturers are packing ever greater functionality on chips. The number of leads required for packaged integrated circuits may reach up to 500 leads for single-chip packages by the late 1990s. With multichip modules (MCMs) offering even higher interconnection densities, the leads accommodated may be even higher.

Currently, about 500 leads can be handled with a 0.3-mm (12.5-mil) outer lead pitch, a 0.1-mm (4-mil) inner lead pitch, and a package size of almost 16 $cm^2$ (2.5 square inches). With bump bonding, the dimensional trend is even smaller. In large packages, molding compound flow-induced leadframe deformation has to be reduced by using a shorter lead length. Both the bond pad spacing on the device and the wirebond spacing in the inner lead interconnection will have to decrease to reduce the probability of paddle shift. A larger paddle for large dies shortens the lead length and increases the spacing between leads at the lead tips. In some package styles, oversized paddles provide a leadframe design that is manufacturable by punching or etching.

Various approaches have been studied and used in special situations to handle large I/O counts from individual devices. Among them, flip-chip technology, where the bonding pads are laid out over the entire top surface of the die, holds particular promise for the future. Another innovation in peripheral bonding pad configuration is Micro SMT (MSMT) packaging for microwave devices [Iscoff 1992]. Unlike conventional surface-mount technology (SMT) packages, the MSMT is not wirebonded or tape bonded; rather it is made with wafer fabrication equipment. The number of leads the package can offer, although theoretically unlimited, is restricted by the interconnect technology of the circuit board. The deposited and defined metal forms an interconnect from the bonding pad to an area in the scribe line. Exposed silicon in some of the scribe-line area is then photomasked and etched with metal interconnect, flexibly bridging the integrated circuits and a silicon post that later becomes the external package backside interconnect (after backside thinning, gold/nickel plating, and tin-lead finishing).

## 10.2 TRENDS IN MATERIALS, DESIGN, AND FABRICATION

Trends in encapsulation materials, package design, and manufacturing methods are presented in this section.

### 10.2.1 Encapsulation Material Trends

Although all materials used in the fabrication of plastic packages contribute to the functional requirements of plastic-encapsulated microelectronics (PEMs), the encapsulating material (molding compound) has the primary impact on the performance and reliability of PEMs. Table 10.2 shows the history of molding compounds from Nitto Denko, and Tables 10.3a and b show Nitto Denko's roadmap for encapsulating methods.

**Table 10.2 History of Nitto Denko molding compound**

| Period | Main use | Nitto product | Key requirements |
|---|---|---|---|
| | Discrete | MP - 2000 | Moldability |
| 1970s | small DIP | MP - 3000 | moisture resistance<br>moldability |
| | middle DIP | HC - 10 - 2 | moisture resistance<br>moldability |
| | large DIP, SOP | MP - 150 SG | low stress |
| 1980s | PLCC | MP - 180 | low stress |
| | QFP | MP - 190 | low stress<br>processability |
| | SSOP, TSOP | MP - 7000 | soldering resistance |
| | SQFP, TQFP | MP - 7000 | soldering resistance |
| 1990s | SOP, DIP | MP - 80 | antidelamination<br>processability |
| | PLCC, QFP | MP - 8000 | floor life for soldering<br>low stress,<br>moldability |
| | large package | Target | soldering resistance<br>low stress,<br>moldability |
| | ultrathin package | Target | moldability<br>soldering resistance |
| | specific package | Target | adaptability to components |

**Table 10.3a Development roadmap of semiconductor encapsulating material [Nitto Denko 1993]**

<table>
<tr><th></th><th>Category</th><th>1992</th><th>1993</th><th>1994</th><th>1995</th></tr>
<tr><td rowspan="4">Design technology standard MC</td><td>small package</td><td>solder dip & pressure-cooker test</td><td rowspan="2">low temperature mold (for pre-plated frame) longer floor life no post cure</td><td>good processability</td><td></td></tr>
<tr><td>low stress</td><td>fast cure anti-delamination</td><td></td><td></td></tr>
<tr><td>anti-solder (thin package)</td><td>anti-solder & moldability</td><td colspan="2">anti-solder (super thin)</td><td></td></tr>
<tr><td>anti-solder (large package)</td><td colspan="2">anti-solder (thick package)</td><td>anti-solder (large package)</td><td>anti-solder (large & thin)</td></tr>
<tr><td rowspan="3">Special MC</td><td>power device</td><td>good processability</td><td>low stress & anti-abrasiveness</td><td colspan="2">high thermal conductive & anti-solder</td></tr>
<tr><td>large/special package</td><td>MC for heat dissipation package</td><td colspan="3">MC for MCM</td></tr>
<tr><td>TAB</td><td></td><td>MC for small TAB</td><td colspan="2">MC for large TAB</td></tr>
</table>

**Table 10.3a (cont.)**

| | Category | 1992 | 1993 | 1994 | 1995 |
|---|---|---|---|---|---|
| System technology | new polymer | | polyimide material | new inflammable (harmless) | recyclable recovery or cull |
| | new system | | liquid encapsulating | | multifunction encapsulating |
| Manufacturing technology | kneading | high filler content | | | |
| | pelletizing | | dust free pellet | | |
| | new manufacturing system | | | | process for new encapsulated system |
| Evolution technology | moldability | analysis of void/ delamination | analysis of pad-tilt etc. | | |
| | | analysis of flow process | simulation by FEM | application of simulation result | |
| | reliability | adhesion study | | application of the study | |

**Table 10.3b Development roadmap of semiconductor encapsulating material [Nitto Denko 1993]**

| | 1993 | 1994 | 1995 | 1996 | Requirements | Nitto's products |
|---|---|---|---|---|---|---|
| Process innovation | | no PMC | | | 1. PMC free | MC |
| | | PPF | | | 2. low temperature | MC |
| | laser marking | | bottomless markless | | 3. package traceability | micro print system |
| | die bond snap cure | | | | 4. high-performance die attach | elep mount system |
| | | | single device burn-in | wafer burn-in | 5. wafer burn-in | ASMAT |
| | | | clean mold system | | 6. dust free | CT, new encapsulant |
| Package innovation | ball-grid array | | | | 1. no warpage no delamination | MC, elep mount system |
| Fine-pitch | | flip-chip, TAB | | | 2. filling space | new encapsulant |
| | | | dam bar less | | 3. dam bar less | new material |
| | | | increasing bonding pad | | 4. high humidity resistance | MC |

**Table 10.3b (cont.)**

| | 1993 | 1994 | 1995 | 1996 | Requirements | Nitto's products |
|---|---|---|---|---|---|---|
| Thinner package multilayer | 1 mmt | 0.5 mmt | | TCP | 1. good moldability antipopcorn warpage free | MC. elep mount system lamination material |
| | | MCM | | multilayer MCM | 2. buffer, low stress<br>3. casting | MC, new material<br>liquid encapsulant |
| High power High frequency | | | package with large heat spreader<br>high thermal radiation<br>high temperature storage | dielectric loss<br>noise problem<br>ESD problem | 1. new combination<br>2. high lambda<br>3. heat resistance<br>4. low dielectric loss<br>5. shielding<br>6. protect from static electricity | MC, elep mount<br>MC<br>polyimide encapsulant<br>new encapsulant<br>new system<br>(multilayer encapsulant) |
| Local stress soft error | | gel coat | cavity mold | | 1. buffer | MC, die coating |
| | | gel coat less | | | 2. new system | cavity package |
| | | data retention error | | | 3. low stress | MC buffers for local stress |

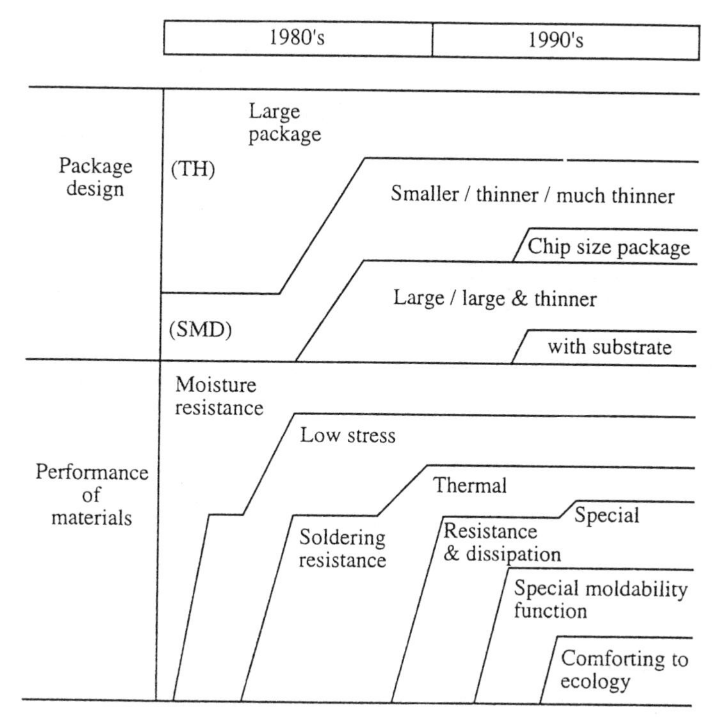


**Figure 10.1** *Technical trend of semiconductor encapsulating technology*

Figure 10.1 depicts the technical trend of semiconductor encapsulating technology in terms of package design and performance of materials. Figure 10.2 shows the recommendations and targets for Nitto's molding compounds following the trend in decreasing package size.

Epoxy molding compounds (EMCs) using transfer molding are expected to remain the major force in molded packaging technology. However, it is unclear whether they will be able to meet the yield requirements of high-pin-count complex packages (such as PQFP) and the heat dissipation needs of future products. The expected drop of supply voltage in future integrated circuits from 5 to 3.3 to 2.5 V will certainly ease the heat dissipation demands on PEMs. Among other uncertainties of enhancements in epoxy molding compound properties are the coefficient of thermal expansion and viscosity tailoring. Figure 10.3 depicts the current development trend of EMCs, with their specific intended applications.

The following explains some of the terms used in the tables and figures above:

- *Anti-solder:* The properties of molding compounds that make the components robust against potential failures during soldering (eg. delamination and package cracking).

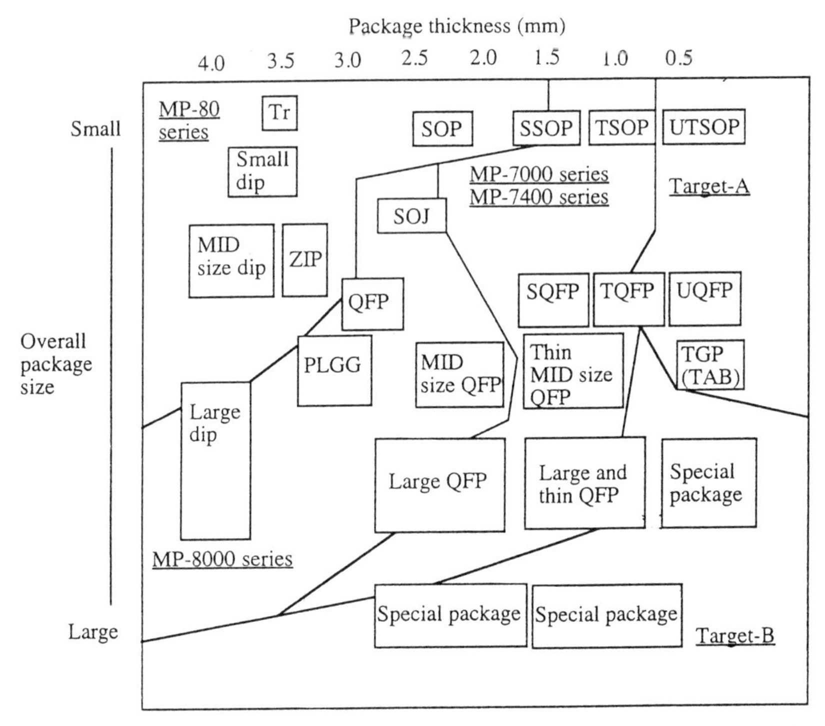


**Figure 10.2** *Recommendation and targets of Nitto molding compounds*

- *Recyclable recovery:* Nitto Denko is in the process of undertaking the development of molding compounds that can be recycled so that the molding compound remaining after the molding process (eg. material left in the runners) can be remelted by the customer and used for the next molding.

- *Bottomless markless:* Plastic package markings on the bottom surface are impossible to read unless the package is stripped of the board. Bottomless-markless technology uses a microprint system (being developed by Nitto Denko) to eliminate bottom-side markings.

- *Single-device burn-in:* Burn-in screens for single devices, rather than manufactured lots, is becoming a necessity as customers look for known-good dies (KGD). Nitto Denko is developing application specific materials (ASMATs) which are unique in that (1) they have polyimide substrates bonded directly to the copper leadframes without any intervening adhesive, (2) can be used for extremely fine-pitch applications, and (3) allow bumps to be made in the leadframe/polyimide substrate to increase the reliability of the die-leadframe bonding during burn-in.

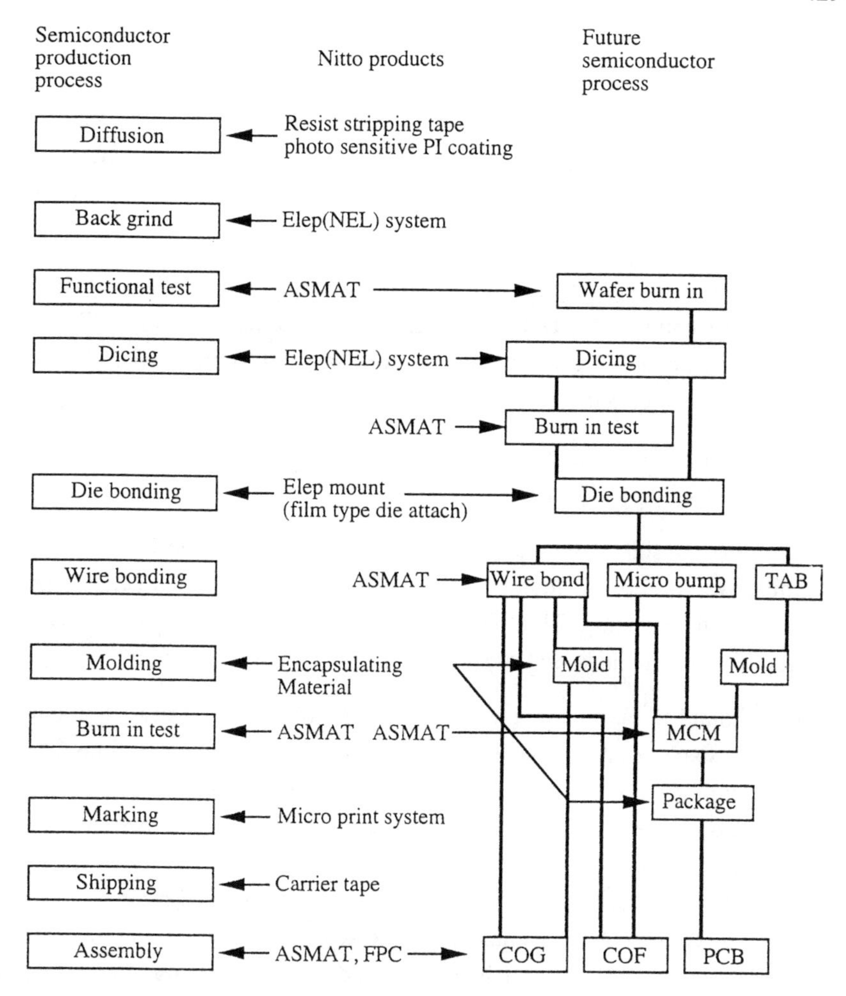


**Figure 10.3** *Nitto's technical development plan of semiconductor related materials*

- *Dam-bar less technology:* The leadframe is usually provided with a dam-bar to constrain the molding compound from overflowing during the molding process. Nitto Denko is developing thin-films which when applied to the leadframe, would serve the same purpose while reducing the leadframe's overall material thickness. These new materials are being developed for fine-pitch thin-package applications where the leadframes are thin (and may deform upon removal of the dam bar).

- *Local-stress soft error:* The die is susceptible to local stresses due to CTE mismatch effects between the die-surface and the molding compound. Because of this, the die functionality can be affected, leading to data retention errors (soft errors) in memory dies. Presently, Nitto Denko provides a gel coat on the die surface to act as a buffer. The company is in the process of developing cavity molds which would not require gel coats. Later Nitto Denko plans to develop low-modular materials for their molding compounds to provide buffers for local stress in non-cavity packages.

- *Kneading:* A process used for forming a mixture from the elements of the molding compound. During kneading, the molding reaction is allowed to occur to about 10% completion to form a solid continuous mass of the molding compound.
- *Pelletizing:* The kneaded mold is compressed into tablets to be used during the transfer molding process.
- *Pad-tilt:* During molding, the die-paddle (along with the die) sometimes tilts with the flow of molding compound, leading to the die being exposed outside the package. Nitto Denko plans to simulate this process to develop molding compounds that are robust against this type of failure.

For postmolded plastic packaging, injection-molded liquid-crystal polymers (LCP) have some promising potential for the future. With a cycle time of only 20 to 30 sec, they can offer several times higher production rates than epoxies. The base resin is very pure, with low viscosity; even with filler particles added for high thermal conductivity, it has a very low coefficient of thermal expansion. Liquid-crystal polymers also have exceptional solvent resistance and are amenable to optical fiber attachment.

Liquid injection and reaction injection molding processes use low-viscosity monomers or oligomers; the former is one component system, the latter is two. Chemically these materials are silicones, epoxies, or epoxy-silicone hybrids. Even with high filler loadings, they have very low viscosities and their properties can be specially formulated.

Polyimides with low coefficients of thermal expansion have been developed with values as low as $4 \times 10^{-7}$/°C. They may be especially useful in very high-precision electronics applications, such as insulation films and plastic-encapsulation underlayers. Most polyimide precursors interact with copper (leadframe), necessitating the use of passivation barrier coatings. However, the polyimide precursor based on ester technology is compatible with copper metallurgy. Its unique isomer ratio enables the development of high solid formulations that allow thick coatings of up to 20 μm.

### 10.2.2 Package Design Trends

Thin, small-outline packages (TSOPs), with their compact profiles, became a key product of the 1990s. By most estimates, they will also be used increasingly in surface-mount small-outline packages in the future, especially for memories, as demand for space-saving packages grows [Iscoff 1992]. Other thin package types include the thin quad flatpack (TQFP), thin sealed small-outline (TSSOP) and thin-body plastic dual in-line (PDIP) packages. Integrated circuit packages with width and/or length shrink include the fine-pitch quad flatpack (FPQFP), quartersize small-outline package (QSSOP), shrink dip (SDIP), shrink quad flatpack (SQFP), shrink small-outline package (SSOP), and very small-outline package (VSOP). With these or new shrunken packages, suppliers will continue to provide more functions in less board space and boards will become smaller. This trend to skinnier housings will challenge die thinning, wafer transportation, chip-pad mounting, leadframe design, lead bonding, board interconnection, molding, and soldering.

At less than 300 pins, present-day packaging issues depend on different uses for plastic quad flatpacks in Japanese and U.S. manufacturing. A Joint Electronic Devices Engineering Council (JEDEC) group is addressing the problems of bond pad pitch requirements finer than 0.65 mm (26 mils). A few manufacturers have tooled for 0.5-mm (20-mil)-pitch parts with plans for small pitch in various stages of development. The Electronic Industry Association of Japan PQFP has received wider acceptance by ASIC users and manufacturers, since greater volumes have been generated at lower costs than for the U.S. JEDEC version.

The quad flat pack (QFP) package has become the surface-mount technology package of choice for microprocessors and peripheral devices and is well suited for hand-held computer application. To address this market segment, Intel Corporation has developed a new package with a thin (2.0-mm) body for light weight; low standoff (0.25 mm maximum) for overall lower profile; in a 32-mm x 32-mm body format [Jain 1994]. Tape automated bonding (TAB) interconnect from die to the leadframe is used to overcome wirebond interconnect limitations. This package provides a more manufacturable package system by using flat handling during test operation. Devices operating at 100 MHz clock/50 MHz bus frequency at 3.3 V voltage and dissipating less than 3.0 W of power in static air can be successfully assembled in this package. The package is trimmed and formed subsequent to test operation and shipped in standard JEDEC thin matrix plastic trays, ready to use for SMT operation with good lead coplanarity. Package reliability is equivalent to other plastic packages. Sample packages were supplied to selected customers to help develop the 0.4-mm pin pitch SMT process, equipment, and materials. The results indicate that the current base of installed SMT equipment is capable of assembling such 0.4-mm pitch packages with very little or no modification and at acceptable quality levels. Figure 10.4 shows an isometric view of the package.

JEDEC has standardized a family of 1.4-mm-thick thin quad flatpack (TQFP) packages that range in size from 5 to 28 mm square and 32 to 256 leads. TQFP usage is driven by applications that require small, thin, and lightweight semiconductor chip packaging, including notebook and subnotebook computers, personal digital assistants, portable consumer products, PCMCIA cards, and disk drives. These applications call for chips with larger sizes, greater power dissipation, or higher electrical performance [Hoffman et al. 1994]. The trend is toward even larger 1.4-mm-thick TQFP packages than currently covered by JEDEC, such as the 32-mm and 40-mm packages, to handle both higher leadcounts and larger chip sizes. However, the molding process gets increasingly complicated when package sizes increase at the same thickness. Olin Interconnect Technologies has developed a high thermal and electrical performance TQFP in which the plastic mold compound is replaced by an anodized aluminum base and lid adhesively sealed to the leadframe [Hoffman et. al 1994]. The package uses the same infrared or vapor phase reflow board mounting profile as a plastic package, weighs the same as a plastic package, and is dimensionally equivalent to a plastic package.

ASAT Inc. has developed a new family of packages, called EDQUAD, to meet the high thermal and electrical performance demands not fulfilled by standard

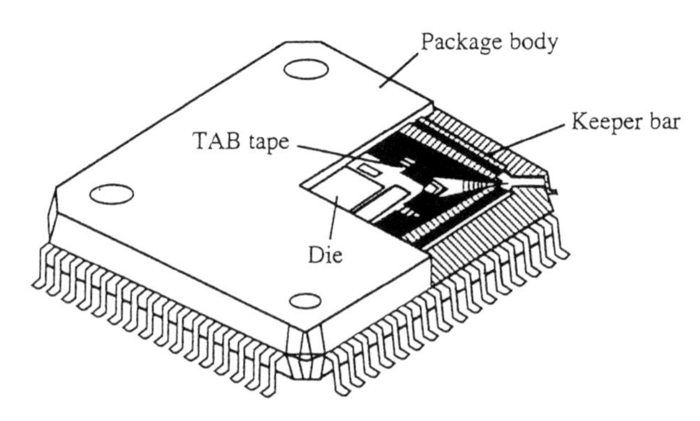


**Figure 10.4** *Isometric view of Intel 296 lead FQFP showing in the quarter portion of the package the internal structure of the TAB interconnect from die to the leadframe*

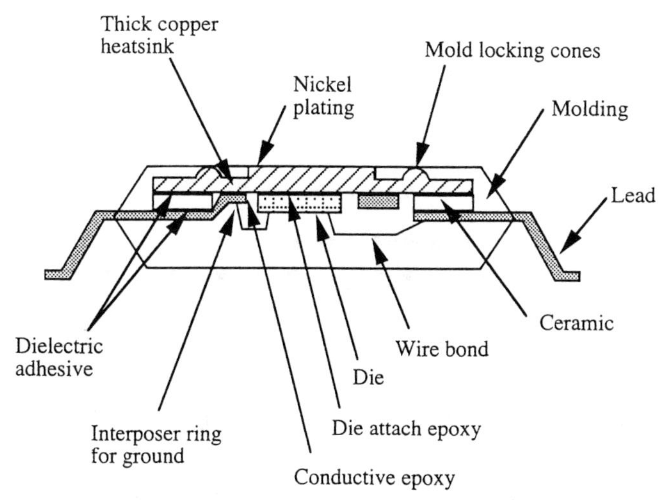


**Figure 10.5** *An EDQUAD package cross section [Karnezos et al. 1994]*

plastic packages [Karnezos et al. 1994]. EDQUAD includes a standard metric QFP, 1.4-mm-thick TQFP and PLCC packages ranging from 7-mm to 40-mm bodies and leadcounts in the 20 to 304 range. The packages are designed with an integral heatspreader exposed to air, used as a ground plane as well. The die is attached directly to the heatspreader to minimize the thermal resistance. The power and ground pads of the die can be wire bonded directly on separate internal rings to minimize the inductance. Measurements on a 28 x 28-mm, 208-lead EDQUAD show that the power dissipation is doubled to over 4 W/chip for most applications. The lead impedance can be controlled to 50 Ω and the inductance of ground/power connections is significantly reduced. The package, a cross section of which is shown in Figure 10.5, meets the commonly accepted standards.

Figure 10.6 shows how ASIC packaging demands are likely to be met by new package design rules [Singer 1993]. Partial yield losses at these finer bond-pad pitches have made most manufacturers unwilling to compromise for a workable standard at these levels; obtaining multiple suppliers for each package has thus become difficult.

Plastic quad flatpacks, the most popular high-lead-count integrated circuit package, and less than 20-mil-thick TAB, could be replaced by the overmolded pad-array carrier (OMPAC) (or ball-grid array) from Motorola [Tuck 1994]. An overmolded pad-array carrier uses an array of solder balls for board attachment, eliminating worries about lead skew and coplanarity. OMPAC can be handled with the same pick-and-place and soldering equipment normally used for low-lead-count components.

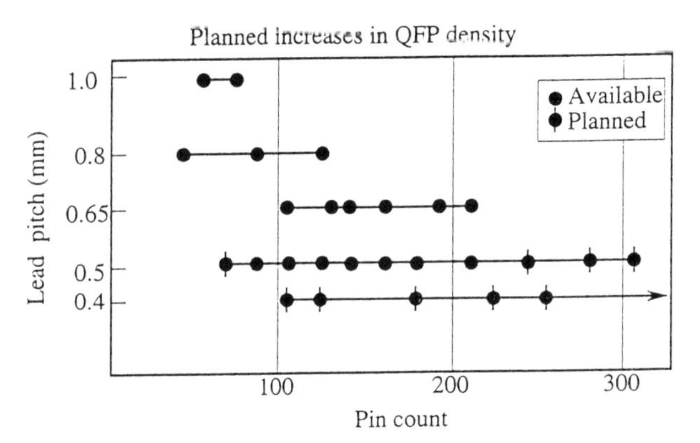


**Figure 10.6** *ASIC packaging demands*

The OMPAC consists of a thin copper-clad FR-4-type printed wiring board; the top side metallization carries a die flag and wire bond pads. The bond pads extend through plated through-holes to the backside of the substrate, where copper traces complete the electrical continuity to solder-pad termination sites. All metal features and solder masks are photodefined and electroplated with copper, nickel, and gold. After conventional epoxy encapsulation, the packages are bumped with 62PB/36Sn/2Ag. Thermal vias under the die facilitate heat dissipation to lands placed on the printed wiring board. With a 49% smaller footprint and a 41% thinner profile than plastic quad flatpacks, OMPAC leads to denser packing on the printed wiring board. The solder-pad spacing in OMPAC is 2.3 times the lead spacing of plastic quad flatpack, so critical circuit timing traces can be routed directly under the package. The soldered board mounting is almost self-aligning. However, the issue of good (inspectable) solder joints is still unresolved.

Aptix Corporation has developed a 1024-pin ball-grid array (BGA) package that contains an area array die flip-chip mounted to a silicon transposer, to reduce the cost of its 1024-pin field programmable interconnect device (FPIC), which has the die flip-chip mounted to a ceramic land-grid array (LGA) package and housed in a molded plastic screw-mounted socket fitted with "fuzz button" contacts [Switky et al. 1994]. The package's contacts are arranged in a 32 x 32 array on 40-mil pitch. The transposer, which fans the area array to two rows of pads on its periphery, is wirebonded to a printed circuit board substrate.

In the area of ultrahigh heat-dissipation plastic packages, Power Quad 2, a 160- and 208-lead plastic quad flatpack from Amkor Electronics, is a significant development in the 1990s [Iscoff 1992]. In this package, thermal management attributes include a large-area solid copper heat sink under the die. For improved electrical and thermal performance, an internal ground plane could be used. With a 8.75 x 8.75-mm (350 x 350-mil) die, theta junction-to-case is less than 0.4°C/W; with a new line of external heat sink and air flow, a junction-to-ambient thermal resistance of 8.0°C/W will result. Power Quad 2 is a similar part with superior electrical performance.

Finally, various design tools for both single and multichip packages exist on the market. These tools typically address only some aspects of the IC production flow. For instance, some tools simulate mainly the electrical performance of the packages for various configurations and operations frequencies. Others concentrate only on the thermal management side by portioning the functions on either the die or the module, while still others focus only on the mechanical aspect. Such a gap is well recognized by the CAD and modeling tool vendors, and attempts have been made at integrating the disjointed modules into one single package. Understandably, most of the development efforts have been devoted to the high value-added packages. All the major CAD vendors now profess to have design tools that can partition, route, and simulate the performance of the complex packages, such as multichip modules, in terms of electrical characteristics, thermal dissipation, and mechanical stresses.

Software tools are also available to ensure that reliability, quality, and yield issues are addressed during package design. One such tool for reliability is the Computer Aided Design of Microelectronic Packages (CADMP II) software developed by the CALCE Electronic Package Research Center at the University of Maryland.

### 10.2.3 Package Fabrication Trends

Evolutionary directions in package fabrication include new cleaning methods for package components and precision, improved manufacturing process control, fine-pitch interconnections, and new leadframe fabrication.

Manufacturing technology and the cost of the advanced materials required for high-quality and reliable plastic packaging determine both acquisition and lifetime cost, and drive the technology for the widespread use of PEMs in all markets. Performance criteria naturally adapt to the cost constraints of the market. At near-perfect yield, the present molding process productivity of the most common PLCC packages is about 800 packaged devices per hour. High-yield plastic packages of the future using thinner leadframes, fine-pitch wirebonding, and flip-chip or TAB structures will need low-viscosity molding compounds at a low molding temperature to reduce shear-rate-induced yield losses. Higher production rates will be required to compete with PGAs and premolded packages. Smaller molds with fewer cavities and a total cycle time (including in-mold cure) of 1 to 2 min in an automated mode will be needed. A high level of automation will result in lower cost, uniform-quality packages with less damage to fragile high-I/O-count assemblies. A clean manufacturing environment will reduce contamination-related failures.

Automation will play an increasingly important role in PEM manufacture. One approach is partial automation of various labor-intensive aspects of the process, such as preform heating and handling. A robotic arm may place the heated preform into the transfer pot and starts the process sequence of transfer, curing, and ejection of the molded leadframes. In this automation approach, a single operator can handle four or five transfer presses. In the total automation approach, the process runs without any operator assistance, although typically 10 to 20% of an operator's time is spent in moving different cassettes and checking equipment malfunctions. The use of a smaller molding tool with fewer cavities in conjunction with faster-curing molding compounds will further increase productivity. Such a packaging system is usually totally enclosed to maximize process cleanliness and personnel safety. The cassettes of molded leadframes feed an automated trim-and-form press, and then move on to a code-marking station. The partial or total automation of single-pot systems has been superseded by multiplunger technology in multiproduct production environments. In automated multiplunger systems, six to twelve pots feed six to twenty four cavities, with one to four cavities per transfer pot. Because of very short flow length, they are very effective in minimizing voids and promoting high molding compound density. State-of-the art encapsulants — very fast-curing molding compounds with a total cycle time of less than 2 min in the mold — will be used in these manufacturing systems.

As discussed in Chapters 3 and 4, stringent contamination control of package assembly components during fabrication of PEMs is necessary for quality assurance and long-term reliability. Over and above the practice of using semiconductor fabrication type class 100,000 or better, clean room operations, high-technology, integrated-circuit wafer-cleaning processes will increasingly spread into packaging process technology. Plasma cleaning of the pad and leadframe assembly for improved bonding with plastics may become widespread. As a function of oxygen/carbon tetrafluoride/argon plasma treatment, lap-shear strength of epoxy adhesives should increase. Ultrahigh purity chemicals are sure to find their way into PEM fabrication lines as initial cleaning agents for particulate contamination control.

The increasing use of analytical techniques and microsensors to fine tune materials and control process-induced defects falls into the category of enhanced process control and quality assurance. X-ray radiography and C-mode scanning acoustic microscopy/tomography (see Chapter 9) are finding increased use as nondestructive techniques for evaluating plastic delamination; die metal, die-attach, and bonding wire deformation; die metal and wirebond voiding; leadframe, die passivation, die-attach, wire, wirebond and case brittle fracture; and dendritic growth under bias. Mercury porosimetry has been successfully employed to find the number and size of epoxy and epoxy-metal pores in epoxy-encapsulated packages. Epoxy pores are less than about 0.2 μm in diameter, epoxy-leadframe voids are about 1 μm in diameter, and surface pores range from 5 to 500 μm in diameter. Piezoresistive strain gauges integrated into test chips will continue to be used to directly measure the mechanical stress induced inside a PEM by encapsulation, die bonding, and other factors, either during fabrication or under environmental stress testing. Solid-state moisture microsensors will also be used to measure the moisture content at any specific location inside a PEM.

Challenged by higher integration levels (particularly in ASICs) with tighter bond-pad pitches at the die level, leading edge pitches will be at 0.060 to 0.075 mm (2.5 to 3 mils) by the late 1990s. Beyond 0.1-mm (4-mil) pitch, bonding with gold wire may be done by wedge bonding. The combination of tight pad pitches and shrink packages demands not only low wire looping (0.09 to 0.18 mm), but also different loop shapes. This requires modifying the dopants in the wire to offer the right set of mechanical properties. With tight pitches, the physical limits of leadframe technology will also force the placement of lead contacts at ever-increasing distances from the package center to control wire sweep. Low looping also promotes wire-to-die edge shorting when the die has inbound bonding pads to avoid encapsulation stress concentration zones. Edge shorting can be prevented by the use of new wirebonder software, that allows an extra bend in the wire at a predetermined site.

The demands of tightly spaced inner-lead bonding (0.05 to 0.075 mm) and outer-lead bonding (0.15 to 0.35 mm), high pin counts, high-end performance-driven applications, and high-volume production could also move bumped-tape automated bonding (TAB) to the forefront of the packaging industry [Iscoff 1992]; a potential for which Japan already has the capability and infrastructure. Limiting factors for TAB applications have been the expense of tape, the lead time needed to obtain it, and the capital needed for the bonding equipment. Bump fabrication has also been an obstacle. Better tape metallurgy, currently available, now permits burn-in and longer shelf-life before encapsulation. Laser bonding of fine pitch devices is particularly suited to TAB.

Trends in leadframe technology have important implications for the future of molded plastic packaging. Copper alloys are replacing 42Fe/58Ni (Alloy 42) as the leadframe material for moderate to high heat dissipating devices, such as processors and logics. But their high coefficient of thermal expansion and average mechanical properties could prohibit their use in other applications. It is also becoming increasingly difficult to manufacture thin (≤ 0.15-mm) leadframes with close lead tips for very high lead-count packages. Although chemical etching has replaced mechanical punching in these situations, as the leads and lead spacing get smaller, maintaining a lead aspect ratio near unity for the leads by leadframe isotropic etching is becoming extremely difficult. Consistent with the need for a mechanically stable lead geometry, leadframe thickness has decreased from 2.5 mm (100 mils) for small pin-count devices to under 0.15 mm (6 mils) for high lead-count (~200) fine-pitch plastic quad flatpacks. Several different types of lead tip layouts offer different pad sizes, lead-tip spacings, and wirebond lengths [Jahsman 1989]. Thinner leadframes, larger die pads (for short, close-spaced lead tips), and longer wirebond spans pose such molding problems as flow-induced deformation, paddle shift, and wire sweep. There are also problems with handling the delicate leadframe in terms of both inner lead wirebonding yields and outer lead soldering yields. Thus, it appears that leadframe manufacturing limitations could hinder the development of a very fine pitch and high lead-count molded plastic packaging.

Development trends in molding compounds have mirrored the advances made in the memory ICs, since large research efforts were devoted to that substantial market share of the semiconductor business [Nguyen et al. 1993]. Indeed, the evolution of the compound technology, from the generation of standard novolac-based resins to the current third generation of ultralow-stress, silicone-modified resins, follows the migration of package form factors for the memory ICs. However, memory ICs have turned into low-priced commodity items, barring short term price fluctuations from material shortage or trade policies. As a result, current efforts attempt to satisfy the functional requirements of the packages rather than the different categories of devices. For instance, ultrathin packages have different reliability challenges than DIPs or SOs.

The demands of high-density circuit boards have provided research and development impetus for new board-level interconnection materials and processes. Integrated circuit metallization materials and processes (subtractive and additive) will increasingly be tried on larger substrates. Large-area pattern generation and transfer on not-so-planar substrates are the major impediments. However, non-mask-limited metallization processes, such as laser direct-write and selective electroplating, could produce high-density second-level interconnection patterns on inorganic ($SiO_2$) or organic (polyimide) substrates. For example, Lawrence Livermore Laboratory has used a 400-Å sputtered titanium-tungsten

**Table 10.4 PCB versus semiconductor shipments - dollar growth % (1992-1993) [Custer 1994]**

| Location | PCB | Semiconductor |
|---|---|---|
| North America | +6.5% | +34.9% |
| Western Europe | +1.2% | +26.9% |
| Japan | -5.1% | +18.0% |
| Rest of the world | +5.0% | +36.0% |
| Total | +1.8% | +28.0% |

adhesive layer under 2 to 10 kÅ of electrodeposited gold and 0.5 to 6 kÅ of sputtered amorphous silicon to argon- laser direct write (at about 11 cm/sec) a gold-silicon alloy interconnect pattern. It then is thickened by rapid electroplating and the multilayered matrix not exposed to the laser is successively removed by sulfur hexafluoride dry etching of the silicon, wet etching of the gold, and dry etching of the titanium-tungsten. This process could become particularly valuable for chip-on-board (COB) die interconnections and has the added advantage of excellent board-level alignment.

## 10.3 TRENDS AND CHALLENGES IN CIRCUIT CARD ASSEMBLIES

Key trends in circuit cards are associated with new card materials, solder, and solderless technologies, especially in terms of how they affect system reliability and impact the environment.

### 10.3.1 Printed Circuit Boards

It is known historically that printed circuit board demand varies with the industrial economy. In response to heavy electronic equipment demand, the U.S. market peaked in 1984 and again in 1988. In 1992, an estimated $17.1 billion of rigid printed circuit boards and an additional $1.1 billion of flexible circuits were produced worldwide [Custer 1994]. Approximately 29% of the total rigid printed circuit boards (PCBs) were produced in the United States in 1992. In 1993, preliminary estimates from IPC indicated that world printed circuit board production amounted to $20.2 billion. Table 10.4 shows the estimated 1992-1993 dollar growth of printed circuit boards versus semiconductor shipments by geographical area. The much higher dollar growth rates of semiconductors versus bare PCBs in 1993 can be attributed to the increase in the semiconductor content of assembled PCBs and price increases in semiconductors (especially memory and microprocessors) compared with the price attrition in PCBs [Custer 1994].

The trends in printed wiring board (PWB) materials and fabrication technology are likely to include relatively new materials with lower coefficients of thermal expansion, lower dielectric constants, and greater dimensional stability, which can be inexpensively machined. High temperature capability will make board distortion a nonissue and matched in-plane coefficient of thermal expansion will help reduce solder-joint fatigue.

### 10.3.2 Solder Technologies

The consumer electronic products industry with its short product lifetimes and rapid market introductions of personal and portable equipment (stereos, televisions, cellular telephones, calculators, and audio/video players), has been the prime mover in the rapid adaptation of surface-mount technology. Surface-mount technology offers the industry significantly better interconnection density than through-hole technology due to smaller package size and the potential for two-sided assembly.

Fine-pitch technology has been a challenge for surface-mount soldering. Good soldering requires a minimum of solder bridges, icicles, and unsoldered joints. To avoid rework, manufacturers are aiming for zero-defect soldering, which demands good solderability of the joining metals, a controlled temperature profile over the board, and minimal variation in environmental conditions at the time of soldering. There is a trend toward decreasing the use of flux in soldering operations to avoid cleaning residues later. Soldering performed in an inert atmosphere, in which no oxygen is present to oxidize the molten solder, is one solution because it obviates the need for a flux. The amount of oxides present on the components and board before the board comes in for soldering is small and can easily be removed.

***Equipment and assembly processes.*** While most surface-mount soldering of fine-pitch packages is accomplished using solder reflow methods, modifications in the fluxing, preheat, and solder-wave technology for fine-pitch applications, may enable reliable wave soldering. Dynamic management of the solder wave in the wave soldering process is being developed as a technique to allow complete wetting of the vias. One a method of altering the flow around the soldering sites is by the use of barriers around the vias in the shape of U's or V's [Cummings 1992]. Ultrasonic wave soldering, which uses a high-frequency blast of ultrasonic energy to form the solder joint, is also being developed as a process whereby a molten solder alloy in contact with a compatible base material is ultrasonically agitated to generate an interface that results in metallurgical or mechanical bonding [Fuchs 1991]. Metallurgical bonding of the two metals occurs with the removal of dirt and oxides from the surface of the base metal by the ultrasonic wave. The oxides leave the surface of the base metal exposed to the solder, and ionic attraction between them causes a strong bond to form.

Solder screening and reflowing are the most popular methods of soldering. However, at pitches of 0.4 mm (16 mils) to 0.3 mm (12 mils), the technology is challenged. For instance, metal stencils screening solder paste are only 0.15 to 0.2 mm (6 to 8 mils) thick for a fine-pitch pattern, compared with 0.2 to 0.25 mm (8 to 10 mils) thick for 1.25-mm (50-mil) pitch ASICs and other devices. Therefore, better equipment with greater control over the process parameters and the solder paste characteristics will be required for fine-pitch applications.

One trends in solder dispensing is the use of positive displacement pumps with programmable shot size for dispensing solder-paste in different sized dots. Solder dots have been dispensed at 0.625-mm (25-mil) pitch footprints with these dispensers [Cavallaro 1991]. To achieve such accuracy, rotary positive-displacement pumps are used with constant pressures to keep the paste flowing. An Archimedes-style auger screw dispenses the solder. Control of the amount of liquid dispensed is precise enough to measure 5-msec increments at speeds of over 16,000 dots/hr when dot sizes range from 0.3 to 0.6 mm (12 to 50 mils). Robotic pick-and-place machines, incorporated with computer vision systems, will be employed in automated dispensing equipment for accurate positioning of the dispensing needle. Speeds of up to 25,000 dots/hr can be achieved with such systems.

Trends in manual soldering and rework processes include the use of microflame-reflow soldering, which combines the soldering iron and laser-soldering approaches, to reduce the processing time, obtain consistent soldering quality, obviate the need for cleaning between soldering cycles, require less maintenance and servicing, and improve process reliability. Another emerging technique of reflow soldering is optical fiber-reflow soldering, which uses optical fibers to direct heat from a radiant source, such as a lamp, to specific solder sites so that components and the surrounding board area are not affected [Kobayashi 1991]. The technique prevents variation in soldering quality and improves process control because of the short response time of the xenon lamp.

The use of solder for connecting PEMs to circuit cards still has reliability challenges. Over and above popcorning, solder fatigue/stress, solder defects, potential flux contamination, and other high-temperature component degradation problems, at or below pitches of 0.4 mm (16 mils), can cause solder bridging and opens. An additional issue with large plastic packages (particularly J-leads) is that the coefficients of thermal expansion thermal mismatch between the PEM and the printed wiring board, which can cause solder-joint fatigue failures.

When using infrared reflow soldering, the entire package is subjected to near-reflow temperatures (215 to 260°C). This assembly process has aggravated popcorn package cracking, particularly in thin and ultrathin package styles (see Chapter 4). This trend toward improved molding compounds that resist delamination at the paddle interface, minimize deformation at the reflow solder temperature, and deter cracking when deformed by the vapor expansion.

***Environmental impact trends.*** Because of the ban on chlorofluorocarbons, environmentally acceptable solvent cleansers are being found for assembled-board cleaning operations. In addition to mild detergents used in industrial dishwashers, nonflammable cleaning solvents are being developed for these operations. Molded plastic package material should thus have good solvent resistance. Due to the smaller standoff distances in Japanese products, flux cleaning becomes quite costly. Ultrasonic cleaning is becoming increasingly popular for its superior cleaning ability when clearance is minimal.

Until inert-atmosphere soldering is perfected, reliable soldering operations will need acidic fluxes to clean soldering surfaces. Although most molding compounds are impervious to these corrosive fluxing agents at the required soldering times and temperatures, the interfacial paths along the leads are particularly vulnerable to ingress by these corrosive materials. Consequently, good sealing and adhesion along these plastic-metal interfaces is needed to prevent flux penetration and package contamination during fluxing. If fluxes are not removed from the assembled board — including the underside — by cleaning operations and effluent testing, temperature cycling eventually opens up more of these interface ingress routes and premature corrosive failure of PEMs results. The current trend is to use very mild fluxes, such as those made from lemon juice (citric acid), that can be easily washed out during board cleaning.

### 10.3.3 Solderless Technologies

Solderless interconnections include the use of conductive surface-mount adhesives (CSMAs) or conducting polymers and fuzz buttons (springs). CSMAs are isotropically or anisotropically conductive thermoplastic or thermoset resins filled with silver or gold particles 10 to 20 μm in diameter. Like solder paste, CSMAs can be printed using stencils and screens and then be cured using heat and catalysts. The anisotropic z-axis CSMAs (mostly elastomers) are properly aligned so that they bridge only the smaller z-axis dimension. These materials can be used for interconnection between the component and the circuit board. Isotropic CSMAs are more advanced as replacements for solder. However, CSMAs are harder to rework than solders and their adhesive characteristics on different metallic surfaces are not yet clear. Silver or gold depletion due to ionic migration in the presence of moisture and electric bias also poses a potential stumbling block to widespread use of CSMAs. The advantages of CSMAs over solders are that (1) they facilitate fine-line resolutions for fine-pitch applications, (2) they have high flexibility, creep resistance, and stress dampening, and are therefore fatigue resistant, (3) they require lower operational temperatures (120 to 150°C), and (4) they pose no known environmental problems. Therefore, eliminating solder using low-curing temperature, low-viscosity conductive epoxies has major implications for the wider application of PEM-populated circuit boards. Both the popcorn and solder stress/fatigue-related reliability limitations would then be nonissues.

Another emerging solderless interconnection technology, fuzz buttons, consists of buttons populated in a molded engineering carrier (called vectra), with a hole pattern matching the electrical pad pattern on both the bottom of the chip carrier and the top of the printed wiring board. The typical button is a continuous gold-plated beryllium copper wire randomly woven into a cylindrical shape. The buttons are typically 0.05 mm in diameter and 0.1 mm in length; they protrude approximately 0.006 mm from both sides of the carrier. The electrical interconnection is formed by sandwiching the button carrier between the chip carrier and the printed wiring board.

Button technology readily permits the use of pad-area-array component packages. Like pin-grid-array packages without the pins, the pad-area-array package can use the entire bottom area of the package for I/Os, thus significantly increasing the packaging density over comparable perimeter I/O packages. Solderless button technology also factors out solder process requirements such as high temperatures, solvents, and cleaning.

Westinghouse Electric Company has patented a technology that sandwiches fuzz buttons between the chip carrier and the printed circuit board. The fuzz buttons are fixed in a nonconducting button carrier, according to the layout of the components on the board. Component cavities are carved out in the module assembly unit and the button carriers are dropped into them. The components are dropped into the cavities next, and a thermal slug is tightened over the components to conduct heat away from them and provide the necessary compression for proper contact formation between the components and the pads below. The button carrier acts as a fixed stop, preventing further compression of the buttons. (See Harris [1993] for more information.)

## 10.4 STANDARDS AND REQUIREMENTS

Methods to ensure high quality and long-term reliability will be established by short-term correlated and tailorable tests. In many cases, qualification-by-design will replace qualification tests. Instead, statistical process control (SPC) will be used. In terms of quality assurance methods, screening will virtually disappear as a procedure for eliminating defective parts. Military qualification and reliability test standards will be discarded, and critical-use environment related-test conditions will be modified to reflect the physics-of-failure processes to be uncovered. Users will probably still demand manufacturer's qualification data for their particular application environments, but they will maintain supplier partnerships with only those vendors that can satisfy parts-reliability criteria from field-failure data, failure analysis, applicable environmental tests, quality assurance techniques, and part-specific design analysis. Qualification requirement matrices will be drawn up to guide suppliers, relating all elements of design, manufacture, and part assembly variables with the required needs.

The future will also have to bring together differences between Japanese and American product standards [Wolfe 1990]. For example, the standoff distance between a circuit board and the most popular PEM (PQFP) is 0.25 to 0.37 mm (10 to 15 mils) in the U.S.-manufactured product, compared with near-zero in its Japanese counterpart. In addition, Japanese lead length requirements on higher pin-count, finer-pitch packages have been shrinking; some lead lengths are as short as 1.0 mm (0.5 mm from the edge of the package to the vertical portion and 0.5 mm for the foot itself). U.S. manufacturers have reached only 0.8 mm for the lead from the package end to the vertical bend. Differences in lead length standards between U.S. and Japanese manufacturers have had a direct adverse effect on board design, as designers have attempted to accommodate both types of parts. This has led to lower board yield and increased rework. Interestingly, JEDEC standards for package outlines are looser than assembly tolerance requirements, resulting in noninterchangeability of even different U.S. vendor parts made to the same specifications.

## REFERENCES

Cavallaro, K.J. Alternate Techniques for Fine-Pitch Dispensing. *Circuits Assembly* 2, 5 (May 1991).

Cummings, M. Dynamic Solder Management Reduces Defects in Wave Soldering. *Electronic Packaging & Production* 32,7 (July 1992), 74-78.

Fuchs, F. J. *Blackstone Ultrasonics, Inc.,* Ultrasonic Soldering in the Electronics Industry. Jamestown, NY (Internal publication, 1991).

Harris, D.B. *Fuzz Button Design Variation and Performance Effects*. Thesis, University of Maryland, College Park, MD (1993).

Hoffman, P., Liang, D., Mahulikar, D., and Parthasarathi, A. Development of a High Performance TQFP Package. (1994) 57-62.

Isagawa, M., Iwasaki, Y., and Sutoh, T. Deformation of Al Metallization in Plastic-encapsulated Semiconductor Devices Caused by Thermal Shock. *IEEE Transactions on Components, Hybrids, Manufacturing Technology,* 3 (1980) 171-180.

Iscoff, R. Amkor Develops Competitor to Multilayer Ceramic, MQUAD Packages. *Semiconductor International* (March 1992) 34.

Iscoff, R. Thin Outline Packages: Handle With Care! *Semiconductor International* (June 1992) 78-82.

Iscoff, R. Micro SMT Package Avoids Traditional Bonding Methods. *Semiconductor International* (November 1992) 40.

Jahsman, W.E. Lead Frame and Wire Bond Length Limitations to Bond Densification. *Journal Electronic Packaging* 111 (1989) 289-94.

Jain, P. 296 Lead Fine Pitch (0.4 mm) Thin Plastic QFP Package with TAB Interconnect. (1994) 50-56.

Karnezos, M., Chang, S.C., Chidambaram, N. (Raj), Tak, C.M., Kyu, A.S., and Combs, E. EDQUAD: An Enhanced Performance Plastic Package. (1994) 63-66.

Kobayashi, M. Local Soldering with Light Beam Using Optical Fiber. *Proceedings of the Technical Program NEPCON West* (1991).

McCall, D.W. Materials for High Density Electronic Packages and Interconnections. *Materials Research Society Meeting* (April 1989).

Nguyen, L.T., Lo, R.H.Y., and Belani, J.G. Molding Compound Trends in a Denser Packaging World, I: Technology Evolution. *Proceedings of Japan International Electronic Manufacturing Technology Symposium* (1993) 204-212.

Nitto Denko Corporation. Personal communication during Japan Technical Evaluation Center visit (Summer 1993).

Powell, D.J. Fabricating PWBs and MCM-Ls with a New Non-woven Aramid Reinforcement. *Proceedings of the Technical Program, Surface Mount International*, San Jose, CA (August September 1993) 299-309.

Semiconductor Industry Association Semiconductor Technology Workshop Conclusions (1993).

Singer, P. Trends in ASICs: 1993 and Beyond. *Semiconductor International* (January 1993) 44-48.

Switky, A., Sajja, V., Darnauer, J., and Dai, W.W.M. A 1024-Pin Plastic Ball Grid Array for Flip Chip Die. (1994) 32-38.

Tuck, J. Next Generation B.G.A. *Circuits Assembly,* 5, 5 (May 1994) 22-25.

Watson, G.F. Interconnections and Packaging. *IEEE Spectrum* (September 1992) 69-71.

Wolfe, G. Electronic Packaging Issues in the 1990s. Electronic Packaging & Production (October 1990) 76-80.

# GLOSSARY

**ACCELERATED TESTING:** A test performed on materials or assemblies that is meant to produce failures caused by the same wearout failure mechanism as expected in field operation but in significantly shorter time. The wearout mechanism is accelerated by increasing one or more of the controlling test parameters significantly above use values.

**ACCELERATOR:** A chemical additive that hastens a chemical reaction under specific conditions. An example is a catalyst.

**ACCEPTANCE TESTS:** Those tests deemed necessary for determining acceptability of a product and agreed to by both purchaser and vendor.

**ACTIVE DEVICES OR COMPONENTS:** Discrete devices, such as diodes or transistors, or integrated devices, such as analog or digital circuits in monolithic or hybrid form, that can operate on an applied electrical signal so as to change its basic characters; e.g., rectification, amplification, switching.

**ADHESION:** The property of one material to remain attached to another, a measure of the bonding strength of the interface between two discrete surfaces.

**ADHESION PROMOTION:** The chemical process of preparing a surface for a uniform, well-bonded interface.

**ALLOY:** A combination of two or more metallic elements in the form of a solid solution.

**ASSEMBLY:** A film circuit or circuits to which discrete components have been attached; also, process steps involved in putting together a functional plastic-encapsulated microcircuit.

**BALL BOND:** A bond formed when a ball-shaped interconnecting wire end is deformed by temperature and pressure against a metallized pad. The bond is also designated a nail-head bond from the appearance of a flattened ball.

**BALL-GRID ARRAY (BGA):** Leadless area array packages. Developed jointly by Motorola and Citizen, the package was initially introduced in 1992 for pager and hand-held radio applications.

**BALL LIFT:** A category of ball-bond failure in which the ball lifts from the surface of the integrated circuit die bond pad metallization or lifts the metallization from the surface of the underlying oxide or silicon.

**BOND:** An interconnection that performs a permanent electrical and/or mechanical function.

**BONDABILITY:** Those surface characteristics and conditions of cleanliness of a bonding area that must exist to provide a capability for successfully bonding an interconnection material by one of several methods, such as ultrasonic or thermocompression wirebonding.

**BONDING:** The joining together of two materials. For example, the attachment of wires to an integrated circuit or the mounting of an integrated circuit to a substrate.

**BONDING PAD:** A metallized area to which an external connection is to be made.

**BONDING WIRE:** Fine gold or aluminum wire for making electrical connections between various bonding pads on the semiconductor device substrate and device terminals or substrate lands.

**BOND INTERFACE:** The interface between the lead and the material to which it was bonded on to the substrate.

**BOND LIFTOFF:** The failure mode whereby the bonded lead separates from the surface to which it was bonded.

**BOND SEPARATION:** The distance between the attachment points of the first and second bonds of a wirebond.

**BOND SITE:** The portion of the bonding areas where the actual bonding took place.

**BOND STRENGTH:** In wirebonding, the pull force to separate wire from the bond site, measured in the unit gram-force.

**BOW:** The deviation from flatness of a wafer, or board, or leadframe strip, characterized by a roughly cylindrical or spherical curvature such that, if the board is rectangular, its four corners are in the same plane.

**B-STAGED RESIN (B-STAGE):** A partially cured resin which is used in the lamination or encapsulation process.

**BUMP:** A means of providing connections to the terminal areas of a device. A small mound is formed on the device (or substrate) pads and is used as a contact for facedown bonding.

**BURN-IN:** The process of electrically stressing 100% of a group of devices (usually at an elevated temperature environment) for an adequate period of time to cause failure of marginal devices.

**CAMBER:** A term that describes the amount of overall warpage present in a substrate; camber $= C/D$, where $C = z$ placement, $D =$ diagonal (longest distance) of the substrate surface. The out-of-plane deflection of a flat cable or flexible laminate of specified length.

**CAPILLARY:** A hollow ceramic-based bonding tool used to guide the bonding wire and to apply pressure to the wire during the bonding cycle.

**CASE TEMPERATURE:** The temperature measured at a specified location on the case of a device.

**CERAMIC:** Inorganic nonmetallic material such as alumina, beryllia, steatite, or forsterite, whose final characteristics are produced by treatment at high temperatures (and sometimes also under applied pressure), often used in microelectronics as parts of components, substrate, or package.

**CERAMIC PACKAGE:** See *CERPACK.*

**CERDIP:** A dual-in-line package composed of a ceramic header and lid, a stamped-metal leadframe, and frit glass that is used to secure the structure.

**CERPACK:** A flatpack composed of a ceramic base and lid, a stamped-metal leadframe, and frit glass that is used to secure the structure.

**CHIP:** The uncased and normally leadless form of an electronic component part, either passive or active, discrete or integrated.

**CHIP CARRIER:** A special type of enclosure or package to house a semiconductor device or a hybrid microcircuit and which has metallized castellations as usually electrical terminations around its perimeter, as well as solder pads on its underside, rather than an extended leadframe or plug-in pins.

**CHIP-OUT:** See *CRATERING.*

**CIRCUIT:** The interconnection of a number of electrical elements and/or devices, performing a desired electrical function. A circuit must contain one or more active elements (devices) in order to distinguish it from a network.

**COEFFICIENT OF THERMAL EXPANSION:** The ratio of the change in length to the change in temperature.

**COINING:** Rounding of leadframe edges by etching.

**COMPLIANT LEAD:** A component lead of sufficiently low stiffness to reduce the stress levels at the termination (solder joints) significantly during thermal cycling or vibration loading.

**COMPONENT:** An individual functional element in a physically independent body that cannot be further reduced or divided without destroying its stated function, for example, a resistor, capacitor, diode, or transistor.

**COMPONENT DENSITY:** The number of components per unit area.

**COMPOUND:** Designating an insulating material made by mixing two or more different materials to make one material.

**CONDUCTIVITY:** The ability of a material to conduct electricity, the reciprocal of resistivity.

**CONFORMAL COATING:** An insulating protective coating that conforms to the configuration of the object coated. It is applied to the completed board assembly or to components.

**CONTACT AREA:** The common area between two mating conductors permitting the flow of electricity.

**CONTACT ANGLE:** The angle made between the bonding material and the bonding pad.

**CONTACT RESISTANCE:** In electronic elements, such as capacitors or resistors, the apparent resistance between the terminating electrode and the body of the device. The electrical resistance between two mating conductors.

**CONTINUITY:** The uninterrupted path of current flow in an electrical circuit.

**COPLANARITY:** Property of lying in the same plane.

**COPLANAR LEADS (FLAT LEADS):** Ribbon-type leads extending from the sides of the circuit package, all lying in the same plane.

**CRACKING:** Breaks in the metallic and/or nonmetallic layers.

**CRATERING:** A category of ball-bond failure or defect in which the ball lifts from the surface of the integrated circuit die-bond pad, taking with it a portion of bond-pad metallization and the underlying oxide or silicon. Also called "CHIP-OUT UNDER BOND."

**CREEP:** The dimensional change with time of a material under load.

**CTE:** Coefficient of Thermal Expansion

**CTE MISMATCH:** The difference in the coefficients of thermal expansion (CTEs) of two materials or components in contact, producing strains and stresses at the joining interfaces or in the attachment structures (solder joints, leads, and so on).

**CYCLIC STRESS:** A completed circuit subjected to stress by cycling temperature and load over a period of time to cause premature failure.

**DEFECT:** A characteristic that does not conform to applicable specification requirements and adversely affects or potentially can affect the quality of a device.

**DEGRADATION:** A gradual deterioration in performance as a function of time.

**DEIONIZED WATER:** Water that has been purified by removal of ionizable materials. Commercial deionized (DI) water has a typical resistance of 18 MΩ. Further purification (e.g, for wafer processing steps) can raise that resistance even further.

**DELAMINATION:** A separation between the laminated layers of base material and/or base material and conductive foil. For PEMs, delamination can occur at interfaces between the molding compound and the die or the leadframe.

**DENDRITIC GROWTH:** Growth of metallic filaments between conductors due to condensed moisture and electrical bias.

**DERATING:** The practice of subjecting parts or components to lesser electrical or mechanical stresses than they can withstand in order to increase the life expectancy of the part or component.

**DESICCANT:** An absorbent material used to maintain a low relative humidity in moisture barrier bags for dry packing moisture-sensitive plastic-packaged devices. Desiccant may be a silica gel or an absorbent clay packed in dustless pouches.

**DEVICE:** A single discrete electronic element such as a transistor or resistor or a number of elements integrated within one die, which cannot be further reduced or divided without eliminating its stated function. Preferred usage is die or dice, bare or prepackaged.

**DEW POINT:** The temperature at which liquid first condenses when a vapor is cooled.

**DEWETTING:** The condition in a soldered area in which liquid solder has not adhered intimately and has pulled back from the conductor area.

**DIE:** (plural, dies or dice): The individual semiconductor element or integrated circuit after it has been cut or separated out of the processed semiconductor wafer, distinct from a completely packaged or encapsulated integrated circuit with leads attached. A semiconductor chip. See also *chip.*

**DIE ATTACH:** The technique of mounting chip to a substrate. Methods include AuSi eutectic bonding, various solders, and conductive (and nonconductive) epoxies, polyimides, or thermoplastics.

**DIE BOND:** The interface of the die to some other entity (substrate, header, etc.)

**DIE PAD, DIE-ATTACH PAD, DIE-MOUNT PAD, OR PADDLE:** The portion of the leadframe on which the IC die is mounted during the assembly process.

**DIELECTRIC LAYER:** A layer of dielectric material between two conductor plates.

**DIELECTRIC:** (1) Any insulating medium which intervenes, between two conductors. (2) A material with the property that energy required to establish an electric field is recoverable in whole or in part as electric energy.

**DIELECTRIC PROPERTIES:** The electrical properties of a material, such as insulation resistance, and breakdown voltage.

**DIFFUSION:** The phenomenon of movement of matter at the atomic level from regions of high concentration to regions of low concentration. A thermal process by which minute amounts of impurities are deliberately impregnated and distributed into semiconductor material.

**DIFFUSION CONSTANT:** The rate at which diffusion takes place under specific conditions.

**DIMENSIONAL STABILITY:** A measurement of dimensional change of a material due to various environmental factors.

**DIP SOLDERING:** Soldering process for populated printed boards accomplished by dipping the exposed leads and conductive pattern in a bath of static molten solder.

**DRY PACKING:** Protecting moisture-sensitive plastic packages in metallized plastic bags/containers to eliminate moisture absorption, which may induce cracking during solder reflow. See also POPCORN.

**DUAL IN-LINE PACKAGE (DIP):** A package having two rows of leads extending at right angles from the base and having standard spacings between leads and between rows of leads.

**DUCTILITY:** That property which permits a material to deform plastically without fracture. Property that allows a material to absorb large overloads.

**DUTY CYCLE:** A statement of energized and deenergized time in repetitious operation, for example, 2 sec on, 6 sec off.

**ELASTIC MODULUS:** A constant of proportionality which is indicative of the stiffness or rigidity of the material.

**ELASTOMER:** Any elastic, rubber-like substance, such as natural or synthetic rubber.

**ELECTRICAL ISOLATION:** Two conductors isolated from each other electrically by an insulating layer.

**ELECTRODEPOSITION:** The deposition of conductive material by applying an electric current in a plating bath.

**ELECTRODES:** The conductor or conductor lands of an electronic device. Also the metallic material portions of a capacitor structure.

**ELECTROLESS DEPOSITION:** The deposition of conductive material from an autocatalytic plating solution without the application of electrical current.

**ELECTROLESS PLATING:** Deposit of a metallic material on a surface by chemical deposition as opposed to the use of an electric current.

**ELECTROLYTE:** Current-conducting solution (liquid or solid) between two electrodes or plates of a capacitor.

**ELECTROLYTIC CORROSION:** Corrosion by means of electrochemical action.

**ELECTROMIGRATION:** The electrical current-induced transfer of metal along a metallic conductor path.

**ELECTRONIC PACKAGING:** A protective enclosure for an electronic circuit so that it will both survive and perform under a plurality of environmental conditions.

**ELECTRONIC SHIELDING:** A physical barrier, usually electrically conductive, designed to reduce the interaction of electric or magnetic fields upon devices, circuits, or portions of circuits.

**ELECTROSTATIC DISCHARGE (ESD):** The instantaneous transfer of charges accumulated on a nonconductor along a conductor into ground. ESD is either caused by direct contact or induced by an electrostatic field.

**ELECTROSTATIC SHIELD:** A metallic shield or foil, usually grounded, between reed switches or other types of contacts and coil, or between adjacent relays, to minimize crosstalk effects.

**ELONGATION:** The ratio of the increase in wire length, in a tensile test, to the initial length, given as a percentage.

**ENCAPSULATE:** Sealing or covering an element or circuit for mechanical and environmental protection.

**ENCAPSULATION:** The process of completely enclosing an article in an envelope of dielectric material.

**ENTRAPPED MATERIAL:** Gas or particles (inclusions) bound up in a solid or in an electrical package.

**ENVIRONMENT:** The physical conditions, including climate, mechanical, and electrical conditions, to which a product may be exposed during manufacture, storage, or operation.

**ENVIRONMENTAL TEST:** A test or series of tests used to determine the sum of external influences affecting the structural, mechanical, and functional integrity of any given package or assembly.

**FAILURE:** The termination of the capability of an item to perform a required function.

**FAILURE ANALYSIS:** The analysis of a component from a circuit to locate the reason for the failure to the specified level.

**FAILURE CRITERIA:** The limiting conditions relating to the admissibility of the deviation from the characteristic value due to changes after stress.

**FATIGUE LIFE:** The number of loading cycles of a specified nature that a design element can sustain before failing.

**FAILURE MODE:** The cause for rejection of any failed device as defined in terms of the specific electrical or physical requirement that it failed to meet.

**FAILURE MECHANISM:** The physical or chemical process by which a device proceeds to the point of failure.

**FATIGUE:** Used to describe a failure of any structure caused by repeated application of stress over a period of time.

**FILLER:** A substance, usually dry and powdery or granular, used to add desirable properties to fluids or polymers.

**FIRST BOND:** The first bond in a sequence of two or more bonds made to form a conductive connection.

**FLAME RETARDANT:** Additives incorporated in trace amounts into the molding compound to reduce its flammability.

**FLIP-CHIP:** A leadless monolithic structure, containing circuit elements, which is designed to electrically and mechanically interconnect to a substrate by means of an appropriate number of bumps located on its face which are covered with a conductive bonding agent.

**FLOOR LIFE:** The allowable time period for a moisture-sensitive plastic-packaged device to be exposed to normal room environment humidity, after removal from a moisture barrier bag and before a solder reflow process.

**FLUX:** In soldering, a material that chemically attacks surface oxides and tarnishes so that molten solder can wet the surface to be soldered.

**FOOTPRINT:** The area needed on a substrate for a component or element. Usually refers to specific geometric dimensions of a chip.

**GANG BONDING:** The act of bonding a plurality of mechanical and/or electrical connections through a single act or stroke or a bonding tool.

**GATE ARRAY:** A semicustom product, implemented from a fully diffused or ion-implanted semiconductor wafer carrying a matrix of identical primary cells arranged into columns with routing channels between them in the x and y directions.

**GEL TIME:** The time (in seconds) required for a molten thermoset resin to reach infinite viscosity.

**GLASS TRANSITION TEMPERATURE ($T_g$):** The temperature at which a vitreous material changes from a hard and relatively brittle condition to a viscous or rubbery condition. In this temperature region, many physical properties, such as hardness, brittleness, thermal expansion, and specific heat, undergo significant, rapid changes.

**GLASSIVATION:** An inert, transparent, glasslike thin layer of pyrolytic insulation material that covers (passivates) the active device areas, including metal conductors, but excluding bonding pads, bumps, and beam leads.

**GLOB-TOP:** A usually organic material dispensed over a wire bonded die mounted to a substrate for environmental protection.

**HARDNESS:** A property of solids, plastics, and viscous liquids that is indicated by their solidity and firmness; resistance of a material to indentation by an indentor of fixed shape and size under a static load or to scratching; and the cohesion of the particles on the surface of a mineral as determined by its capacity to scratch another or be itself scratched. Indentation hardness may be measured by various hardness tests, such as Brinell, Rockwell, and Vickers.

**HEEL (OF THE BOND):** The part of the lead adjacent to the bond that has been deformed by the edge of the bonding tool used in making the bond. The back edge of the bond.

**HEEL BREAK:** The rupture of the lead at the heel of the bond.

**HEAT SINK:** The supporting member to which electronic components or their substrate or their package bottom is attached. This is usually a heat conductive metal with the ability to rapidly transmit heat from the generating source (component).

**HEEL CRACK:** A crack across the width of the bond in the heel region.

**HERMETIC:** A package sealed so that the object is gastight. The test for hermeticity is to fill the object with a test gas, often helium, and observe leak rates when the object is placed in a vacuum. Plastic encapsulation is not hermetic because it allows permeation by gases.

**HERMETICITY:** The ability of a package to prevent exchange of its internal gas with the external atmosphere. The figure of merit is the gaseous leak rate of the package measured in Pa-$m^3$-$sec^{-1}$.

**HIGHLY ACCELERATED STRESS TEST (HAST):** Accelerated test done at high temperature and relative humidity to induce rapid failure of plastic-encapsulated devices. Test conditions vary between companies. Current JEDEC standards recommend 135°C and 85% RH for 168 hours (static).

**HOT SPOT:** A small area on a circuit that is unable to dissipate the generated heat and, therefore, operates at an elevated temperature above the surrounding area.

**HUMIDITY INDICATOR CARD (HIC):** A card with color-sensitive dyes to expose the level of relative humidity inside a container. HIC is used for dry packing. Color changes are reversible as long as high-temperature baking is not used.

**INCLUSION:** A foreign object in either the conductive layer, plating, or base material.

**INFRARED SOLDERING (IR):** The soldering process in which unfocused infrared lamps or heater elements are used to bring the printed circuit board and components to the solder reflow temperature.

**INJECTION MOLDING:** Molding of electronic packages by injecting liquefied plastic into a mold.

**INTEGRATED CIRCUIT:** A microcircuit (monolithic) consisting of interconnected elements inseparably associated and formed in situ on or within a single substrate (usually silicon) to perform an electronic circuit function.

**INTERCONNECTION:** The conductive path required to achieve connection from a circuit element to the rest of the circuit.

**INTERFACE:** The boundary between dissimilar materials, such as between a film and substrate.

**INTERMETALLIC BOND:** The ohmic contact made when two metal conductors are welded or fused together.

**INTERMETALLIC COMPOUND:** A definite composition of two or more metals that has a characteristic crystal structure.

**INTERNAL LAYER:** A conductive layer internal to a multilayer printed board.

**ION MIGRATION:** The movement of free ions within a material or across the boundary between two materials under the influence of an applied electric field.

**IONIC CONTAMINANT:** Any contaminant that exists as ions and, when in solution, increases electrical conductivity.

**ISOLATION:** A technique for electrically separating circuit elements. In dielectric isolation, components are isolated by means of insulating layers. In diode isolation components are isolated by means of reverse-biased *pn* junctions.

**J-LEAD:** A surface-mount device that has leads (external) formed into a J pattern, folding under the device body.

**JUNCTION TEMPERATURE:** The temperature of the region of transition between the *p* and *n* type semiconductor material in a transistor or diode element.

**JUNCTION:** (1) In solid-state materials, a region of transition between *p*- and *n*-type semiconductor materials as in a transistor or diode; (2) A contact between two dissimilar metals or materials, (3) A connection between two or more conductors or two or more sections of a transmission line.

**KIRKENDALL VOIDS:** The formation of voids by diffusion across the interface between two different materials, in the material having the greater diffusion rate into the other. Formed by vacancy collapse.

**KOVAR:** An alloy of 53% iron, 17% cobalt, 29% nickel, and trace elements, with thermal expansion matching alumina substrate and certain sealing glasses. Most commonly used leadframe and pin material in hermetic parts. Conforms to the ASTM designation F15.

**LAMINATE:** A layered sandwich of sheets of substances bonded together under heat and pressure to form a single structure.

**LASER BONDING:** Forming of a metal-to-metal bond of two conductors by welding the two materials together using a laser beam for a heat source.

**LEAD:** A conductive path, usually self-supporting. That portion of an electrical component used to connect it to the outside world. See also *COMPLIANT LEAD.*

**LEADLESS DEVICE:** A chip device using no input/output leads.

**LEADLESS CHIP CARRIER:** A chip carrier with integral metallized terminations and no compliant external leads.

**LEADLESS SURFACE MOUNTING:** The surface mounting of components directly to a substrate by means of solder joints.

**LEADFINGERS:** The interior ends of the leadframe leads, to which the bond wires are connected to complete the circuit from the IC bond pads.

**LEAD CAPACITANCE:** Often is broken into two components. The self-capacitance is the capacitance of the lead with respect to a ground conductor or a ground plane. It is dependent on the lead geometry and the dielectric material. The mutual capacitance is the lead-to-lead capacitance.

**LEAD INDUCTANCE:** Often is broken into two components. The self-inductance is the inductance of the loop formed in a subsystem comprising of the wirebond, package lead, and a ground conductor or ground plane. The self-inductance is a function of the loop length. The mutual inductance is the inductance that exists between two signal leads as a result of magnetic coupling between the leads.

**LEAD RESISTANCE:** DC resistance and the cause of voltage drops in a package.

**LEADFRAME:** The metallic portion of the device package that completes the electrical connection path from the die to ancillary hybrid circuit elements of the card assembly.

**LEAKAGE CURRENT:** An undesirable small stray current that flows through or across an insulator between two or more electrodes, or across a back-biased junction.

**LIFE TEST:** Test of a component or circuit under electrical load and high temperature to assure reliability device.

**LOOP:** The curve or arc of the wire between the attachment points at each end of a wirebond.

**LOOP HEIGHT:** A measure of the deviation of the wire loop from the straight line between the attachment points of a wirebond. Usually, it is the maximum perpendicular distance from this line to the top of the wire loop.

**MARKING:** A method of identifying printed boards or components with part number, revision letter, manufacturer code, and so on.

**METALLIZATION:** A deposited or plated thin metallic film used for its protective or electrical properties. The process of applying a conductive metal film to the surface of the integrated circuit chip. This metallization is used both for interconnection on the chip and to define a place for the attachment of bond wires.

**MICROCRACKS:** A thin crack in a substrate or chip, or in thick-film trim-kerf walls, that can only be seen under magnification and which can contribute to latent failure phenomena.

**MICROSECTIONING:** The preparation of a specimen for the metallographic examination of the material (usually by cutting out a cross section, followed by polishing, etching, staining, and mounting).

**MIL:** An unit of length that is equal to 0.001 in or 0.0254 mm.

**MISREGISTRATION:** Improper alignment of successively produced features or patterns.

**MODULE:** A generic term referring to assembled units in electronic equipments/assemblies.

**MODULUS OF ELASTICITY:** The ratio of stress to strain in a material that is elastically deformed.

**MOISTURE BARRIER BAG (MBB):** A plastic bag designed to restrict the transmission of water vapor and used to pack moisture-sensitive plastic-encapsulated components.

**MOISTURE STABILITY:** The functional stability of a circuit under high-humidity conditions.

**MOLDING COMPOUND:** The plastic formulation used to package ICs through a molding process.

**MONOMER:** A precursor of low molecular weight used as a starting material for polymerization to produce molecules of larger molecular weight, called polymers.

**MULTICHIP PACKAGE:** An electronic package that carries a number of chips and interconnects them through several layers of conductive patterns separated by insulative layers and interconnected through holes.

**MULTILAYER CIRCUITS:** A composite circuit formed by alternate layers of conductive circuit and insulating materials (ceramic or dielectric compositions) bonded together with the conductive layers interconnected as required.

**MULTILAYER PRINTED BOARD:** Printed circuit or printed wiring configuration that consists of more than one conductive layer bonded together to form a multiple conductive layer assembly. The term applies to both rigid and flexible multilayer boards.

**NONWETTING:** (1) The lack of metallurgical wetting between molten solder and a metallic surface due to the presence of a physical barrier on the metallic surface, (2) A condition in which a surface has contacted molten solder, but the solder has not adhered to all of the surface.

**OUTGASSING:** Gaseous emission from a material when exposed to reduced pressure and/or heat.

**PACKAGE:** In the electronics/microelectronics industry, an enclosure for a single element, an integrated circuit, or a hybrid circuit. It provides environmental protection, determines the form factor, and serves as external interconnection for the device by means of package terminals.

**PACKAGE CRACKING::** Cracks caused in an IC package due to stresses in the case.

**PACKAGING DENSITY:** The amount of electronic function per unit volume, often defined qualitatively as high, medium, or low.

**PAD:** The metallized area on a substrate or on the face of an integrated circuit used for making electrical contact or connection.

**PASSIVATION:** The formation of an insulating layer directly over a circuit or circuit element to protect the surface from contaminants, moisture, or particles.

**PEEL STRENGTH (PEEL TEST):** A measure of adhesion between a conductor and the substrate. The test is performed by pulling or peeling the conductor off the substrate and observing the force required. Preferred unit is $g\text{-}mm^{-1}$ or $kg\text{-}mm^{-1}$ of conductor width.

**PERCENT DEFECTIVE ALLOWABLE (PDA):** The maximum observed percent defective that will permit a lot of devices to be accepted after the specified 100% test.

**PERMEABILITY:** The property of a material, such as solid plastic, that allows penetration by a liquid or gas.

**PHASE (MATERIAL):** (As in glassy phase or metal phase). Refers to the part or portion of a material system that is metallic or glassy in nature. In general, a phase is a structurally and chemically homogeneous physically distinct portion of a substance or a group of substances that are in equilibrium with each other.

**PIN:** Round, cross-sectional electrical terminal and/or mechanical support. Used in plug-in type packages, either straight or modified as nail head, upset, pierced, or bent varieties. The primary functions of a pin are, internally, to support a wire bond or other joint and, externally, to plug into a second-level package connector.

**PIN CONTACT:** A male-type contact, usually designed to mate with a socket or female contact. It is normally connected to the "dead" side of a circuit.

**PIN DENSITY:** The number of pins per unit area on a printed board.

**PIN-GRID ARRAY:** A package or interconnect scheme, featuring a multiplicity of plug-in-type-electrical terminals arranged in a prescribed matrix format or array. Used with high-input/output-count devices.

**PINHOLE:** Small holes occurring as imperfections that penetrate entirely through film elements, such as metallization, dielectric, or passivation films.

**PITCH:** The nominal centerline-to-centerline dimension between adjacent conductors.

**PLASTIC:** A polymeric material, either organic, such as epoxy and polyimide, or inorganic, such as silicone, used for conformal coating, encapsulation, or overcoating.

**PLASTIC DEVICE:** A device wherein the package, or the encapsulant material for the semiconductor die, is plastic. Such materials as epoxies, phenolics, silicones, etc. are included.

**PLASTIC ENCAPSULATION:** Environmental protection of a completed circuit by embedding it in a plastic such as epoxy or silicone.

**PLASTIC LEADED CHIP CARRIER (PLCC):** A JEDEC standard surface mount family of IC packages, with leads exiting all four (4) sides of the package and formed into a "J" lead formation, with 50 ml lead-to-lead pitch.

**PLASTIC QUAD FLATPACK (PQFP):** A JEDEC standard surface mount family of IC packages, with leads exiting all four (4) sides of the package and formed into "gull wing" lead formation, with 25-mil lead-to-lead pitch.

**PLASTIC SURFACE MOUNT COMPONENTS (PSMCs):** Includes the family of components which are attached to the surface of a printed circuit board. such as PLCC, PQFP, SOIC, etc.

**PLASTICIZER:** A chemical agent added in compounding plastics to make them softer and more flexible.

**PLATED THROUGH-HOLE:** A hole in which an electrical connection is made between internal or external conductive patterns, or both, by the plating of metal on the wall of the hole.

**POISSON'S RATIO:** The proportionality ratio of a lateral strain to an axial strain when a material is placed in tension.

**POLARIZATION:** The elimination of inplane symmetry so that parts can be engaged in only one way.

**POLYMERIZATION:** A chemical reaction in which the molecules of a monomer are cross-linked to form large molecules whose molecular weight is a large multiple of that of the original substance.

**POPCORN:** Package cracking during solder reflow caused by high moisture content in the package.

**POROSITY:** The ratio of solid matter to voids in a material.

**POT-LIFE:** The length of time that a two-part epoxy system remains useful, usually measured in hours; it can be extended by refrigeration.

**POTTING:** Encapsulating a circuit in a plastic by applying it in the liquid form on the mounted surface.

**POWER DENSITY:** The amount of power dissipated from a functioning electrical device and is measured, in $W\text{-}cm^{-2}$ of the substrate.

**POWER DISSIPATION:** Heat generated from a circuit when a current flows through it.

**PRECONDITIONING:** Process to simulate surface mounting such that the plastic packages are subjected to moisture prior to solder reflow.

**PRINTED WIRING ASSEMBLY:** One or more populated printed wiring boards that perform a specific function in a system.

**PRINTED WIRING BOARD:** A printed board containing a conductive pattern that consists only of printed wiring.

**PRINTED CIRCUIT BOARD:** A printed board that consists of conductive pattern which includes printed components and/or printed wiring.

**PROBE:** A pointed conductor used in making electrical contact to a circuit pad for testing.

**PROCURING ACTIVITY:** The organizational element (equipment manufacturer, government, contractor, subcontractor, or other responsible organization) that contracts for articles, supplies, or services, and has the authority to grant waivers, deviations, or exceptions to the procurement documents.

**PULL STRENGTH:** The values of the pressure achieved in a test where a pulling stress is applied to determine breaking strength of a lead or bond.

**PULL TEST:** A test for bond strength of a lead, interconnecting wire, or conductor.

**PULSE SOLDERING:** Soldering a connection by melting the solder in the joint area by pulsing current through a high-resistance point applied to the joint area and the solder.

**PURPLE PLAGUE:** The term used for the formation of one of several gold-aluminum intermetallic compounds formed when bonding gold to aluminum and accentuated by exposure to moisture and high temperatures. Purple plague is purplish in color and is very brittle, potentially leading to time-based failure of the bonds.

**QUAD FLATPACK:** Generic term for surface mount technology packages with leads on all four sides. Commonly used to describe chip carrier-like devices with gull-wing leads.

**QUALITY ASSURANCE:** A technique or science that uses all the known methods of quality control and quality engineering to ensure the manufacture of a product of uniform acceptable quality standards.

**REFLOW:** Wetting and melting of the solder to join the package leads onto a printed-circuit board. Heating of the solder can be done in an infrared oven or in a vapor-phase reflow oven.

**REFLOW SOLDERING:** A method of soldering involving application of solder prior to the actual joining. To solder, the parts are joined and heated, causing the solder to remelt, or reflow.

**RELIABILITY:** A collective name for those measures of quality that reflect the effect of time in storage or use of a product, as distinct from those measures that show the state of the product at the time of delivery. Generally, it is the capability of an item to perform a required function under stated conditions for a stated period of time.

**RESIN:** An organic polymer that cross-links to form a thermosetting plastic when mixed with a curing agent. It can also refer to a thermoplastic in the solid form.

**RESISTANCE SOLDERING:** A method of soldering in which a current is passed through and heats the soldering area by contact with one or more electrodes.

**RESOLUTION:** The physical integrity of a printed pattern. The ability of photographic materials to reproduce fine dimensions.

**REVERSE BIAS:** The bias that tends to produce current flow in the reverse direction.

**REVERSION:** A chemical reaction resulting in a polymerized material that degenerates to a lower polymeric state or to the original monomer.

**REWORK:** An operation performed on a nonconforming part or assembly that restores all nonconforming characteristics to the requirements in the contract, specification, drawing, or other approved product description.

**RHEOLOGY:** The science dealing with deformation and flow of fluids.

**RISERS:** The conductive paths that run vertically from one level of conductors to another in a multilayer substrate or screen-printed film circuit.

**RUPTURE:** In breaking-strength or creep tests, the point at which a material physically comes apart.

**SCRATCH:** In optical observations, surface mark with a large length-to-width ratio.

**SCREEN:** A network of metal or fabric strands, mounted snugly on a frame, and upon which the film circuit patterns and configurations are superimposed by photographic means.

**SCREENING:** The process whereby the desired film circuit patterns and configurations are transferred to the surface of the substrate during manufacture by forcing a material through the open areas of the screen using the wiping action of a soft squeegee. The term is also used for the process of separation of electronic parts with defects from those without according to a specific specification.

**SEMICONDUCTORS:** Solid materials such as silicon that have a variable (tailorable) resistivity midway between those of a conductor and an insulator. These materials are used for the fabrication of semiconductor devices, such as transistors, diodes, and integrated circuits.

**SHEAR RATE:** With regard to viscous fluids, the relative rate of flow or movement under lateral forces.

**SHEAR STRENGTH:** The limiting stress of a material determined by measuring a strain resulting from applied forces that cause or tend to cause bonded contiguous parts of a body to slide relative to each other in a direction parallel to their plane of contact; the value of the force achieved when shearing stress is applied to the bond (normally parallel to the substrate) to determine the breaking load; strength to withstand shearing of a material.

**SMALL-OUTLINE INTEGRATED CIRCUIT (SOIC):** A standard surface-mount family of IC packages with two (2) rows of formed leads, with 50-mil center-to-center spacing. Lead formation may be "J" or "gull-wing."

**SOLDER:** A low melting point alloy, usually of lead (Pb)-tin (Sn), that can wet copper, conduct current, and mechanically join conductors.

**SOLDER BUMPS:** The round solder balls bonded to a contact pad and used to make connection to a conductor.

**SOLDER BRIDGING:** A short between two or more conductors due to solder.

**SOLDER CONNECTION:** An electrical or mechanical connection between two metal parts formed by solder.

* Disturbed solder connection - a cold solder joint resulting from movement of one or more of the parts during solidification.
* Excess solder connection - a solder joint containing so much solder that the joined part contours are not visible. Also refers to solder that has flowed beyond the designated solder area.
* Insufficient solder connection - a solder joint characterized by one or more of the parts incompletely covered by solder.
* Overheated solder connection - a solder joint characterized by a dull, chalky, grainy, and porous/pitted appearance, which is the result of excess heat application during the solder process.
* Preferred solder connection - a smooth, bright, and well-wetted solder connection with no exposed bare lead material and no sharp protrusions or embedded foreign material with the lead contours visible.
* Rosin solder connection - a solder connection that has a cold solder joint appearance but is further characterized by rosin flux entrapment separating the surfaces to be joined.

**SOLDER IMMERSION TEST:** A test that immerses the electronic package leads into a solder bath to check resistance to soldering temperatures or solderability.

**SOLDER PASTE:** A composition of metal (e.g., Sn/Pb) power, flux, and other organic vehicles.

**SOLDER REFLOW:** The solder attachment process in which previously applied solder paste is melted to solder a component to the printed circuit board.

**SOLDER WICKING:** The rising of solder between individual strands of wire due to capillary action.

**SOLDERABILITY:** The ability of a conductor to be wetted by solder and to form a strong bond with the solder.

**SOLDERING:** A process of joining metallic surfaces with solder, without melting the base material.

**SOLVENT:** A material that has the ability to dissolve other materials.

**SPECIFIC HEAT:** The ratio of the heat capacities of a body and water at a reference temperature; i.e., the quantity of heat required to raise the temperature of 1 g of a substance by 1°C.

**STITCH BOND:** A bond in which a capillary tube is used for feeding the wire and forming the bond sequentially in a stitch pattern. The wire is not formed into a ball prior to bonding.

**STORAGE TEMPERATURE:** The temperature at which a device, without any power applied, is stored.

**STRESS:** Caused by thermal mismatch between the various materials of construction in the device. In a plastic-encapsulated device, part of the stress is also due to the curing of the epoxy polymer network which shrinks during the polymerization. Also often referred to as packaging stress, shrinkage stress, molding stress, or encapsulating stress.

**STRESS RELAXATION:** The time-dependent decrease in stress in a solid under given constraint conditions.

**SUBSTRATE:** A supporting platform for an active or passive electrical or electronic component.

**SURFACE-MOUNT TECHNOLOGY (SMT):** The general category of expertise for mounting surface mount components onto substrates.

**SURFACE RESISTIVITY:** The resistance to a current flow along the surface of a material.

**TAPE AUTOMATED BONDING (TAB):** The utilization of a metal tape material as a support and carrier of a microelectronic component in a gang bonding process.

**TEMPERATURE CYCLING:** An environmental test in which the specimen is subjected to several changes from one temperature to another over a period of time.

**TENSILE STRENGTH:** The pulling stress that has to be applied to a material to break it, usually measured in Pa.

**THERMAL CONDUCTIVITY:** The amount of heat per unit time per unit area that can be conducted through a unit thickness of a material.

**THERMAL EXPANSION:** The expansion of a material when subjected to temperature change (usually a temperature increase).

**THERMAL GRADIENT:** The plot of temperature change across the surface or the bulk thickness of a material being heated.

**THERMAL MISMATCH:** Difference of thermal coefficients of expansion of materials that are bonded together.

**THERMAL SHOCK:** A condition whereby devices are subjected alternately to extreme heat and cold liquids. Used to screen out defective parts.

**THERMOCOMPRESSION BONDING:** A process involving the use of pressure and temperature to join two materials by plastic flow and interdiffusion across the interface.

**THERMOPLASTIC:** A substance that becomes plastic (malleable) on being heated; a plastic material that can be repeatedly melted or softened by heat without change of properties.

**THERMOSET:** A resin that is cured, set, or hardened, usually by heating, into a permanent shape through an irreversible polymerization reaction (cross-linking). Once set, a thermosetting plastic cannot be remelted, although most soften with the application of heat.

**THERMOSONIC BONDING:** The bonding of wires to metal pads on an integrated circuit by means of heat and ultrasonic scrubbing of the wire into the pad to create a metallurgical bond.

**THROUGH-HOLE MOUNTING:** The electrical interconnection of components with leads that are inserted through the holes of a printed circuit board and soldered to the board on the protrusion side.

**TRANSFER MOLDING:** Molding circuit modules by transferring molten plastic into a cavity holding the circuit.

**TRANSFER SOLDERING:** A soldering process using a hand soldering iron and a measured amount of solder in the form of a ball, chip, or disk.

**ULTRASONIC BONDING:** A process involving the use of ultrasonic energy and pressure to join two materials.

**USEFUL LIFE:** The length of time a product functions with a failure rate that is considered to be satisfactory.

**VAPOR-PHASE REFLOW:** A technique for solder reflow to form package interconnections. The solder joint is heated by the heat of condensation of an inert vapor.

**VIA:** An opening in the dielectric layer(s) through which a riser passes and whose walls are made conductive.

**VIA HOLE:** A plated through hole providing electrical interconnection for two or more conductive layers, but not intended to have a component lead inserted through it.

**VISCOSITY:**: The resistance that a gaseous or liquid system offers to flow when it is subjected to a shear stress. The unit of viscosity measurement is poise, more commonly centipoise. Viscosity varies inversely with temperature.

**VISUAL EXAMINATION:** Qualitative examination of physical characteristics using the unaided eye or defined levels of magnification.

**VOID:** In visual inspection of solid materials, a space not filled with the specific solid material, such as a gap or opening that is an unintentional defect in the material.

**WAFER:** A slice of semiconductor crystal ingot used as a host for transistors, diodes, and monolithic integrated circuits.

**WARPAGE:** The distortion of a substrate from a flat plane.

**WAVE SOLDERING:** A soldering process in which populated printed boards are soldered by passing them over a continuous flowing wave of molten solder.

**WEDGE BOND:** A wedge-shaped wirebond formed by a wedge-shaped pressing tool, usually ultrasonically.

**WETTING:** The spreading of molten solder on a metallic surface, with proper application of heat and flux.

**WHISKERS:** Needle-shaped metallic growths between conductors.

**WIRE SWEEP:** Flow-induced deformation of the wirebonds during molding.

**WIREBOND:** A completed wire connection that includes all its constituents and provides electrical continuity between the semiconductor die (pad) and a terminal.

**YIELD:** The ratio of the number of acceptable items produced in a production run to the total number that were attempted to be produced.

**YIELD STRENGTH:** The stress at which a material exhibits a specified strain deviation from purely elastic stress-strain behavior.

# INDEX